Undergraduate Topics in Computer Science

Series Editor

Ian Mackie, University of Sussex, Brighton, France

Advisory Editors

Samson Abramsky, Department of Computer Science, University of Oxford, Oxford, UK

Chris Hankin, Department of Computing, Imperial College London, London, UK

Mike Hinchey, Lero—The Irish Software Research Centre, University of Limerick, Limerick, Ireland

Dexter C. Kozen, Department of Computer Science, Cornell University, Ithaca, USA

Hanne Riis Nielson, Department of Applied Mathematics and Computer Science, Technical University of Denmark, Kongens Lyngby, Denmark

Steven S. Skiena, Department of Computer Science, Stony Brook University, Stony Brook, USA

Iain Stewart, Department of Computer Science, Durham University, Durham, UK

Joseph Migga Kizza, Engineering and Computer Science, University of Tennessee at Chattanooga, Chattanooga, USA

Roy Crole, School of Computing and Mathematics Sciences, University of Leicester, Leicester, UK

Elizabeth Scott, Department of Computer Science, Royal Holloway University of London, Egham, UK

'Undergraduate Topics in Computer Science' (UTiCS) delivers high-quality instructional content for undergraduates studying in all areas of computing and information science. From core foundational and theoretical material to final-year topics and applications, UTiCS books take a fresh, concise, and modern approach and are ideal for self-study or for a one- or two-semester course. The texts are authored by established experts in their fields, reviewed by an international advisory board, and contain numerous examples and problems, many of which include fully worked solutions.

The UTiCS concept centers on high-quality, ideally and generally quite concise books in softback format. For advanced undergraduate textbooks that are likely to be longer and more expository, Springer continues to offer the highly regarded *Texts in Computer Science* series, to which we refer potential authors.

Bernard Zygelman

A First Introduction to Quantum Computing and Information

Second Edition

Bernard Zygelman
Department of Physics and Astronomy
University of Nevada Las Vegas
Las Vegas, NV, USA

ISSN 1863-7310　　　　　ISSN 2197-1781　(electronic)
Undergraduate Topics in Computer Science
ISBN 978-3-031-66424-3　　　ISBN 978-3-031-66425-0　(eBook)
https://doi.org/10.1007/978-3-031-66425-0

1st edition: © Springer Nature Switzerland AG 2018
2nd edition: © The Editor(s) (if applicable) and The Author(s), under exclusive license to Springer Nature Switzerland AG 2025

This work is subject to copyright. All rights are solely and exclusively licensed by the Publisher, whether the whole or part of the material is concerned, specifically the rights of translation, reprinting, reuse of illustrations, recitation, broadcasting, reproduction on microfilms or in any other physical way, and transmission or information storage and retrieval, electronic adaptation, computer software, or by similar or dissimilar methodology now known or hereafter developed.
The use of general descriptive names, registered names, trademarks, service marks, etc. in this publication does not imply, even in the absence of a specific statement, that such names are exempt from the relevant protective laws and regulations and therefore free for general use.
The publisher, the authors and the editors are safe to assume that the advice and information in this book are believed to be true and accurate at the date of publication. Neither the publisher nor the authors or the editors give a warranty, expressed or implied, with respect to the material contained herein or for any errors or omissions that may have been made. The publisher remains neutral with regard to jurisdictional claims in published maps and institutional affiliations.

This Springer imprint is published by the registered company Springer Nature Switzerland AG
The registered company address is: Gewerbestrasse 11, 6330 Cham, Switzerland

If disposing of this product, please recycle the paper.

For Judith

Foreword

For over 40 years, I have been involved in various research activities involving "classical" computations. I have used nearly all computing platforms, including punch-card, serial, scalar, vector, parallel, and massively parallel processors. With these tools and the computational power they facilitate, my collaborators and I have tackled some of the most challenging problems in my field of aerospace engineering.

With the noticeable decline in the rate of progress of high-performance classical computing over the past decade, it is now widely believed that the longevity of Moore's law is limited. Many believe quantum computing (QC) is poised to become the new standard, supplanting or augmenting classical computing methods. In 2012, encouraged by some of the leading program managers of the US-AFOSR, I took the initiative to learn about QC and evaluate its prospects for aerospace applications. At that time, the primary viable source to learn QC was the classic text, the masterpiece of Mike and Ike (Nielsen and Chuang, *Quantum Computation and Quantum Information*, Cambridge U. Press, 2011). This text is widely considered the "bible" and the standard reference in quantum computing and information. However, because of its mathematical elegance, its contents are challenging to people like myself who are not trained as quantum physicists or computer scientists. Fortunately, within the past decade or so, many books have appeared on the subject, which are easier and more accessible to non-experts. The level of difficulty and also the quality of these books vary significantly. Some of these books are good, and some of them are excellent. The first edition of the present text by Bernard Zygelman, published in 2018, belongs to the latter category.

I came across this book in my search for a more gentle introduction to QC and am very excited to have found it. It helped me understand QC. Professor Zygelman does an incredible job of making this difficult topic understandable by anyone with some basic mathematics background (Calculus 2 and linear algebra). The approach and flow in this book are, indeed, like no other! The text starts with the simple definition of qubits and then covers just enough linear algebra as needed for the remainder of the chapters. The circuit model of QC follows this and then the journey to the killer applications of QC and more advanced topics. Many of my colleagues and our students enjoyed the first edition. So, we invited

Prof. Zygelman to be our guest as a visiting scholar at our Pittsburgh Quantum Institute (PQI.org). In his sabbatical here, Prof. Zygelman completed the present Second Edition. We were lucky to have a first peek at the additional two chapters as included and also to listen to his lectures on these chapters. In this second edit, links to more working examples and exercises are provided in the chapters. Some additional figures are provided to help me understand the algorithms better, and all of the minor typos are fixed (to the extent that I can detect them). Professor Zygelman is a gifted teacher with a knack for explaining some of the most difficult topics very simply. Being lucky to have the opportunity to attend his lectures at PQI, I also learned to become a better teacher, as I learned a lot about QC.

Some of the brightest minds in the world believe that QC has the potential to be a disruptive technology that will revolutionize many current computational tasks in science and engineering. The QC industry aims to build a general-purpose *digital universal quantum computer*. Google, IBM, Intel, Microsoft, Amazon, Honeywell, and many start-ups in the US and abroad are developing such machines. Almost all research universities worldwide now offer new courses in QC at both undergraduate and graduate levels. This book will be an excellent primary text or a companion for all these courses. It will also be a useful source for researchers in other fields who want to learn QC and potentially use it in future computations.

June 2024

Peyman Givi
Distinguished Professor and James T.
MacLeod Chair in Engineering
Co-director: Ph.D. Program
in Computational Modeling
and Simulation
University of Pittsburgh
Pittsburgh, PA, USA

Preface

This monograph is an outgrowth of courseware developed and offered at UNLV to a predominantly undergraduate audience. The motivation for the course and, by extension, for this textbook is two-fold. First, it responds to the explosive growth and transformative technological promise of quantum computing and information science (QIS). As outlined in the treatise *"Law and Policy for the Quantum Age"* by Chris Jay Hoofnagle and Simon L. Garfinkel, Cambridge University Press 2022, https://doi.org/10.1017/9781108883719, developments in this field will have a reach far beyond the confines of science and engineering.

Second, the material in this textbook should be accessible to undergraduate students with some background in mathematical-oriented analysis, typically obtained during a first-year calculus class. Thus, quantum concepts can be delivered at a much earlier stage in a student's career than those available through traditional curricula. A standard undergraduate quantum curriculum usually requires some knowledge of the theory of partial differential equations, complex analysis, linear algebra, and advanced vector calculus, which is why most universities introduce quantum theory to students in their third and fourth years. As recent advances in quantum science will underpin twenty-first-century technology, quantum information science (QIS) should be available to a broader and more diverse audience rather than being limited to advanced offerings for physics or computer science majors. This textbook aims to ignite curiosity and allow students to gain the necessary background for advanced studies, such as that presented in the classic treatise Quantum Computation and Quantum Information, Cambridge U. Press (2011), by M. A. Nielsen and I. L. Chuang. For this reason, the emphasis is on accessibility rather than rigor, and more advanced material is avoided, such as a detailed discussion of computational complexity, universality, and measurement theory. Throughout, we have included links to Mathematica® notebooks that offer numerical demonstrations and exercises that help illuminate difficult and counterintuitive concepts. The latter helps guide the student, via an interactive approach, through the written material on the page.

In Chaps. 1 and 2, I introduce the foundational framework for the Copenhagen interpretation of quantum mechanics as it applies to qubit systems. We discuss the Dirac Bra-ket notation and the tools necessary to manipulate objects in a multiqubit Hilbert space. Chapter 3 reviews the circuit model of computation and introduces the first quantum gates and circuits. Chapter 4 provides a detailed discussion of the Shor and Grover algorithms. Chapters 5 and 6 are concerned with aspects of quantum information theory. With Chap. 9, that deals with the theory of quantum error correction; the chapters mentioned above constitute a reasonably paced 14-week semester introduction to QCI. I included Chaps. 7 and 8 which cover developments in trapped ion, quantum cavity electrodynamics (CQED), and quantum circuit (cQED) computers. That discussion proved challenging because a comprehensive treatment requires an advanced understanding of atomic physics, quantum optics, and many-body physics. However, I believe that I succeeded in conveying the essential features of these paradigms by using models that require only a passing familiarity with differential equations and complex analysis. In an introductory course, these chapters could be bypassed.

Since the 2018 edition, significant advancements in quantum computing have occurred, necessitating this updated edition. Notable developments include the advent of NISQ (Noisy Intermediate-Scale Quantum) era machines boasting dozens to hundreds of high-fidelity qubits and new error-correcting codes. Furthermore, the 2018 National Quantum Initiative Act and the CHIPS Act underscore the importance of a quantum-smart workforce and the need for inclusive quantum education that spans the physical sciences, engineering, and mathematics. This edition introduced an extended discussion of quantum error correction, including an introduction to surface codes. A new chapter on adiabatic quantum computing highlights the latest research and practical applications in this promising area. This last chapter can be ignored in an undergraduate offering, as it is on a level more attuned to the background of a graduate student. This edition will have a dedicated website https://bernardzygelman.github.io/ to allow downloads of Mathematica notebooks, exercises, problems, and solution manuals.

On Notation

I use a convention standard among physicists for labeling states in multi-qubit kets that differs from that used in many QCI texts. Given ket $|abc\rangle$, I refer to the entry c as the first qubit, b as the second, and a as the third qubit. That order (from right to left) is inverted for the corresponding bra vectors. This convention also leads to a notational adjustment in wire diagrams. With the convention used in this monograph, the first qubit c is placed on the lowest rung of a wiring diagram, with the second and third, etc., qubit states stacked on it. I employ arrow notation to label vector quantities and use boldface for most operators. I used the regular typeface for common, such as the Pauli, operators where there is little chance of misunderstanding. In most QCI literature, operators are synonymous with their matrix representations, so I make little effort to distinguish them. However, in some instances where there is a possibility for ambiguity, I explicitly use an under-bar notation to stress the matrix character of an operator.

Las Vegas, NV, USA
July 2024

Bernard Zygelman

Acknowledgements

I am deeply grateful to Prof. Peyman Givi for his invitation and support of my sabbatical leave at the University of Pittsburgh and the Pittsburgh Quantum Institute (PQI). Most of the new material in this second edition was written during my time at PQI, enriched by the invaluable interaction with Prof. Givi, his students, and his colleagues. Their insights and feedback were instrumental in refining the content of this edition. I am particularly thankful to "Chief" Peyman, Hirad Aliphana, and Masoud Baranti for their valuable comments and suggestions. I also extend my appreciation to Prof. Dr. Bernd Baumann (HAW Hamburg), Prof. Jed Brody (Emory University), and Dr. Greg Colarch (UNLV) for their review of the first edition, which led to the identification and correction of several typographical and other errors.

I also want to express my appreciation to my UNLV colleagues, Yan Zhou, Joshua Island, and student Maci Kesler, for insightful discussions on all things quantum. I acknowledge the support of the UNLV Faculty Leave and Sabbatical Committee for granting leave from UNLV. I am grateful to Wayne Wheeler of Springer Nature for his encouragement and support in starting this project. Some circuit diagrams were created using the QPIC package that Thomas Draper and Samuel Kutin developed.

Finally and foremost, I would like to thank my wife, Judi, for her constant support and encouragement throughout my career. Along with her editorial contributions, this book would not have been possible without her by my side.

Contents

1	**A Quantum Mechanic's Toolbox**		1
	1.1 Bits and Qubits		1
		1.1.1 Binary Arithmetic	2
	1.2 A Short Introduction to Linear Vector Spaces		3
	1.3 Hilbert Space		5
		1.3.1 Dirac's Bra-Ket Notation	6
		1.3.2 Outer Products and Operators	11
		1.3.3 Direct and Kronecker Products	13
	Problems		20
	References		22
2	**Apples and Oranges: Matrix Representations**		25
	2.1 Matrix Representations		25
		2.1.1 Matrix Operations	27
		2.1.2 The Bloch sphere	29
	2.2 The Pauli Matrices		31
	2.3 Polarization of Light: A Classical Qubit		35
		2.3.1 A Qubit Parable	37
	2.4 Spin		39
		2.4.1 Non-commuting Observables and the Uncertainty Principle	41
	2.5 Direct Products		45
	Problems		47
	Reference		49
3	**Circuit Model of Computation**		51
	3.1 Boole's Logic Tables		51
		3.1.1 Gates as Mappings	55
	3.2 Our First Quantum Circuit		56
		3.2.1 Multi-qubit Gates	59
		3.2.2 Deutsch's Algorithm	63

		3.2.3	Deutsch-Josza Algorithm	69
	3.3	Hamiltonian Evolution		72
	Problems			74
	Reference			76

4 Quantum Killer Apps: Quantum Fourier Transform and Search Algorithms ... 77

	4.1	Introduction		77
	4.2	Fourier Series		78
		4.2.1	Nyquist-Shannon Sampling	79
		4.2.2	Discrete Fourier Transform	80
	4.3	Quantum Fourier Transform		83
		4.3.1	QFT Diagrammatics	84
		4.3.2	Period Finding with the QFT Gate	89
		4.3.3	Shor's Algorithm	93
	4.4	Grover's Search Algorithm		95
	Problems			100
	References			102

5 Quantum Mechanics According to Martians: Density Matrix Theory ... 103

	5.1	Introduction		103
	5.2	Density Operators and Matrices		104
	5.3	Pure and Mixed States		109
	5.4	Reduced Density Operators		110
		5.4.1	Entangled States	111
	5.5	Schmidt Decomposition		114
	5.6	von Neumann Entropy		117
	Problems			118
	References			121

6 No-Cloning Theorem, Quantum Teleportation and Spooky Correlations ... 123

	6.1	Introduction		123
	6.2	On Quantum Measurements		124
	6.3	The No-Cloning Theorem		125
	6.4	Quantum Teleportation		126
	6.5	EPR and Bell Inequalities		129
		6.5.1	Bertlmann's Socks	133
		6.5.2	Bell's Theorem	136
	6.6	Applications		138
		6.6.1	BB84 Protocol	139
		6.6.2	Ekert Protocol	141

		6.6.3 Quantum Dense Coding	142
	6.7	GHZ Entaglements	143
	Problems		144
	References		144
7	**Quantum Hardware I: Ion Trap Qubits**		147
	7.1	Introduction	147
		7.1.1 The DiVincenzo Criteria	148
	7.2	Lagrangian and Hamiltonian Dynamics in a Nutshell	149
		7.2.1 Dynamics of a Translating Rotor	150
	7.3	Quantum Mechanics of a Free Rotor: A Poor Person's Atomic Model	151
		7.3.1 Rotor Dynamics and the Hadamard Gate	155
		7.3.2 Two-Qubit Gates	158
	7.4	The Cirac-Zoller Mechanism	160
		7.4.1 Quantum Theory of Simple Harmonic Motion	161
		7.4.2 A Phonon—Qubit Pair Hamiltonian	163
		7.4.3 Light-Induced Rotor-Phonon Interactions	164
	7.5	Trapped Ion Qubits	170
		7.5.1 Mølmer-Sørenson Coupling	175
	Problems		177
	References		178
8	**Quantum Hardware II: cQED and cirQED**		181
	8.1	Introduction	181
	8.2	Cavity Quantum Electrodynamics (cQED)	184
		8.2.1 Eigenstates of the Jaynes-Cummings Hamiltonian	188
	8.3	Circuit QED (cirQED)	190
		8.3.1 Quantum *LC* Circuits	190
		8.3.2 Artificial Atoms	194
		8.3.3 Superconducting Qubits	195
	Problems		199
	References		201
9	**Computare Errare Est: Quantum Error Correction**		203
	9.1	Introduction	203
	9.2	Quantum Error Correction	206
		9.2.1 Phase Flip Errors	209
	9.3	The Shor Code	211
	9.4	Stabilizers	213
		9.4.1 A Short Introduction to the Pauli Group	214
		9.4.2 Stabilizer Analysis of the Shor Code	218
	9.5	Stabilizers II	220
		9.5.1 Review of Coding Terminology	220
		9.5.2 Single-Qubit System	221
		9.5.3 Two-Qubit System	222

		9.5.4	Three-Qubit System	222
		9.5.5	The Five-Qubit Code	224
		9.5.6	The Seven-Qubit Steane Code	227
		9.5.7	Surface Codes	229
		9.5.8	Fault Tolerant Computing and the Threshold Theorem	234
	Problems			239
	References			239
10	**Adiabatic Quantum Computing**			241
	10.1	Introduction		241
		10.1.1	The Quantum Adiabatic Theorem	242
		10.1.2	The AQS Strategy	243
		10.1.3	Model System	244
		10.1.4	Avoided Crossings	249
		10.1.5	The Grover Algorithm	252
		10.1.6	Summary	256
	10.2	What About the Phase? Exploring Holonomic Quantum Computing		256
		10.2.1	The Rabi Model	259
		10.2.2	Geometric Phase as a Holonomy	260
		10.2.3	Berry's Phase	262
		10.2.4	Non-Abelian Phases	264
		10.2.5	From Phases to Forces	264
		10.2.6	A Topological Rabi Model	269
		10.2.7	Outlook	271
	Problems			272
	References			272

Index 275

A Quantum Mechanic's Toolbox

Abstract

The concepts of bit and qubit, the fundamental units of information in classical and quantum computing respectively, are introduced. We discuss features of the binary number system, linear vector, and Hilbert spaces. We learn how to manipulate qubits and introduce Dirac's bra-ket formalism to facilitate operations in Hilbert space. Scalar, direct, and outer products of bra-kets in multi-qubit systems are introduced. We define Hermitian and unitary operators in a 2^n-dimensional Hilbert space and use the bra-ket formalism to construct them. I summarize the foundational postulates of quantum mechanics, as espoused by the Copenhagen interpretation, for a finite set of qubits.

1.1 Bits and Qubits

The basic unit of information is called a *bit*. It represents a binary-valued quantity such as a yes/no answer to a question, the position of the toggle for an on/off switch, or a stop/go decision. For example, a light bulb can either be in the on or off state and thus serve as a storage device for a single bit of information. All digital computing machines, no matter how large and complex, are constructed from indivisible bits. Typically, the integers 0 and 1 denote the value of a bit. The *qubit* is a similar but distinct concept. I will elaborate on the difference between the two in subsequent discussions, but for now, we differentiate qubits from bits by a simple change in notation so that

$$0 \to |0\rangle \tag{1.1}$$
$$1 \to |1\rangle$$

where $|0\rangle, |1\rangle$ are the two possible states of the qubit.

© The Author(s), under exclusive license to Springer Nature Switzerland AG 2025
B. Zygelman, *A First Introduction to Quantum Computing and Information*,
Undergraduate Topics in Computer Science,
https://doi.org/10.1007/978-3-031-66425-0_1

1.1.1 Binary Arithmetic

The values 0, 1 of a bit form the *letters* of an alphabet that we call the binary number system. Just as words in the English language are concatenations of the 26 letters that comprise the western alphabet, so can a two-character bit alphabet form "words". The Morse code, where the "dash" and "dot" intervals are the two letters of the alphabet, is a familiar example of that proposition. The "book" of life written on a DNA polymer has four letters in its alphabet.

Bits are also used to represent numbers. Obviously, the integer 0 is represented by the bit whose state is 0 and the number 1 by the bit whose value is 1. How about the integers 2, 3 … ? Consider the string

$$1100010$$

which is interpreted by the following *algorithm*,

$$1100010 \rightarrow 1 \times 2^6 + 1 \times 2^5 + 0 \times 2^4 + 0 \times 2^3 + 0 \times 2^2 + 1 \times 2^1 + 0 \times 2^0$$

and has the value 98 in the base ten system. Each entry in the string represents a coefficient of consecutive powers of 2^n, where n is an integer. Note that

$$98 = 0 \times 10^2 + 9 \times 10^1 + 8 \times 10^0$$

conforms to this algorithm but powers of 10^n replace 2^n, and the coefficients of each term are the symbols 0, 1, 2…9.

Imagine a set, or *register*, of five light bulbs. As each light bulb can store a bit of information, a little thought shows that this register can store one of $2^5 = 32$ integers at any single instant of time. For example, reverting to qubit notation, the number 26 is represented by the physical state

$$|11010\rangle.$$

In this notation, the first light bulb, starting from the right-hand side (r.h.s), is off, the second one is on, the third is off, while the fourth and fifth bulbs are in the on position. Consider the following expression

$$|11010\rangle + |00101\rangle. \tag{1.2}$$

What should the + symbol signify here and how can we interpret this construct? At first sight, it might seem reasonable to define it as the arithmetic operation of addition so that $|11010\rangle + |00101\rangle$ is equal to $|11111\rangle$ which represents the integer 31 in binary form. It is apparent, under scrutiny, that this definition is unsatisfactory for the following reason. We agreed that $|11010\rangle$ represents a physical state in which the light bulbs have the corresponding on-off values at a single instance of time. Thus the expression $|11010\rangle + |01001\rangle$ suggests a register of light bulbs simultaneously in two different configurations at the same time, an absurd proposition and so construct (1.2) appears to be meaningless.

Let's posit a five-qubit register comprised of the quantum mechanical analog of a light bulb, that I call a $q$bulb. Because atoms/ions obey the laws of quantum mechanics (QM), it is plausible to define a $q$bulb register as an array of five atoms, each of which can be toggled on and off, in analogy with a light bulb.[1] A more precise definition is forthcoming in later chapters. As a quantum system, this array of atoms/ions must obey the postulates of the quantum theory. Within this theory, we will show how to make sense of expression (1.2). Before elaborating on this statement, let's first embark on a short mathematical detour.

1.2 A Short Introduction to Linear Vector Spaces

Consider a set of objects $\alpha, \beta, \gamma \cdots$. We say that these objects belong to a *linear vector space V* provided that

(i) There exists an operation, which we denote by the $+$ sign, so that if α, γ are any two members of the vector space V then so is the quantity $\alpha + \gamma$.
(ii) For scalar c, there exists a scalar multiplication operation defined so that if β is a vector in V then so is $c\beta = \beta c$. If a, b are scalars $a b \beta = a(b\beta)$.
(iii) Scalar multiplication is distributive, i.e. $c(\alpha + \beta) = c\alpha + c\beta$, also for scalar a, b, $(a+b)\alpha = a\alpha + b\alpha$.
(iv) The $+$ operation is associative, i.e. $\alpha + (\beta + \gamma) = (\alpha + \beta) + \gamma$.
(v) The $+$ operation is commutative, i.e. $\alpha + \beta = \beta + \alpha$.
(vi) There exists a null vector 0 which has the property $0 + \alpha = \alpha$ for very vector α in V.
(vii) For every α in V there exists an inverse vector $-\alpha$ that has the property

$$\alpha + -\alpha = 0.$$

Numerous mathematical structures form vector spaces. Perhaps the most familiar are the vectors that define a direction in space. Convince yourself that the set of numbers on the real number line form a vector space. Do the set of all integers constitute a vector space?

Armed with these definitions, we are ready to tackle, in the context of the five-$q$bulb register, the postulates of quantum mechanics. Because our discussion here is limited to the five-$q$bulb register, we stress the tentative nature of these postulates by labeling them as meta-postulates or m-postulates for short.

m-Postulate I | Following a measurement (observation) of this quantum register, we observe only one out of the 32 possible $q$bulb-on/off configurations. Immediately after that measurement the register is found in the state corresponding to the values measured.

---

[1] David Wineland [7], shows that such a notion is not as fanciful as might first appear.

For example, if the measurement resulted in the first q bulb on, the second off, the last three on, the state corresponding to this measurement is $|11101\rangle$, and the postulate asserts that the system occupies this state immediately after measurement. The verity of this postulate appears to be self-evident. It almost seems not worthwhile stating as it surely applies to the classical counterpart of this system. After introducing the second postulate we will appreciate its significance and implications.

> m-Postulate II a The possible 32 states of our qubit register are vectors in a linear vector space.

According to this postulate, and property (i) of the itemized list above, expression (1.2) must also be a vector in this space. Indeed so is

$$|\Phi\rangle \equiv$$
$$|00000\rangle + |00001\rangle + |00010\rangle + |00011\rangle + |00100\rangle + |00101\rangle + |00110\rangle + |00111\rangle$$
$$|01000\rangle + |01001\rangle + |01010\rangle + |01011\rangle + |01100\rangle + |01101\rangle + |01110\rangle + |01111\rangle$$
$$|10000\rangle + |10001\rangle + |10010\rangle + |10011\rangle + |10100\rangle + |10101\rangle + |10110\rangle + |10111\rangle$$
$$|11000\rangle + |11001\rangle + |11010\rangle + |11011\rangle + |11100\rangle + |11101\rangle + |11110\rangle + |11111\rangle,$$
(1.3)

as is any combination defined by the + operation. This property, that a quantum state $|\Psi\rangle$ can be expressed as a linear combination of other quantum states, is sometimes called the *superposition principle* (Fig. 1.1).

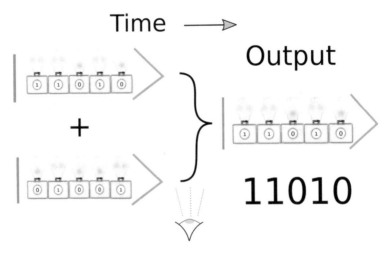

Fig. 1.1 The $|11010\rangle + |01001\rangle$ superposition state prior to measurement (left panel). The measured register value 11010 results in the collapse of the amplitude into state $|11010\rangle$. The normalization constant is not shown in this figure. The yellow bulbs are in the on-state, denoted by the binary digit 1

> **m-Postulate II b** A complete physical description of this quantum register is encapsulated by a vector $|\Psi\rangle$ in this vector space.

What is a complete physical description? A classical register requires an itemization of bulbs that are on/off to characterize its physical state. For the quantum version of this system, m-Postulate II states that an abstract quantity, a vector $|\Psi\rangle$, that is a linear combination of states in which the system finds itself after a measurement is made, defines the system. For example, suppose the complete physical description of our quantum register is given by the vector (state) (1.3), i.e. $|\Phi\rangle$. Construct $|\Phi\rangle$ suggests that all possible on/off configurations exist at the same time, a strange proposition that counters everyday experience. If we are going to make sense of this theory, the necessity for m-Postulate I becomes apparent. If $|\Phi\rangle$ does exist, m-Postulate I insures that we observe only one of the possible $q$bulb configurations. In addition, immediately after that observation $|\Phi\rangle$ "collapses" into that particular state. The above scenario should stimulate lots of questions, perhaps even healthy skepticism. m-Postulate II argues for a "reality" in which all possible outcomes exist at the same time. m-Postulate I grounds us because it prevents direct observation of this "reality". The theory as expressed by these postulates reminds us of the child's conundrum; is the moon there when we are not watching it? [5].

We have not yet addressed the question; which of the possible 32 configurations do we observe after a measurement on a system described by $|\Phi\rangle$? m-Postulate I guarantees that one of the configurations is found but says nothing about which one. To answer this question we look toward another postulate, sometimes called the *Born rule*, after *Max Born* one of the founders of the quantum theory. But first we need to provide additional structure to our linear vector space. That discussion leads us to consider a special type of vector space called a Hilbert space.

> Mathematica Notebook 1.1: Why Superposition? https://bernardzygelman.github.io

1.3 Hilbert Space

We pointed out that if $|\alpha\rangle$ is a vector so is $c|\alpha\rangle$ where c is a scalar quantity. In Hilbert space the scalar c is generally a complex number. Consider the following linear combination of n vectors

$$c_1|\alpha_1\rangle + c_2|\alpha_2\rangle + \ldots c_n|\alpha_n\rangle. \quad (1.4)$$

If this sum equates to the null vector 0 only if $c_1 = c_2 \cdots = c_n = 0$ then the set of vectors $|\alpha_1\rangle, |\alpha_2\rangle, \ldots |\alpha_n\rangle$ are said to be *linearly independent*. A space that admits n linear independent vectors, but not $n+1$, is called an n-dimensional space. In general, a Hilbert space allows infinite dimension [2,4] but we are primarily concerned, in this text, with Hilbert spaces spanned by a finite and denumerable set of basis vectors. The Hilbert space of the five-$q$bulb system has dimension 32.

1.3.1 Dirac's Bra-Ket Notation

The *inner, or scalar product* is a Hilbert space structure that provides a measure of the degree of "overlap" between two vectors. The dot product of two vectors in ordinary three-dimensional Euclidian space is an example of the latter. Here we will be a little bit more abstract and introduce the bra-ket notation to define an inner product. Indeed, we already made use of this notation in the previous sections where we employed the symbol $|...\rangle$ to denote a vector in Hilbert space. It is called a ket and was introduced by the brilliant twentieth century physicist *Paul Dirac*, an architect of the modern quantum theory. Throughout this text, we will employ ket notation to describe quantum states. For example, ket $|\Psi\rangle$ is a vector in Hilbert space where the symbol Ψ nested within the ket is an identification parameter.

Having established ket notation, we introduce a new linear vector space called *dual space*. For our purposes, it is convenient to think of dual space as a mirror image of the vectors (or kets) in Hilbert space. For example, the ket $|\alpha\rangle$ has a mirror counterpart in dual space. The label α should also parameterize it, but we cannot use ket notation as this vector "lives" in a different linear vector space. Dirac suggested the notation $\langle\alpha|$ to denote it and called the symbol $\langle...|$ a bra. Keeping in mind that every ket $|\beta\rangle$ has a corresponding bra $\langle\beta|$, we require that the ket

$$|\Psi\rangle = c_1|\alpha_1\rangle + c_2|\alpha_2\rangle + \ldots c_n|\alpha_n\rangle \quad (1.5)$$

has a bra counterpart. The rule for generating a bra $\langle\Psi|$ from ket $|\Psi\rangle$ is

$$\langle\Psi| = c_1^*\langle\alpha_1| + c_2^*\langle\alpha_2|, \ldots c_n^*\langle\alpha_n|. \quad (1.6)$$

The expansion coefficients for the bra vector are complex conjugates of the corresponding coefficients in ket space. Kets and bras are added to each of their kind, so the following expression

$$|\Psi\rangle + \langle\Phi|$$

is meaningless, but we are allowed to build additional structures by certain combinations of kets and bras. Two constructs which use kets and bras as its building blocks are expressions that have the form

$$\langle\Phi|\Psi\rangle \quad (1.7)$$

and

$$|\Psi\rangle\langle\Phi|. \quad (1.8)$$

The former (1.7), denotes an inner product and evaluates to a complex number. The latter expression (1.8) is called an *outer product*, and it is neither a scalar or vector. Later we show how it can be interpreted as an *operator*. Both structures enable various

1.3 Hilbert Space

transformations and manipulations of vectors in Hilbert space. Let's first discuss the inner, or scalar, product. The scalar quantity defined by expression (1.7) provides a measure of overlap between the vectors $|\Phi\rangle$, $|\Psi\rangle$.[2] According to Dirac notation, the inner product of two vectors in Hilbert space $|\Phi\rangle$, $|\Psi\rangle$ is formed by taking the dual of $|\Phi\rangle$, i.e. the bra $\langle\Phi|$, and placing it side-by-side, as if a Lego® block, with the ket $|\Psi\rangle$. We also require that

$$\langle\Phi|\Psi\rangle = \langle\Psi|\Phi\rangle^*, \quad (1.9)$$

i.e. they are complex conjugates of each other. According to condition (1.9), we recognize that $\langle\Psi|\Psi\rangle = \langle\Psi|\Psi\rangle^*$ and so the inner product of a vector with itself must be a real number. An additional postulate requires

$$\langle\Psi|\Psi\rangle \geq 0. \quad (1.10)$$

Property (1.10) allows us to affix a "length" $\sqrt{\langle\Psi|\Psi\rangle}$ to any vector, in particular, if $\langle\Psi|\Psi\rangle = 1$ that vector is said to be of unit length or *normalized*.

We now state Dirac's distributive axiom for inner products. Given two vectors $|\Psi\rangle = c_1|\alpha_1\rangle + c_2|\alpha_2\rangle$ and $|\Phi\rangle = d_1|\alpha_1\rangle + d_2|\alpha_2\rangle$ the inner product of $|\Psi\rangle$ with $|\Phi\rangle$ is

Dirac's Distributive Axiom for Scalar Products

$$\langle\Psi|\Phi\rangle = \left(c_1^*\langle\alpha_1| + c_2^*\langle\alpha_2|\right)\left(d_1|\alpha_1\rangle + d_2|\alpha_2\rangle\right) =$$
$$c_1^*d_1\langle\alpha_1|\alpha_1\rangle + c_1^*d_2\langle\alpha_1|\alpha_2\rangle + c_2^*d_1\langle\alpha_2|\alpha_1\rangle + c_2^*d_2\langle\alpha_2|\alpha_2\rangle. \quad (1.11)$$

The axiom shows how to distribute the inner product of a compound expression with all its components. Two vectors $|\alpha_1\rangle$ and $|\alpha_2\rangle$ are said to be *orthogonal* if they have a null inner product, i.e.

$$\langle\alpha_1|\alpha_2\rangle = \langle\alpha_2|\alpha_1\rangle = 0.$$

Suppose we have a set, in an n-dimensional Hilbert space, of n normalized linear independent vectors $|\alpha_1\rangle, |\alpha_2\rangle, ...|\alpha_n\rangle$ that are mutually orthogonal. That is if for each $|\alpha_i\rangle$

$$\langle\alpha_i|\alpha_j\rangle = \delta_{ij} \quad (1.12)$$

for all i, j (Here, δ_{ij} is the *Kronecker delta* which has the property $\delta_{ij} = 1$ if $i = j$, otherwise $\delta_{ij} = 0$). The set of vectors $|\alpha_i\rangle$ that satisfies those properties are said to form a *basis* for the vector space. You are probably familiar with the unit vectors, $\hat{\mathbf{i}}, \hat{\mathbf{j}}, \hat{\mathbf{k}}$ that constitute a basis for three-dimensional Euclidean space.

---

[2] In older textbooks the definition for the inner product is given by the alternative notation $(|\Phi\rangle, |\Psi\rangle)$ and does not rely on the use of dual space.

m-Postulate III a   The 32 vectors itemized on the right-hand side of (1.3) form a basis that spans the Hilbert space of the five-qubit register.

According to this postulate any of these vectors, for example, $|10110\rangle$, is orthogonal to all other vectors itemized on the r.h.s of (1.3) and each vector has unit length. Basis vectors are *orthonormal*, meaning that they are both orthogonal and have unit length. According to m-Postulate II b

$$|\Psi\rangle = \sum_{i=1}^{n} c_i |\alpha_i\rangle \qquad (1.13)$$

and if we require that $|\Psi\rangle$ has unit length, i.e. $\langle\Psi|\Psi\rangle = 1$, we find that

$$\langle\Psi|\Psi\rangle = \sum_{i=1}^{n}\sum_{j=1}^{n} c_i^* c_j \langle\alpha_i|\alpha_j\rangle = \sum_{i=1}^{n}\sum_{j=1}^{n} c_i^* c_j \delta_{ij} = \sum_{i=1}^{n} |c_i|^2 = 1. \qquad (1.14)$$

Taking the inner product of $|\alpha_m\rangle$ with $|\Psi\rangle$, we obtain

$$\langle\alpha_m|\Psi\rangle = \sum_{j=1}^{n} c_j \langle\alpha_m|\alpha_j\rangle = \sum_{j=1}^{n} c_j \delta_{mj} = c_m \qquad (1.15)$$

where we used the orthonormality condition (1.12).

Before proceeding further it's useful to consolidate our understanding of this formalism with more familiar examples from vector calculus. We are all accustomed to expressions like,

$$\vec{A} = A_x \hat{\mathbf{i}} + A_y \hat{\mathbf{j}} + A_z \hat{\mathbf{k}}.$$

It is a vector in 3-d Euclidian space expressed in terms of the unit basis vectors $\hat{\mathbf{i}}, \hat{\mathbf{j}}, \hat{\mathbf{k}}$. The latter are a set of orthonormal vectors since

$$\hat{\mathbf{i}} \cdot \hat{\mathbf{j}} = \hat{\mathbf{i}} \cdot \hat{\mathbf{k}} = \hat{\mathbf{j}} \cdot \hat{\mathbf{k}} = 0$$

and

$$\hat{\mathbf{i}} \cdot \hat{\mathbf{i}} = \hat{\mathbf{j}} \cdot \hat{\mathbf{j}} = \hat{\mathbf{k}} \cdot \hat{\mathbf{k}} = 1.$$

The scalars A_x, A_y, A_z are the components of vector $\vec{A}$ so that, using the orthonormality properties of the basis,

$$A_x = \vec{A} \cdot \hat{\mathbf{i}} \quad A_y = \vec{A} \cdot \hat{\mathbf{j}} \quad A_z = \vec{A} \cdot \hat{\mathbf{k}}.$$

Similarly, the Hilbert space vector

$$|\Psi\rangle = c_1 |\alpha_1\rangle + c_2 |\alpha_2\rangle + \cdots c_n |\alpha_n\rangle \qquad (1.16)$$

is a linear combination of orthonormal basis vectors $|\alpha_i\rangle$. Using orthonormality we find,

$$c_1 = \langle\alpha_1|\psi\rangle, \; c_2 = \langle\alpha_2|\psi\rangle, \; \ldots c_n = \langle\alpha_n|\psi\rangle. \qquad (1.17)$$

1.3 Hilbert Space

Table 1.1 Examples and comparison of three-dimensional Euclidean space structures with n-dimensional Hilbert space analogues

Structure	Euclidean space	Hilbert space
Basis expansion	$\vec{A} = A_x \hat{\mathbf{i}} + A_y \hat{\mathbf{j}} + A_z \hat{\mathbf{k}}$	$|\Psi\rangle = c_1|\alpha_1\rangle + c_2|\alpha_2\rangle + \cdots c_n|\alpha_n\rangle$
Inner product	$\vec{A} \cdot \vec{B}$	$\langle \Phi | \Psi \rangle$
Basis components	$A_x = \vec{A} \cdot \hat{\mathbf{i}}$, ...etc.	$c_i = \langle \alpha_i | \Psi \rangle \quad i = 1, 2 \ldots n$
Outer product	(dyadic) $\hat{\mathbf{j}}\hat{\mathbf{k}}$, ...etc.	$|\alpha_i\rangle \langle \alpha_j |$

The similarity of expressions (1.16), (1.17) with the corresponding relations for vector $\vec{A}$ should be apparent. In the next section we discuss and elaborate on the Hilbert space outer product (1.8), but before going there let's review an analogous construct, the *dyadic* in Euclidean space. For the space covered by the basis vectors $\hat{\mathbf{i}}, \hat{\mathbf{j}}, \hat{\mathbf{k}}$, a dyadic is a bilinear expression such as $\hat{\mathbf{i}}\hat{\mathbf{j}}$ or any of the other eight combinations of basis vectors. A dyadic is positioned either before, or after, a vector $\vec{A}$ so that

$$\hat{\mathbf{j}}\hat{\mathbf{k}}\vec{A} \equiv \hat{\mathbf{j}}(\hat{\mathbf{k}} \cdot \vec{A}) = \hat{\mathbf{j}} A_z$$
$$\vec{A}\hat{\mathbf{j}}\hat{\mathbf{k}} \equiv (\vec{A} \cdot \hat{\mathbf{j}})\hat{\mathbf{k}} = \hat{\mathbf{k}} A_y. \qquad (1.18)$$

Table 1.1 summarizes some similarities between vector calculus constructs in Euclidean space and the Hilbert space of a n-qubit system.

At this stage of the narrative, it is useful to introduce a new, more economical, notation to represent basis vectors. In this alternative notation $|00000\rangle \equiv |0\rangle_5$, $|00010\rangle \equiv |2\rangle_5$, $|01001\rangle \equiv |9\rangle_5$, etc., i.e., we label each basis ket by the Arabic numeral value of the binary number label. The subscript 5 reminds us that we are labeling a five-qubit register. Therefore $|\Phi\rangle = \sum_{i=0}^{31} |i\rangle_5$ and $\langle \Phi | \Phi \rangle = \sum_{i=0}^{31} \sum_{j=0}^{31} \langle i | j \rangle_5 = 32$ where the last identity follows from the fact that basis vectors are orthonormal. Because vectors in Hilbert space have a length, we attach an addendum to m-Postulate II b and require physical states to be of unit length, i.e. $\langle \Psi | \Psi \rangle = 1$. We are now in a position to state the Born rule for our five-qubit register.

> **m-Postulate III b (the Born rule)** If the five-qubit register is in state $|\Psi\rangle = \sum_{i=0}^{31} c_i |i\rangle_5$ then a measurement yields the $q$bulb configuration corresponding to one of the 32 states $|i\rangle_5$, with probability $p_i = |c_i|^2 = |\langle i | \Psi \rangle_5|^2$. The condition $\langle \Psi | \Psi \rangle = 1$, insures that $\sum_i p_i = 1$.

The Born rule addresses the question not answered by m-Postulate I. To explore and appreciate its import we invoke a hypothetical *gedanken experiment*.[3] Imagine 1000 experimenters who each have a five-$q$bulb register in their respective laboratories. Furthermore, assume that each system is described by an identical state vector $|\Psi\rangle$. Each $q$bulb configuration is measured and m-Postulate I asserts that a measurement finds one of 32 possible configurations. The scientists tabulate the results of their

---

[3] Thought experiment.

Table 1.2 Tally of results for the gedanken experiment

Measurement	00000	01000	10001	11000
# of observations	101	209	321	369

measurements in lab books and convene a meeting later in the day to compare the data. A possible tally might look like that given in Table 1.2. For this trial, the scientists observed only four configurations out of the possible 32. In summary, 101 physicists found all q bulbs (atoms/ions) in the off position corresponding to collapsed state $|0\rangle_5$, 209 found that only the second q bulb (from the left) is on, corresponding to collapsed state $|8\rangle_5$ and 321 scientists found a configuration in which the first and last bulbs are on, corresponding to collapsed state $|17\rangle_5$. Finally, 369 found that the two left-most q bulbs are on, corresponding to the state $|24\rangle_5$. The frequency interpretation of probability tells us that, after a large number N of trials, the probability of obtaining the i'th result p_i = (# of trials resulting in choice i)/N. Clearly, $\sum_i p_i = 1$.

> Mathematica Notebook 1.2: An Introduction to Probability Theory http://www.physics.unlv.edu/%7Ebernard/MATH_book/Chap1/chap1_link.html; See also https://bernardzygelman.github.io

With this tabulated data we venture an educated guess for state $|\Psi\rangle$. One reasonable hypothesis is

$$|\Psi\rangle = \sqrt{\frac{101}{1000}}|0\rangle_5 + \sqrt{\frac{209}{1000}}|8\rangle_5 + \sqrt{\frac{321}{1000}}|17\rangle_5 + \sqrt{\frac{369}{1000}}|24\rangle_5, \quad (1.19)$$

as it is consistent with the Born rule. However, it is not a unique choice. The expansion coefficients c_i are complex numbers and $|c_i|^2$, the probability measure, is un-changed if the coefficients are altered by an arbitrary phase β, i.e., $c_i \to \exp(i\beta)c_i$. For example, if the coefficient $c_0 = \sqrt{101/1000}$ of the first term in sum (1.19) is replaced by $-i\sqrt{101/1000}$ the probability distribution is unchanged. Also, the frequency interpretation does not guarantee those outcomes unless $N \to \infty$. So repeating the experiment with another set of 1000 trials, a small number could, because of statistical fluctuations, instantiate configurations that do not appear in Table (1.2).

m-Postulate-III b informs us that quantum mechanics is a probabilistic theory where full knowledge of the system, i.e., $|\Psi\rangle$, does not guarantee a definite outcome for a measurement. It does offer a probability distribution of outcomes for an ensemble of measurements. Knowing $|\Psi\rangle$ we can predict the relative outcome of any configuration, let's say 11100, over the other 31 possibilities. Born's rule and (1.15) tells us that evaluation of $|\langle 11100|\Psi\rangle|^2$ provides us with that information.

I live and work in Las Vegas, a city whose existence attests to the predictive power of probabilistic inference. With the Copenhagen interpretation of quantum mechanics, Born's rule illuminates the mystery behind expressions like $|10001\rangle + |10011\rangle$.

1.3 Hilbert Space

In that interpretation, there is no reality[4] where configurations 10001 and 10011 are simultaneously in "existence". Born's rule simply provides an operational algorithm for calculating probabilities of possible outcomes in an experiment. However, once a measurement is made, m-Postulate I requires that the system "collapses" into the basis state corresponding to that measurement outcome. Immediately after that measurement, since all $c_i = 0$, save the state j in which the system "collapses" to, a subsequent measurement has probability 1, or certainty, for that outcome.

1.3.2 Outer Products and Operators

In this section, we use Dirac's bra-ket formalism to construct an outer product. We first state Dirac's distributive axiom for outer products. Given states $|\Psi\rangle = c_1|\alpha_1\rangle + c_2|\alpha_2\rangle$ and $|\Phi\rangle = d_1|\alpha_1\rangle + d_2|\alpha_2\rangle$ the outer product of $|\Psi\rangle$ with $|\Phi\rangle$ is

$\boxed{\text{Dirac's Distributive Axiom for Outer Products}}$

$$|\Psi\rangle\langle\Phi| = \Big(c_1|\alpha_1\rangle + c_2|\alpha_2\rangle\Big)\Big(d_1^*\langle\alpha_1| + d_2^*\langle\alpha_2|\Big) = \qquad (1.20)$$
$$c_1 d_1^* |\alpha_1\rangle\langle\alpha_1| + c_1 d_2^* |\alpha_1\rangle\langle\alpha_2| + c_2 d_1^* |\alpha_2\rangle\langle\alpha_1| + c_2 d_2^* |\alpha_2\rangle\langle\alpha_2|.$$

Consider the outer product $\mathbf{X} \equiv |\Phi\rangle\langle\Psi|$. According to Dirac notation we are allowed to place it in front of a ket $|\Gamma\rangle$, or to the right of a bra $\langle\Gamma|$, i.e. the constructs $\mathbf{X}|\Gamma\rangle$ and $\langle\Gamma|\mathbf{X}$ are valid expressions in Hilbert and dual space respectively. However, the expressions $|\Gamma\rangle\mathbf{X}$, $\mathbf{X}\langle\Gamma|$ are illegal.

Dirac's associative axiom for outer products states that

$\boxed{\text{Dirac's Associative Axiom for Outer Products}}$ For an outer product $\mathbf{X}$ and ket $|\Gamma\rangle$

$$\mathbf{X}|\Gamma\rangle = \Big(|\Phi\rangle\langle\Psi|\Big)|\Gamma\rangle = |\Phi\rangle\Big(\langle\Psi|\Gamma\rangle\Big) = c|\Phi\rangle$$
where $c \equiv \langle\Psi|\Gamma\rangle$, Also
$$\langle\Gamma|\mathbf{X} = \langle\Gamma|\Big(|\Phi\rangle\langle\Psi|\Big) = \Big(\langle\Gamma|\Phi\rangle\Big)\langle\Psi| = \langle\Psi|d$$
$$d \equiv \langle\Gamma|\Phi\rangle. \qquad (1.21)$$

According to (1.21), placing an outer product to the left of a ket, "un-hinges" the bra vector $\langle\Psi|$ from $\mathbf{X}$ and "locks" it onto the right-most ket $|\Gamma\rangle$ forming the scalar product, i.e., the complex number $\langle\Psi|\Gamma\rangle$. The result is a new ket $c|\Phi\rangle$. In short, outer product $\mathbf{X}$ "operates" on vector $|\Gamma\rangle$ and transforms it to vector $c|\Phi\rangle$ in Hilbert space. When operating on bra vectors, it plays a similar role in dual space. Evidently, outer

---

[4] There exist alternative descriptions, including the many-worlds interpretation of QM [1,6] and the consistent histories approach [3,6], but in this monograph we adhere to orthodox dogma.

products play an essential role in facilitating transformations of vectors in Hilbert space. Thus the outer product allows us to construct *operators* in Hilbert space. We can think of operators as objects that map vectors to other vectors in Hilbert space.

According to (1.21) we find that the dual of the transformed vector $\mathbf{X}\,|\Gamma\rangle$ is $\langle\Phi|\,c^*$ where $c^* = \langle\Gamma|\Psi\rangle$. So the dual of $\mathbf{X}|\Gamma\rangle \neq \langle\Gamma|\,\mathbf{X}$. There is an outer product called $\mathbf{X}^\dagger$ that has the following property.

Definition 1.1 The adjoint operation For operator $\mathbf{X}$ and ket $|\Phi\rangle$, the dual of $\mathbf{X}\,|\Phi\rangle$ is given by the expression $\langle\Phi|\,\mathbf{X}^\dagger$ for all $|\Phi\rangle$. $\mathbf{X}^\dagger$ is called the *adjoint*, or conjugate transpose, operator to $\mathbf{X}$.

Definition 1.2 Hermitian operators Operators $\mathbf{X}$ that have the property

$$\mathbf{X} = \mathbf{X}^\dagger \tag{1.22}$$

are called Hermitian, or self-adjoint, operators.

Definition 1.3 Unitary operators Operator $\mathbf{U}$ is a unitary operator, if

$$\mathbf{U}\,\mathbf{U}^\dagger = \mathbf{U}^\dagger\,\mathbf{U} = \mathbb{1} \tag{1.23}$$

where $\mathbb{1}$ is a unit operator, i.e. $\mathbb{1}|\Psi\rangle = |\Psi\rangle$ for all $|\Psi\rangle$ in Hilbert space.

Both Hermitian and unitary operators play a central role in quantum computing and information (QCI) applications.

We know that operators map, or transform, a vector in Hilbert space to another vector in that same space. A special class of mappings, generated by operator $\mathbf{X}$, have the following property. For some vectors $|\Phi\rangle$,

$$\mathbf{X}\,|\Phi\rangle = \phi\,|\Phi\rangle \tag{1.24}$$

where ϕ is a scalar. An equation of this type is called an *eigenvalue* equation. The vector $|\Phi\rangle$ is called an *eigenvector*, and the constant ϕ is called the eigenvalue associated with that eigenvector. Two theorems [10] that concern important properties of Hermitian operators, are

Theorem 1.1 *The eigenvalues of a Hermitian operator are real numbers.*

Theorem 1.2 *If the eigenvalues of Hermitian operator are distinct, then the corresponding eigenvectors are mutually orthogonal. If some of the eigenvalues are not distinct, or* degenerate, *then a linear combination of that subset of eigenvectors can be made to be mutually orthogonal.*

Let's investigate how some of these concepts relate to our five-qubit register. Because we are performing all operations in 32-dimensional Hilbert space,

1.3 Hilbert Space

kets $|j\rangle_5$ will simply be replaced with the symbol $|j\rangle$. Consider operator $\mathbf{N}_6 \equiv |00110\rangle\langle 01100|$, or in the alternate notation $|6\rangle\langle 6|$. First, note that $\mathbf{N}_6$ is Hermitian, and according to Theorem 1.1, its eigenvalues are real numbers. Indeed, the eigenvalues for $\mathbf{N}_6$ are 1 and 0, associated with eigenvectors $|6\rangle$ and kets $|j\rangle$ for $j \neq 6$ respectively.

Proof Using Dirac's associative axiom

$$\mathbf{N}_6 |6\rangle = (|6\rangle\langle 6|) |6\rangle = |6\rangle (\langle 6|6\rangle) = 1 |6\rangle$$

and

$$\mathbf{N}_6 |j\rangle = |6\rangle (\langle 6|j\rangle) = 0 |j\rangle \quad j \neq 6$$

$\square$

where we have used the orthonormality property of the vectors $|j\rangle$. Note that the eigenvalue 0, is associated with each of the $|j \neq 6\rangle$ eigenvectors. It is 31-fold degenerate because each $|j \neq 6\rangle$ share the same eigenvalue. Convince yourself that the eigenvectors of $\mathbf{N}_6$ also satisfy Theorem 1.2.

Let's define yet another operator, $\mathbf{N} \equiv \sum_{j=0}^{31} j \, (|j\rangle\langle j|)$. Here, outer product $(|j\rangle\langle j|)$ is multiplied by the integer j that labels each ket. Now

$$\mathbf{N}^\dagger = \sum_{j=0}^{31} j \, (|j\rangle\langle j|)^\dagger = \sum_{j=0}^{31} j \, (|j\rangle\langle j|) = \mathbf{N} \qquad (1.25)$$

and is therefore Hermitian.

Its eigenvectors $|j\rangle$ are labeled by j which happens to be an eigenvalue of $\mathbf{N}$. According to the Born rule a measurement, with device $\mathbf{N}$, of the $q$bulb register in state $|\Psi\rangle$, yields the probability $p_j = |\langle j|\Psi\rangle|^2$ that the register is found in the j'th configuration. In other words, p_j is calculated by taking the inner product of the ket that is an eigenvector of $\mathbf{N}$, that is $|j\rangle$ with the system state vector $|\Psi\rangle$. The eigenvalue associated with this ket is the measured configuration index. For this reason, operator $\mathbf{N}$ is a configuration, or *occupation number, measurement operator*, as its eigenvalues identify each possible configuration label revealed by a measurement.

1.3.3 Direct and Kronecker Products

Up to this point, we introduced and discussed almost all foundational postulates, theorems, and definitions that are needed to describe a five-$q$bulb quantum register. To explore Hilbert spaces beyond the five-qubit example, we need to introduce an additional tool, that of the *direct, or tensor*, product. When we think of points in a plane as an ordered pair (x, y), we recognize it as the pairing of a x-axis coordinate with a coordinate on the y-axis number line. The pair is called a *Cartesian product* of two lower dimensional vector spaces, i.e., the real x and y-axis number lines. In the same way, we build higher-dimensional Hilbert spaces from direct products of single qubit Hilbert spaces. Below, we will show how to construct a register that contains an arbitrary number of qubits from a constituent, lower dimensional space via direct products.

First, let's discuss in more detail a single qubit system of Hilbert space dimension 2. The kets $|0\rangle_1$ and $|1\rangle_1$ are possible basis vectors as we require them to be linearly independent and orthonormal. The subscript reminds us that ket $|0\rangle_1$ differs from ket $|0\rangle_5$. The latter represents a five-qubit register in which all $q$bulbs are in the off-position, whereas the former denotes the off-position of a single $q$bulb. In the subsequent discussion, we will assume, unless otherwise stated, that vectors $|0\rangle, |1\rangle$ are single-qubit vectors and ignore the subscript. Any state vector $|\Psi\rangle$ in this Hilbert space can be expressed as a linear combination of those two basis kets. A single $q$bulb (e.g., a two-state atom) serves as a physical realization of this Hilbert space.

Definition 1.4 | The direct product. | Given kets $|a\rangle, |b\rangle$ their direct product is given by the expression

$$|a\rangle \otimes |b\rangle.$$

Suppose $a, b \in 0, 1$, i.e. both $|a\rangle, |b\rangle$ are single qubit basis kets. Itemize all distinct combinations of indices a, b to get

$$|0\rangle \otimes |0\rangle, \quad |0\rangle \otimes |1\rangle, \quad |1\rangle \otimes |0\rangle, \quad |1\rangle \otimes |1\rangle. \tag{1.26}$$

Note that the direct product of $|a\rangle$ with $|b\rangle$ is not the same as the direct product of $|b\rangle$ with $|a\rangle$, i.e.

$$|a\rangle \otimes |b\rangle \neq |b\rangle \otimes |a\rangle.$$

Without proof we state the following.

Theorem 1.3 *Given a Hilbert space of dimension d that is spanned by basis vectors $|b\rangle$, and the single qubit vector $|a\rangle$ for $a \in 0, 1$, the direct products $|a\rangle \otimes |b\rangle$ for all a, b are basis vectors in a Hilbert space of dimension $2d$. The dual vector for*

$$c |a\rangle \otimes |b\rangle$$

is

$$c^* \langle b| \otimes \langle a|$$

for all values of a, b.

Note the ordering of the bra vectors from that of the corresponding kets. According to this theorem vectors $|a\rangle \otimes |b\rangle$ allow inner products with all vectors in the direct product Hilbert space. Let's assume that $|a\rangle, |b\rangle, |c\rangle, |d\rangle$ are single qubit states. Given $|\Psi\rangle = c_1|a\rangle \otimes |b\rangle + c_2|c\rangle \otimes |d\rangle$, and $|\Phi\rangle = d_1|a\rangle \otimes |b\rangle + d_2|c\rangle \otimes |d\rangle$ the inner product $\langle\Psi|\Phi\rangle$ is obtained by application of Dirac's axiom in the following manner

$$\left(c_1^*\langle b| \otimes \langle a| + c_2^*\langle d| \otimes \langle c|\right)\left(d_1|a\rangle \otimes |b\rangle + d_2|c\rangle \otimes |d\rangle\right) =$$
$$c_1^*d_1\langle b|b\rangle\langle a|a\rangle + c_1^*d_2\langle b|d\rangle\langle a|c\rangle + c_2^*d_1\langle c|a\rangle\langle d|b\rangle + c_2^*d_2\langle c|c\rangle\langle d|d\rangle =$$
$$c_1^*d_1 + c_2^*d_2. \tag{1.27}$$

1.3 Hilbert Space

where the last identity follows from the orthonormality property of the single-qubit kets. Using (1.27) we find that the four vectors itemized in (1.26) are also mutually orthonormal, e.g. $(\langle 0| \otimes \langle 0|)(|0\rangle \otimes |0\rangle) = \langle 0|0\rangle \langle 0|0\rangle = 1$, $(\langle 0| \otimes \langle 0|)(|0\rangle \otimes |1\rangle) = \langle 0|0\rangle \langle 0|1\rangle = 0$, etc. Because direct products are a common feature in our subsequent discussions, use of the $\otimes$ symbol becomes somewhat cumbersome. Let's drop the $\otimes$ symbol and re-express the four kets in (1.26) as follows

$$|00\rangle, \quad |01\rangle, \quad |10\rangle, \quad |11\rangle. \tag{1.28}$$

There should be no ambiguity with this notation as long as we accept an implicit understanding that it is shorthand for the direct product of two qubits. In this notation $|\Psi\rangle = c_1|ab\rangle + c_2|cd\rangle$, and $|\Phi\rangle = d_1|ab\rangle + d_2|cd\rangle$. The inner product $\langle \Psi|\Phi\rangle$ is given by the following expression

$$\left(c_1^* \langle ba| + c_2^* \langle dc| \right) \left(d_1|ab\rangle + d_2|cd\rangle \right)$$

and, when expanded, evaluates to $c_1^* d_1 + c_2^* d_2$, in harmony with result (1.27).

The kets itemized in (1.28) reminds us of the notation used to describe five-qubit kets. In that case, each ket was labeled by its binary number representation. Kets (1.28) are similarly labeled, except that the dimension of Hilbert space is 2^2 and so the kets are enumerated accordingly. In analogy with the five-qubit notation, we could label the basis vectors in (1.28) as

$$|0\rangle_2, \quad |1\rangle_2, \quad |2\rangle_2, \quad |3\rangle_2. \tag{1.29}$$

The subscript refers to the fact that the number of qubits $n = 2$, and the dimension of Hilbert space is $d = 2^n$. With this notation, it is straightforward to construct a higher dimensional Hilbert space from qubit constituents. For example, consider the direct product of qubit $|a\rangle$, $a \in 0, 1$ with kets (1.26). It should be clear, if we make use of Theorem 1.3, that this product generates the basis kets

$$|0\rangle_3, \quad |1\rangle_3, \quad |2\rangle_3, \quad |3\rangle_3,$$
$$|4\rangle_3, \quad |5\rangle_3, \quad |6\rangle_3, \quad |7\rangle_3 \tag{1.30}$$

in a $d = 2^3$ dimensional Hilbert space. Generalization to direct products of n-qubits follows. We now recognize that the Hilbert space of our five-$q$bulb register can be expressed as a direct product of 5 individual qubits. For a modest number of n-qubits the dimension of Hilbert space $d = 2^n$ is huge.

We can think of direct products as qubit entries in different slots. For example, the register $|100\rangle$, or $|4\rangle_3$, is a direct product (starting from the right and going left) of ket $|0\rangle$ in slot 1, ket $|0\rangle$ in slot 2, and ket $|1\rangle$ in slot 3. According to Theorem 1.3 the bra vector associated with this ket is $\langle 001|$. So now the slot order for bra vectors is reversed from that of ket space, i.e., going from left to right bra $\langle 0|$ is in slot 1, $\langle 0|$ is in slot 2, and $\langle 1|$ is in slot 3. If we take an inner product of this three-qubit vector with $|abc\rangle$ ($a, b, c \in 0, 1$) then, according to (1.27), the inner product

$$\langle 001|abc\rangle = \langle 1|a\rangle \langle 0|b\rangle \langle 0|c\rangle.$$

Each qubit ket in a given slot "hooks" up with the bra in that same slot.

Using the slot analogy we can also express multi-qubit operators, such as $\mathbf{X}_6$ and $\mathbf{N}$, introduced in the previous section, as direct products of single qubit operators. Operators, or outer products, are not vectors and so we should not use the same symbol to define products of the latter. Instead, we employ the symbol $\tilde{\otimes}$ to denote products of operators. Given $|ab\rangle \equiv |a\rangle \otimes |b\rangle$ and $|cd\rangle \equiv |c\rangle \otimes |d\rangle$.

$$|ab\rangle\langle dc| = (|a\rangle \otimes |b\rangle)(\langle d| \otimes \langle c|) \equiv |a\rangle\langle c|\tilde{\otimes}|b\rangle\langle d| \qquad (1.31)$$

The symbol $\tilde{\otimes}$ is called the *Kronecker product*. In this expression, the outer product $|b\rangle\langle d|$ "operates" only on a qubit in slot 1, whereas $|a\rangle\langle c|$ operates on qubits in slot 2. For example, consider the two-qubit state $|\Psi\rangle = c_1|01\rangle + c_2|10\rangle$ then, according to Dirac's axiom,

$$(|ab\rangle\langle dc|)|\Psi\rangle = c_1(|a\rangle\langle c|\tilde{\otimes}|b\rangle\langle d|)|01\rangle + c_2(|a\rangle\langle c|\tilde{\otimes}|b\rangle\langle d|)|10\rangle =$$
$$|ab\rangle\Big(c_1 \langle c|0\rangle\langle d|1\rangle + c_2\langle c|1\rangle\langle d|0\rangle\Big) = |ab\rangle\Big(c_1 \langle cd|10\rangle + c_2 \langle cd|01\rangle\Big). \quad (1.32)$$

The two-qubit operator $|ab\rangle\langle dc|$, when operating on two-qubit state $|\Psi\rangle$, projects the latter into vector $c\,|ab\rangle$ where the constant $c = c_1 \langle cd|10\rangle + c_2 \langle cd|01\rangle$.

In the next chapter, we introduce matrix representations of states and operators and, in that framework, we will find that the operation $\tilde{\otimes}$ is equivalent to the direct product operation $\otimes$. With definitions for direct products of kets and the corresponding Kronecker products of operators, we generalize and re-frame the postulates itemized in the previous sections for an arbitrary n-qubit register.

Postulate I — Kets $|0\rangle, |1\rangle$ constitute a basis for the qubit Hilbert space. An n-qubit register is spanned by basis vectors that are direct products of n-qubits $|a\rangle \otimes |b\rangle \otimes |c\rangle \ldots |n\rangle$, where $a, b, c \cdots n \in 0, 1$.

Postulate II — A full description of the system is encapsulated by a vector $|\Psi\rangle$, of unit length, in this 2^n dimensional Hilbert space.

Postulate III (Born's rule) — The act of measurement associated with Hermitian operator $\mathbf{A}$ results in one of its eigenvalues. The probability for obtaining a nondegenerate eigenvalue a is given by the expression $|\langle a|\Psi\rangle|^2$ where $|a\rangle$ is an eigenvector of $\mathbf{A}$ that corresponds to eigenvalue a. If the eigenvalue a is degenerate, the probability to find that value is $\sum_i |\langle a_i|\Psi\rangle|^2$ where the sum is over all i in which $a_i = a$.

Postulate IV (collapse hypothesis) — Immediately after measurement by $\mathbf{A}$ with result a, the system is described, up to an undetermined phase, by state vector $|a\rangle$. If a is degenerate, the system is in a linear combination of the corresponding eigenvectors.

1.3 Hilbert Space

We now recognize that a five-$q$bulb state, lets say $|22\rangle = |10110\rangle$ is shorthand for the direct product $|1\rangle \otimes |0\rangle \otimes |1\rangle \otimes |1\rangle \otimes |0\rangle$. It is an eigenstate, with eigenvalue $j = 22$, of operator $\mathbf{N}_{22} = 22\,|10110\rangle\langle 01101|$, equivalent to the Kronecker product

$$\mathbf{N}_{22} = 22\,|1\rangle\langle 1|\tilde{\otimes}|0\rangle\langle 0|\tilde{\otimes}|1\rangle\langle 1|\tilde{\otimes}|1\rangle\langle 1|\tilde{\otimes}|0\rangle\langle 0|. \tag{1.33}$$

Let's define the single qubit operators $\mathbf{n} \equiv |1\rangle\langle 1|$ and $\mathbf{1} \equiv |0\rangle\langle 0| + |1\rangle\langle 1|$. Since $\mathbf{n}|0\rangle = 0$, $\mathbf{n}|1\rangle = 1|1\rangle$, $\mathbf{n}$ is an occupation number operator for a single qubit, whereas $\mathbf{1}$ is an identity operator. It has the property $\mathbf{1}|0\rangle = |0\rangle$, $\mathbf{1}|1\rangle = |1\rangle$, and $(\mathbf{1} - \mathbf{n}) = |0\rangle\langle 0|$. With these definitions we re-express (1.33) as

$$\mathbf{N}_{22} = 22\,\mathbf{P}_{22}$$
$$\mathbf{P}_{22} = \mathbf{n}\tilde{\otimes}(\mathbf{1}-\mathbf{n})\tilde{\otimes}\mathbf{n}\tilde{\otimes}\mathbf{n}\tilde{\otimes}(\mathbf{1}-\mathbf{n}) =$$
$$\mathbf{n}\tilde{\otimes}\mathbf{1}\tilde{\otimes}\mathbf{n}\tilde{\otimes}\mathbf{n}\tilde{\otimes}\mathbf{1} - \mathbf{n}\tilde{\otimes}\mathbf{1}\tilde{\otimes}\mathbf{n}\tilde{\otimes}\mathbf{n}\tilde{\otimes}\mathbf{n} - \mathbf{n}\tilde{\otimes}\mathbf{n}\tilde{\otimes}\mathbf{n}\tilde{\otimes}\mathbf{n}\tilde{\otimes}\mathbf{1} + \mathbf{n}\tilde{\otimes}\mathbf{n}\tilde{\otimes}\mathbf{n}\tilde{\otimes}\mathbf{n}\tilde{\otimes}\mathbf{n} \tag{1.34}$$

The operator $\mathbf{P}_{22}$ is called a *projection operator* and has the properties

$$\mathbf{P}_{22}\mathbf{P}_{22} = \mathbf{P}_{22},$$

$$\mathbf{P}_{22}|22\rangle = |22\rangle \quad \text{and for } j \neq 22 \quad \mathbf{P}_{22}|j\rangle = 0.$$

Throughout this chapter, we focused primarily on one type of measurement, that for the configuration occupation number j of a register. Postulate III allows other measurement devices as there are no restrictions on operator $\mathbf{A}$ as long as it is Hermitian. So what other measurement possibilities arise? An obvious choice is a measurement for the occupation number of a $q$bulb in a given slot, disregarding the values for the other $q$bulbs. For example, if we are only interested on the state of the $q$bulb in slot 3 we can define the measurement operator

$$\mathbf{P}_3 \equiv \mathbf{1}\tilde{\otimes}\mathbf{1}\tilde{\otimes}\mathbf{n}\tilde{\otimes}\mathbf{1}\tilde{\otimes}\mathbf{1}$$

which also happens to be a projection operator. For any basis ket $|j\rangle$, $\mathbf{P}_3|j\rangle = |j\rangle$, if slot 3 is in the on position and it is zero otherwise. However, there are many kets with this property including

$$|00100\rangle, |00101\rangle, |00110\rangle, |00111\rangle$$
$$|01100\rangle, |01101\rangle, |01110\rangle, |01111\rangle$$
$$|10100\rangle, |10101\rangle, |10110\rangle, |10111\rangle$$
$$|11100\rangle, |11101\rangle, |11110\rangle, |1111\rangle.$$

Each of these kets is an eigenstate of $\mathbf{P}_3$ with eigenvalue 1. They are 16-fold degenerate and Postulate III tells us that the probability to find the $q$bulb of slot 3 to be in the on position is $\sum_j |\langle j|\Psi\rangle|^2$ where the sum is over all states $|j\rangle$ itemized above. Obviously, because there are multiple qubits in a register, many possibilities for occupation type measurements arise. For example, we could measure the state of only a single $q$bulb in a given slot, or permutations for several $q$bulbs at a time. The various operators associated with all such measurements can ultimately be expressed, as in (1.34), by sums of direct products involving the single qubit operators $\mathbf{1}$, and $\mathbf{n}$. Are there other possibilities not involving occupation type measurements? To investigate this question let's consider the simplest system, the single $q$bulb or qubit. The $q$bulb can either be on or off and one might, incorrectly, conclude that the measurement operator **n** exhausts all possibilities in this Hilbert space. But consider the following operator

$$\mathbf{A} \equiv |0\rangle\langle 1| + |1\rangle\langle 0|.$$

A is Hermitian, i.e., $\mathbf{A} = \mathbf{A}^\dagger$ and so it is a measurement device candidate. But what does **A** measure since it is not expressed solely by **n** and **1**? It turns out that, for a real qubit, there does exist a measurement device associated with this operator. We will discuss it, and others, in detail in the next chapter. That discussion will force us to reconsider the simple $q$bulb analogy employed in this chapter. Instead of a quantum light bulb having two distinct properties, we will learn that a real qubit is multi-faceted and forces us to broaden our conception of what it means for the $q$bulb to be in the on or off state. Nevertheless the $q$bulb analogy is still very useful as the eigenstates $|j\rangle_n$, also called the *computational basis*, of the configuration number operator $\mathbf{N} = \sum_{j=0}^{2^n-1} j\, (|j\rangle\langle j|)_n$ play a special role in quantum information theory.

> Mathematica Notebook 1.3: The Born rule and projective measurements http://www.physics.unlv.edu/%7Ebernard/MATH_book/Chap1/chap1_link.html; See also https://bernardzygelman.github.io

Before proceeding to that discussion in the next chapter, we need address two issues not yet discussed. In addition to the four postulates itemized above, there is a fundamental postulate that tells us how the state vector $|\Psi\rangle$ evolves in time. We defer that discussion to Chap. 3. Now, consider the state

$$|\Psi\rangle = \sum_{j=0}^{2^n-1} c_j |j\rangle \qquad (1.35)$$

for an n-dimensional register. It is expressed as an expansion over the computational basis $|j\rangle$. For the sake of simplicity, we dropped the subscript index that defines the latter as n-dimensional basis kets. As the computational basis vectors are orthonormal, let's take the scalar product of both sides of (1.35) with ket $|m\rangle$, where m is an arbitrary index $0 \leq m \leq 2^n - 1$. We find

$$\langle m|\Psi\rangle = \sum_{j=0}^{2^n-1} c_j \delta_{mj} = c_m \qquad (1.36)$$

where we used the fact $\langle i|j\rangle = \delta_{ij}$. Now relation (1.36) must be true for all values of m and so inserting this identity back into (1.35) we obtain

$$|\Psi\rangle = \sum_{j=0}^{2^n-1} \langle j|\Psi\rangle |j\rangle = \sum_{j=0}^{2^n-1} |j\rangle\langle j|\Psi\rangle. \qquad (1.37)$$

1.3 Hilbert Space

Now according to Dirac's rule for inner products this sum should be the same as

$$|\Psi\rangle = \left(\sum_{j=0}^{2^n-1} |j\rangle\langle j|\right)|\Psi\rangle$$

and which only makes sense if the sum between the parenthesis has no effect on the n-dimensional qubit state $|\Psi\rangle$. In other words (1.37) implies that

$$\sum_{j=0}^{2^n-1} |j\rangle\langle j| = \mathbb{1}\tilde{\otimes}\mathbb{1}\tilde{\otimes}\mathbb{1}\tilde{\otimes}\cdots \equiv \mathbb{1}. \tag{1.38}$$

Relation (1.38) is called the *closure*, or *completeness*, property for any set of n qubits. We will often make use of this property in our subsequent discussions. The symbol $\mathbb{1}$ represents an identity operator in the full n-qubit Hilbert space. In subsequent chapters we will use this symbol to represent the identity without explicit reference to the dimensionality of the Hilbert space.

A Note on Notation.

Quantum Information Science is multidisciplinary, with contributions from the fields of physics, computer science, mathematics, and engineering. It is not surprising that this can lead to clashes in notational convention. In this textbook, I use a notation more aligned with physics convention when describing a multi-qubit vector. In this convention, the expression

$$|a\,b\,c\rangle = |a\rangle \otimes |b\rangle \otimes |c\rangle$$

identifies c as the first qubit, b the second, and a the third. Also, keeping within the spirit that dual space vectors are "mirror" images of ket vectors, the dual to $|a\,b\,c\rangle$, is written as

$$\langle c\,b\,a| = \langle c| \otimes \langle b| \otimes \langle a|.$$

An advantage of this notation is that inner products of kets with bras allow a series of nested contractions as shown below,

$$\langle c\,b\,a|a\,b\,c\rangle = \langle a|a\rangle \langle b|b\rangle \langle c|c\rangle.$$

Unfortunately this notation has pitfalls when translating into bit strings. For example, take the ket

$$|7\rangle_5 = |0\,0\,1\,1\,1\rangle.$$

Our convention dicates that the inner product $\phi = {}_5\langle 7|7\rangle_5$ has the binary string representation
$$\phi = \langle 1 1 1 0 0 | 0 0 1 1 1\rangle.$$
This can cause confusion if the binary string in the bra vector is read in the standard manner (from right to left, leading to the bit string value 28.) Instead, bit strings in dual space should read from left to right (remember that dual space is a Bizzaro world!)

Later, in Chap. 3, we will adhere to the convention described below when we discuss wire diagrams. For example, in a wire diagram, the state $|a\,b\,c\rangle$ is represented as in Fig. 1.2, where the state $|c\rangle$ of the first qubit occupies the lowest rung, while the remaining qubit states occupy the next rungs in progressive order.

Problems

1.1 Evaluate the exercises in Mathematica Notebook 1.1

1.2 Given the set of polynomials of degree 3 in variable x, $P_a = a_0 + a_1 x + a_2 x^2 + a_3 x^3$, where $a_0 \ldots a_3$ are real numbers. Let the binary operation $P_a + P_b$ denote ordinary addition. Show that set P_γ constitutes a linear vector space.

1.3 Answer the exercises in Mathematica Notebook 1.2

1.4 The state
$$|\psi\rangle = \frac{1}{\sqrt{6}}|01101\rangle + \sqrt{\frac{2}{3}}|11111\rangle + \frac{1}{\sqrt{6}}|00001\rangle$$
describes a register of five $q$bulbs. (a) Calculate the probability that the first $q$bulb is in the on position, after making a measurement with device $\mathbb{1}\tilde{\otimes}\mathbb{1}\tilde{\otimes}\mathbb{1}\tilde{\otimes}\mathbb{1}\tilde{\otimes}\mathbf{n}$ (b) What is the probability that all $q$bulbs are in the on position, after a measurement with device $\mathbf{n}\tilde{\otimes}\mathbf{n}\tilde{\otimes}\mathbf{n}\tilde{\otimes}\mathbf{n}\tilde{\otimes}\mathbf{n}$. (c) Calculate the probability that at least three $q$bulbs are found in the off-position. (d) A measurement reveals the occupation configuration 01101. Immediately after that measurement another measurement with $\mathbf{N}$ is made. Calculate the probability that at least three $q$bulbs are in the off-position.

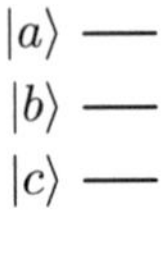

Fig. 1.2 Wire diagram for state $|a\,b\,c\rangle$

$|a\rangle$ ———
$|b\rangle$ ———
$|c\rangle$ ———

1.3 Hilbert Space

1.5 Given the states $|\psi\rangle = \frac{1}{\sqrt{6}}|01101\rangle + \sqrt{\frac{2}{3}}|11111\rangle + \frac{i}{\sqrt{6}}|00001\rangle$ and $|\Phi\rangle = \frac{1}{\sqrt{2}}|01101\rangle + \frac{1}{\sqrt{2}}|01111\rangle$. Evaluate $\langle\Phi|\psi\rangle$, $|\Phi\rangle\langle\psi|$ and $|\psi\rangle\langle\Phi|$.

1.6 Consider operator $\mathbf{X} = |\Phi\rangle\langle\psi|$, show that $\mathbf{X}^\dagger = |\psi\rangle\langle\Phi|$. Hint: use this expression for $\mathbf{X}^\dagger$ and operate it on bra $\langle\Gamma|$. Compare the result with that obtained by $\mathbf{X}|\Gamma\rangle$.

1.7 Re-express the following states $|17\rangle_5, |5\rangle_5, |12\rangle_5$ in binary notation, i.e., $|k_4 k_3 k_2 k_1 k_0\rangle$, $k_i \in \{0, 1\}$.

1.8 Consider the operator $\mathbf{X} \equiv exp(i\alpha)|00110\rangle\langle00100| + |00111\rangle\langle11111|$. Evaluate $\mathbf{X}|\Phi\rangle$ where $|\Phi\rangle$ is given by Eq. (1.3) in the text.

1.9 Show that operator, in the Hilbert space of a single qubit, $\mathbf{X} \equiv |0\rangle\langle 1| + |1\rangle\langle 0|$ is Hermitian. Solve the following equation

$$\mathbf{X}|\psi\rangle = \lambda|\psi\rangle$$

for the state $|\psi\rangle$. Hint: express $|\psi\rangle = c_1|0\rangle + c_2|1\rangle$, and solve for the coefficients c_1, c_2. Show that this equation admits solutions only for select values of parameter λ.

1.10 In the Hilbert space of three qubits, consider the operator

$$\mathbf{A} \equiv |000\rangle\langle 000| + 2|001\rangle\langle 100| - 2|010\rangle\langle 010| + 3|100\rangle\langle 001| + |011\rangle\langle 110| - |101\rangle\langle 101|.$$

Find all the eigenvalues and eigenvectors of $\mathbf{A}$. Identify the degenerate eigenvalues and show that any linear combination of the corresponding eigenvectors are also eigenstates of $\mathbf{A}$.

1.11 In a two-qubit Hilbert space, consider the operator

$$\mathbf{A} = |00\rangle\langle 01| + |10\rangle\langle 00| + |01\rangle\langle 10| + |10\rangle\langle 01|.$$

Find the eigenvalues and eigenstates of $\mathbf{A}$.

1.12 Given the single qubit operators

$$\mathbf{A} = |0\rangle\langle 0| + |1\rangle\langle 1|, \quad \mathbf{B} = i|0\rangle\langle 1| - i|1\rangle\langle 0|,$$

show that $[\mathbf{A}, \mathbf{B}] \equiv \mathbf{AB} - \mathbf{BA} = 0$. Find the eigenstates for operator $\mathbf{B}$ and show that they are also eigenstates of operator $\mathbf{A}$. What are the eigenvalues associated with operator $\mathbf{A}$?

1.13 Prove the following: for two Hilbert space operators $\mathbf{A}, \mathbf{B}$, $[\mathbf{A}, \mathbf{B}] = 0$, show that, if $\mathbf{A}$ has non-degenerate eigenstates $|a_1\rangle, |a_2\rangle, \ldots$, then the kets $|a_i\rangle$ are also eigenstates of operator $\mathbf{B}$.

1.14 Prove Theorems 1.1 and 1.2.

1.15 Given operator $\mathbf{A}$ in an n-dimensional Hilbert space with orthonormal eigenvectors $|a_1\rangle, |a_2\rangle, \ldots, |a_n\rangle$, prove that $\mathbf{A} = \sum_i^n a_i |a_i\rangle \langle a_i|$, where a_i is the eigenvalue associated with $|a_i\rangle$.

1.16 Show that operator

$$\mathbf{U} = \cos\theta\,|0\rangle\langle 0| + \exp(i\phi)\sin\theta\,|0\rangle\langle 1| + \exp(-i\phi)\sin\theta\,|1\rangle\langle 0| - \cos\theta\,|1\rangle\langle 1|,$$

where θ, ϕ are real parameters, is unitary.

1.17 Consider the operator $\mathbf{X} = |0\rangle\langle 1| + |1\rangle\langle 0|$, evaluate

$$\tilde{\mathbf{X}} = \mathbf{U}\mathbf{X}\mathbf{U}^\dagger$$

where $\mathbf{U}$ is given in Problem 1.16.

1.18 Find the eigenvalues and eigenstates for operator $\tilde{\mathbf{X}}$ given in Problem 1.17.

1.19 Evaluate $\tilde{\mathbf{Y}} = \mathbf{U}\mathbf{Y}\mathbf{U}^\dagger$ where, $\mathbf{Y} = -i|0\rangle\langle 1| + i|1\rangle\langle 0|$ and $\mathbf{U}$ is defined in Problem 1.16. Demonstrate that

$$\left[\tilde{\mathbf{X}}, \tilde{\mathbf{Y}}\right] = 2i\tilde{\mathbf{Z}}$$

where $\tilde{\mathbf{Z}} = \mathbf{U}(|0\rangle\langle 0| - |1\rangle\langle 1|)\mathbf{U}^\dagger$, and $\tilde{\mathbf{X}}$ is defined in Problem 1.17.

1.20 Consider the operator

$$\mathbf{P} = 2\,\mathbf{n}\tilde{\otimes}\mathbf{n}\tilde{\otimes}\mathbf{n} + \mathbf{1}\tilde{\otimes}\mathbf{1}\tilde{\otimes}\mathbf{n} - \mathbf{n}\tilde{\otimes}\mathbf{1}\tilde{\otimes}\mathbf{n} - \mathbf{1}\tilde{\otimes}\mathbf{n}\tilde{\otimes}\mathbf{n}.$$

Show that $\mathbf{P}$ is a projection operator.

References

1. D. Deutsch, *The Fabric of Reality* (Allen Lane, London, 1997)
2. C. Cohen-Tannoudji, B. Dieu, F. Laloe, *Quantum Mechanics*, vol. I (Wiley, 1991)
3. R.B. Griffiths, in *Stanford Encyclopedia of Philosophy* (2014). http://plato.stanford.edu/entries/qm-consistent-histories

References

4. F.T. Jordan, *Linear Operators for Quantum Mechanics* (Dover Publications Inc., Mineola New York, 1997)
5. D. Mermin, *Physics Today*, vol. 39. https://doi.org/10.1063/1.2815188
6. J.-M. Schwindt, *Conceptual Basis of Quantum Mechanics* (Springer, 2017)
7. D. Wineland, http://www.youtube.com/watch?v=RFkyvkBV5dM#t=32m50s

Apples and Oranges: Matrix Representations

2

Abstract

I discuss and elaborate on the isomorphism between kets that span the Hilbert space of n-qubits with column matrices of dimension 2^n. The collection of all row matrices is shown to constitute the corresponding dual space. We illustrate how outer products, or operators, are represented by $n \times n$ square matrices. The various matrix operations that provide the inner, outer, direct or Kronecker, products for the corresponding Hilbert space are introduced and discussed. The concepts of spin and the Bloch sphere are introduced. A qubit interpretation of spin, and the polarization properties of light is discussed.

2.1 Matrix Representations

People often use the expression "apples and oranges" to flag false analogies and similes. To a mathematician, apples are very much like oranges in the sense that each apple (orange) can be represented by an integer or a deficit of apples (oranges) by a negative integer. Both sets exhibit the behavior of a mathematical structure called a *group*. So if two apples and three apples add to five apples, the same is true for the oranges. Mathematical structures, in which a member of set A, and operations within that set, can be put into a one-to-one correspondence with members of set B, is called an *isomorphism*. In this chapter, we illustrate how the formal structure introduced in the previous chapter is isomorphic to a vector space whose elements are matrices. We will find that ket vectors can be represented by column matrices and their dual, the bra vectors, by row matrices.

Let's start with the simplest, non-trivial, Hilbert space of a single qubit. Since the kets $|0\rangle, |1\rangle$ span a Hilbert space of dimension $d = 2$, any ket $|\Psi\rangle$ in this space can be expressed as the linear combination $|\Psi\rangle = c_1|0\rangle + c_2|1\rangle$. The constraint $\langle\Psi|\Psi\rangle = 1$ on a physical state imposes the restriction that $|c_1|^2 + |c_2|^2 = 1$.

Consider the following array, also called a column matrix,

$$\begin{pmatrix} c_1 \\ c_2 \end{pmatrix} \tag{2.1}$$

where c_1, c_2 are complex numbers. Matrix addition of two column matrices follows the rule

$$\begin{pmatrix} c_1 \\ c_2 \end{pmatrix} + \begin{pmatrix} d_1 \\ d_2 \end{pmatrix} = \begin{pmatrix} c_1 + d_1 \\ c_2 + d_2 \end{pmatrix}$$

and it is apparent that the set of all possible column matrices form a vector space. This proposition is verified by checking conditions (i-vii) enumerated in the previous chapter. According to the definition of matrix addition condition (i) is obviously satisfied. So is condition (ii), as multiplication of a column matrix by a number (scalar) simply multiplies each entry in the array by that number. Conditions (iii-v) follow from definitions for matrix addition and scalar multiplication. The null vector is given by the array

$$\begin{pmatrix} 0 \\ 0 \end{pmatrix}$$

and each vector (2.1) has a unique inverse

$$\begin{pmatrix} -c_1 \\ -c_2 \end{pmatrix}.$$

Because the equality

$$c_1 \begin{pmatrix} 1 \\ 0 \end{pmatrix} + c_1 \begin{pmatrix} 0 \\ 1 \end{pmatrix} = \begin{pmatrix} 0 \\ 0 \end{pmatrix} \tag{2.2}$$

is satisfied, if and only if, $c_1 = c_2 = 0$, the vectors

$$\begin{pmatrix} 1 \\ 0 \end{pmatrix} \text{ and } \begin{pmatrix} 0 \\ 1 \end{pmatrix}$$

are linearly independent. Also, since this is the largest set of independent vectors, the dimension of this vector space is $d = 2$.

We conclude that there is an isomorphism between the abstract kets $|0\rangle, |1\rangle$ and column matrices. We assert the association

$$|0\rangle \Leftrightarrow \begin{pmatrix} 1 \\ 0 \end{pmatrix}$$

$$|1\rangle \Leftrightarrow \begin{pmatrix} 0 \\ 1 \end{pmatrix}. \tag{2.3}$$

The matrices on right-hand side of (2.3) are said to be a matrix representation of the ket vectors and the arrows suggest that we can always replace a ket with its matrix representation, and vice-versa. This identification is practiced so often that physicists make it a habit to denote column matrices as kets without bothering to acknowledge the implicit isomorphism. As we get more accustomed to working with matrices, we will also fall into this habit and replace $\Leftrightarrow$ with an equality sign.

Having established the isomorphism between the vector space of column matrices with the ket space of a qubit, we now focus on row vectors. Obviously, they also form a vector space distinct from that of column matrices. Nevertheless, we can associate with each column matrix a corresponding row matrix. The row matrices constitute a dual space. Given column matrix

$$\begin{pmatrix} c_1 \\ c_2 \end{pmatrix} = c_1 \begin{pmatrix} 1 \\ 0 \end{pmatrix} + c_2 \begin{pmatrix} 0 \\ 1 \end{pmatrix},$$

we define its dual, the row matrix

$$\begin{pmatrix} c_1^* & c_2^* \end{pmatrix} = c_1^* \begin{pmatrix} 1 & 0 \end{pmatrix} + c_2^* \begin{pmatrix} 0 & 1 \end{pmatrix}. \tag{2.4}$$

It's evident that the row matrices $\begin{pmatrix} 1 & 0 \end{pmatrix}$ and $\begin{pmatrix} 0 & 1 \end{pmatrix}$ are basis vectors for the vector space of all $2d$ row matrices, and it follows that the association

$$\langle 0| \Leftrightarrow \begin{pmatrix} 1 & 0 \end{pmatrix}$$
$$\langle 1| \Leftrightarrow \begin{pmatrix} 0 & 1 \end{pmatrix} \tag{2.5}$$

is appropriate.

> Mathematica Notebook 2.1: Matrix manipulations and operations with Mathematica http://www.physics.unlv.edu/%7Ebernard/MATH_book/Chap2/chap2_link.html; See also https://bernardzygelman.github.io

2.1.1 Matrix Operations

We established that the matrix representation of a qubit ket is a two-dimensional column matrix, whereas its bra dual is the corresponding row matrix obtained using rule (2.5). But we also know that column and row matrices can be multiplied in two different ways. Let's review those rules. Convention tells us that if an n-dimensional row matrix with entries $a_1, a_2, \ldots a_n$ is placed to the left of an n-dimensional column matrix with entries $b_1, b_2, \ldots b_n$, their product is

$$\sum_{i=1}^{n} a_i b_i. \tag{2.6}$$

For the case $n = 2$,

$$\begin{pmatrix} a_1 & a_2 \end{pmatrix} \begin{pmatrix} b_1 \\ b_2 \end{pmatrix} = a_1 b_1 + a_2 b_2. \tag{2.7}$$

The latter is called the scalar product and, as we show below, is identical to the definition discussed in Chap. 1. Let $|\Psi\rangle = c_1|0\rangle + c_2|1\rangle$ and $|\Phi\rangle = d_1|0\rangle + d_2|1\rangle$. According to the discussions of the previous chapter, their inner product is $\langle\Phi|\Psi\rangle = d_1^*c_1 + d_2^*c_2$. If we make the associations

$$|\Psi\rangle \Rightarrow \begin{pmatrix} c_1 \\ c_2 \end{pmatrix}$$

$$|\Phi\rangle \Rightarrow \begin{pmatrix} d_1 \\ d_2 \end{pmatrix} \tag{2.8}$$

then, according to the above definitions,

$$\langle\Phi|\Psi\rangle \rightarrow \begin{pmatrix} d_1^* & d_2^* \end{pmatrix} \begin{pmatrix} c_1 \\ c_2 \end{pmatrix} = d_1^*c_1 + d_2^*c_2. \tag{2.9}$$

In the previous chapter, we arranged bra-ket combinations to construct operators in the following manner

$$\mathbf{X} \equiv |\Psi\rangle\langle\Phi|.$$

Prescription (2.8) leads us to consider the matrix operation

$$\mathbf{X} \Rightarrow \begin{pmatrix} c_1 \\ c_2 \end{pmatrix} \begin{pmatrix} d_1^* & d_2^* \end{pmatrix} \equiv$$

$$\begin{pmatrix} d_1^* & d_2^* \end{pmatrix}$$

$$\begin{pmatrix} c_1 \\ c_2 \end{pmatrix} \begin{pmatrix} c_1 d_1^* & c_1 d_2^* \\ c_2 d_1^* & c_2 d_2^* \end{pmatrix}. \tag{2.10}$$

Thus the outer product $\mathbf{X}$ is isomorphic to a 2×2 square matrix. When we express this as an equality, i.e.

$$\mathbf{X} = \begin{pmatrix} c_1 d_1^* & c_1 d_2^* \\ c_2 d_1^* & c_2 d_2^* \end{pmatrix}, \tag{2.11}$$

it is implicit that $\mathbf{X}$ is a matrix representation of operator $|\Psi\rangle\langle\Phi|$. There is another equivalent way to express this isomorphism. Using the definition $\mathbf{X} = |\Psi\rangle\langle\Phi|$, we form all possible inner products of the two vectors $\mathbf{X}|0\rangle$ and $\mathbf{X}|1\rangle$ with bras $\langle 0|, \langle 1|$. We then organize the resulting scalar quantities in the following tabular, or matrix, form

$$\begin{pmatrix} \langle 0|\mathbf{X}|0\rangle & \langle 0|\mathbf{X}|1\rangle \\ \langle 1|\mathbf{X}|0\rangle & \langle 1|\mathbf{X}|1\rangle \end{pmatrix} = \begin{pmatrix} \langle 0|\Psi\rangle\langle\Phi|0\rangle & \langle 0|\Psi\rangle\langle\Phi|1\rangle \\ \langle 1|\Psi\rangle\langle\Phi|0\rangle & \langle 1|\Psi\rangle\langle\Phi|1\rangle \end{pmatrix} \tag{2.12}$$

and which, when evaluated, agrees with expression (2.11). In general, an n-dimensional column matrix with entries $a_1, a_2, \ldots, a_n$ placed to the left of a n-dimensional row matrix with entries $b_1, b_2, \ldots, b_n$ implies an $n \times n$ table, or square matrix, whose ith row and jth column contain the product $a_i b_j$.

Now $|\Psi\rangle\langle\Phi|$ operates on kets to its right, or bras on its left and engenders a transformation of vectors in Hilbert and dual space respectively. Let's illustrate this

2.1 Matrix Representations

transformation in the matrix representation. Consider the vector $\mathbf{X}|0\rangle$, and using the matrix representation of $\mathbf{X}$ and ket $|0\rangle$ we find

$$\mathbf{X}|0\rangle \implies \begin{pmatrix} c_1 d_1^* & c_1 d_2^* \\ c_2 d_1^* & c_2 d_2^* \end{pmatrix} \begin{pmatrix} 1 \\ 0 \end{pmatrix} = \begin{pmatrix} c_1 d_1^* \\ c_2 d_1^* \end{pmatrix}. \tag{2.13}$$

Because $\mathbf{X}|0\rangle = |\Psi\rangle\langle\Phi|0\rangle$ and $\langle\Phi|0\rangle = d_1^*$

$$\mathbf{X}|0\rangle = d_1^* |\Psi\rangle.$$

The latter agrees with (2.13) if $|\Psi\rangle$ is replaced by its matrix representation. Similarly, taking the conjugate transpose[1] of (2.11)

$$(1\ 0) \begin{pmatrix} c_1^* d_1 & c_2^* d_1 \\ c_1^* d_2 & c_2^* d_2 \end{pmatrix} = \begin{pmatrix} c_1^* d_1 & c_2^* d_1 \end{pmatrix}, \tag{2.14}$$

the matrix representation of the relation $\langle\Psi|d_1 = \langle 0|\mathbf{X}^\dagger$.

2.1.2 The Bloch sphere

We now have the tools that allow us to investigate, in more detail, the properties of the qubit Hilbert space. According to the above discussion the matrix representation for a qubit $|\Psi\rangle$ is

$$|\Psi\rangle = \begin{pmatrix} c_1 \\ c_2 \end{pmatrix} \qquad |c_1|^2 + |c_2|^2 = 1, \tag{2.15}$$

where the equivalence symbol $\Leftrightarrow$ is replaced by an equality. c_1, c_2 are complex numbers which we express in the form $c_1 = x_0 + i x_1$, $c_2 = x_2 + i x_3$, and where x_0, x_1, x_2, x_3 are four independent real parameters. The requirement that $|\Psi\rangle$ is a physical (normalized) state imposes a constraint on the constants c_1, c_2 so that $x_0^2 + x_1^2 + x_2^2 + x_3^2 = 1$. This equation describes a 3-sphere embedded in a four-dimensional space, and whose center is located at the origin. Let's define a *Hopf map* of a point x_0, x_1, x_2, x_3 in this space to a point (x, y, z) in a three-dimensional space where

$$\begin{aligned} x &= 2(x_0 x_2 + x_1 x_3) \\ y &= 2(x_3 x_0 - x_1 x_2) \\ z &= x_0^2 + x_1^2 - x_2^2 - x_3^2. \end{aligned} \tag{2.16}$$

With relation (2.16) we find that

$$\sqrt{x^2 + y^2 + z^2} = \sqrt{x_0^2 + x_1^2 + x_2^2 + x_3^2} = 1,$$

---

[1] The transpose of a matrix in which each element is replaced with it's complex conjugate.

where we used the fact that (x_0, x_1, x_2, x_3) lies on a unit 3-sphere. Therefore, points x, y, z lie on the surface of a three-dimensional sphere, the Bloch Sphere, of unit length. If we parameterize the coordinates

$$\begin{aligned} x_0 &= \cos(\theta/2)\cos(\beta) \\ x_1 &= \cos(\theta/2)\sin(\beta) \\ x_2 &= \sin(\theta/2)\cos(\beta+\phi) \\ x_3 &= \sin(\theta/2)\sin(\beta+\phi) \end{aligned}$$

for $0 \le \theta \le \pi, 0 \le \phi \le 2\pi, 0 \le \beta \le 2\pi$ we find that

$$(x, y, z) = (\sin\theta\cos\phi, \sin\theta\sin\phi, \cos\theta),$$

the standard parameterization of a unit 2-sphere in a spherical coordinate system. Here θ, ϕ are the polar and azimuthal angles respectively. With this parameterization we find that

$$|\Psi\rangle = \exp(i\beta) \begin{pmatrix} \cos\theta/2 \\ \exp(i\phi)\sin\theta/2 \end{pmatrix}. \tag{2.17}$$

Therefore, the state of a qubit (disregarding an overall phase factor, $\exp(i\beta)$) is represented by a point on the surface of the Bloch sphere. The point $\theta = 0$, located on the "north pole" of the Bloch sphere, identifies the computational basis vector $|0\rangle$, whereas the $|1\rangle$ vector is described by the point located on the "south pole" in which $\theta = \pi$ in Fig. 2.1.

> Mathematica Notebook 2.2: Visualizing qubits on the Bloch sphere surface. http://www.physics.unlv.edu/%7Ebernard/MATH_book/Chap2/chap2_link.html; See also https://bernardzygelman.github.io

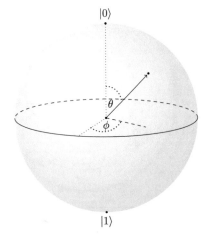

Fig. 2.1 The Bloch sphere. The arrow points to the location on the sphere of the state $|\psi\rangle$ given by 2.17. The points on the north and south poles of the sphere denote the locations of the computational basis states

2.2 The Pauli Matrices

In the previous chapter, I introduced a physical model for a qubit that I called a $q$bulb. In analogy with a light bulb, the $q$bulb is an atom in either an off or on state. Those states are represented by the kets $|0\rangle$, $|1\rangle$ or their corresponding matrix representations (2.3). They are eigenstates of the operator $\mathbf{n} \equiv \sum_{j=0}^{1} j \, |j\rangle\langle j|$, which when expressed in matrix form

$$\mathbf{n} = \begin{pmatrix} 0 & 0 \\ 0 & 1 \end{pmatrix}. \tag{2.18}$$

$\mathbf{n}$ represents a measuring device whose outputs are its eigenvalues 0, 1 and indicates whether the atom or $q$bulb, is in the on or off position. As required by the foundational postulates, $\mathbf{n}$ must be a self-adjoint, or Hermitian, operator. In the previous chapter, we introduced the † operation, so that if $\mathbf{X} = |\Phi\rangle\langle\Psi|$, then $\mathbf{X}^\dagger = |\Psi\rangle\langle\Phi|$. If $\mathbf{X}$ is given by its matrix representation, the † operation on it is equivalent to the complex conjugate of each element of its *transpose matrix*. That is, given matrix $\mathbf{X}$ with entries X_{mn} for the m'th row and n'th column, the matrix $\mathbf{X}^\dagger$ has as it's corresponding entries X_{nm}^*. Using this prescription it is obvious that

$$\mathbf{n}^\dagger = \begin{pmatrix} 0 & 0 \\ 0 & 1 \end{pmatrix}^\dagger = \mathbf{n}. \tag{2.19}$$

The basis states (2.3), as they are eigenstates of the "on-off" measurement operator $\mathbf{n}$, play a special role as the computational basis. Keeping with convention, we agree that all matrix representations, for both vectors and operators, are taken with respect to the computational basis.

Let's define the linear combinations

$$|u\rangle \equiv \frac{1}{\sqrt{2}}|0\rangle + \frac{1}{\sqrt{2}}|1\rangle$$
$$|v\rangle \equiv \frac{1}{\sqrt{2}}|0\rangle - \frac{1}{\sqrt{2}}|1\rangle$$

whose matrix representations are

$$|u\rangle = \frac{1}{\sqrt{2}} \begin{pmatrix} 1 \\ 1 \end{pmatrix}$$
$$|v\rangle = \frac{1}{\sqrt{2}} \begin{pmatrix} 1 \\ -1 \end{pmatrix}. \tag{2.20}$$

Because

$$c_1|u\rangle + c_2|v\rangle = \frac{1}{\sqrt{2}} \begin{pmatrix} c_1 + c_2 \\ c_1 - c_2 \end{pmatrix} = \begin{pmatrix} 0 \\ 0 \end{pmatrix}$$

only if $c_1 = c_2 = 0$, $|u\rangle$, $|v\rangle$ are linearly independent. Furthermore, since $\langle u|v\rangle = 0$, $\langle u|u\rangle = \langle v|v\rangle = 1$, they constitute an alternative basis for a qubit. Let's define the operator

$$\sigma_X \equiv |u\rangle\langle u| - |v\rangle\langle v| =$$

$$\frac{1}{2}\begin{pmatrix} 1 & 1 \\ 1 & 1 \end{pmatrix} - \frac{1}{2}\begin{pmatrix} 1 & -1 \\ -1 & 1 \end{pmatrix} = \begin{pmatrix} 0 & 1 \\ 1 & 0 \end{pmatrix} \quad (2.21)$$

where the second line expresses the outer products by their matrix representations. Matrix σ_X is called the Pauli σ_X-matrix. There are two other Pauli-matrices,

$$\sigma_Y \equiv \begin{pmatrix} 0 & -i \\ i & 0 \end{pmatrix} \quad \sigma_Z \equiv \begin{pmatrix} 1 & 0 \\ 0 & -1 \end{pmatrix}. \quad (2.22)$$

All three Pauli matrices play an important role in generating operations in the Hilbert space of a qubit. Since physical measurement devices are represented by self-adjoint, or Hermitian, operators or matrices it is natural to ask; what is the most general self-adjoint 2×2 matrix ? A little thought suggests

$$\begin{pmatrix} a & b - ic \\ b + ic & d \end{pmatrix} \quad (2.23)$$

where a, b, c, d are arbitrary real numbers. We re-write this self-adjoint matrix in the form

$$b\,\sigma_X + c\,\sigma_Y + \alpha\,\sigma_Z + \beta\,\mathbb{1} \quad (2.24)$$

where $\alpha = (a - d)/2$, $\beta = (a + d)/2$ and $\mathbb{1}$ is the 2×2 identity matrix. Thus, an arbitrary 2×2 Hermitian matrix can be represented by a linear combination of the three Pauli matrices and the identity matrix. The four matrices form a basis for the linear vector space of all 2×2 Hermitian matrices. Pauli matrices also possess interesting algebraic properties. Evaluating the following matrix products

$$\sigma_X\,\sigma_Y - \sigma_Y\,\sigma_X = 2i\sigma_Z$$
$$\sigma_Y\,\sigma_Z - \sigma_Z\,\sigma_Y = 2i\sigma_X$$
$$\sigma_Z\,\sigma_X - \sigma_X\,\sigma_Z = 2i\sigma_Y \quad (2.25)$$

we notice that the three Pauli matrices are closed under the bracket operation $[A, B] \equiv AB - BA$. In other words, given a linear combination of two Pauli matrices A and B, and performing the binary operation $[A, B]$ one always arrives at a linear of combination Pauli matrices. Because matrix multiplication is non-commutative, i.e., AB is not necessarily equal to BA, the bracket operation is non-trivial. It is useful to introduce the shorthand notation

$$[\sigma_i, \sigma_j] = 2i \sum_k \epsilon_{ijk}\sigma_k \quad (2.26)$$

2.2 The Pauli Matrices

where the subscripts $i = 1, 2, 3$ denote X, Y, Z respectively, and ϵ_{ijk} is called the *Levi-Civita symbol*. It has the property that $\epsilon_{123} = \epsilon_{312} = \epsilon_{231} = 1$, $\epsilon_{213} = \epsilon_{321} = \epsilon_{132} = -1$ and $\epsilon_{ijk} = 0$ if any of the two subscript have identical values. In addition to providing a basis for all 2×2 Hermitian matrices, the Pauli matrices serve as generators of all 2×2 unitary matrices. According to the discussion in the previous chapter, a unitary operator **U** has the property $\mathbf{U}^\dagger \mathbf{U} = \mathbf{U}\mathbf{U}^\dagger = \mathbb{1}$. Because operators in qubit Hilbert space are represented by 2×2 matrices and are parameterized by four real parameters, it is convenient to express a general 2×2 unitary matrix as

$$\mathbf{U} = \exp(i\gamma) \begin{pmatrix} \exp(i\phi)\cos\theta & \exp(i\beta)\sin\theta \\ -\exp(-i\beta)\sin\theta & \exp(-i\phi)\cos\theta \end{pmatrix}, \quad (2.27)$$

where $\gamma, \phi, \beta, \theta$ are real numbers. To see how **U** is related to the Pauli matrices we first need to understand how exponentiation of a matrix is defined. We learned in calculus that exponentiation of a number α can be defined by its infinite power series representation, i.e.

$$\exp(\alpha) = 1 + \alpha + \frac{\alpha^2}{2!} + \frac{\alpha^3}{3!} + \ldots$$

So we define the exponentiation of a 2×2 matrix **A** by the expression

$$\exp(\mathbf{A}) \equiv \mathbb{1} + \mathbf{A} + \frac{\mathbf{A}\mathbf{A}}{2!} + \frac{\mathbf{A}\mathbf{A}\mathbf{A}}{3!} + \ldots \quad (2.28)$$

Given $\mathbf{A} = \mathbf{A}^\dagger$, we construct the following

$$\mathbf{U}_A \equiv \exp(i\mathbf{A}) = \mathbb{1} + i\mathbf{A} - \frac{\mathbf{A}\mathbf{A}}{2!} - i\frac{\mathbf{A}\mathbf{A}\mathbf{A}}{3!} + \ldots$$

Taking the conjugate of the r.h.s of this expression we find that $\mathbf{U}_A^\dagger \equiv \exp(-i\mathbf{A}) = (\exp(i\mathbf{A}))^\dagger$. Evaluating $\mathbf{U}_A \mathbf{U}_A^\dagger = \exp(i\mathbf{A})\exp(-i\mathbf{A})$ by multiplying and collecting like terms of the series representation we find

$$\mathbf{U}_A \mathbf{U}_A^\dagger = \mathbf{U}_A^\dagger \mathbf{U}_A = \mathbb{1}. \quad (2.29)$$

Therefore, for any Hermitian operator **A**, $\mathbf{U}_A$ is unitary. Consider

$$\mathbf{U}_X \equiv \exp(i\alpha\sigma_X) = \mathbb{1} + i\alpha\,\sigma_X - \alpha^2 \frac{\sigma_X \sigma_X}{2!} - i\alpha^3 \frac{\sigma_X \sigma_X \sigma_X}{3!} + \alpha^4 \frac{\sigma_X \sigma_X \sigma_X \sigma_X}{4!} + \ldots$$

and which, is guaranteed to be unitary. Using the fact that $\sigma_X \sigma_X = \mathbb{1}$ we simplify this expression so that

$$\mathbf{U}_X = \mathbb{1}\left(1 - \frac{\alpha^2}{2!} + \frac{\alpha^4}{4!} + \cdots\right) + i\,\sigma_X\left(\alpha - \frac{\alpha^3}{3!} + \cdots\right) =$$
$$\mathbb{1}\cos\alpha + i\,\sigma_X \sin\alpha \quad (2.30)$$

where we have replaced the power series in α by their trigonometric representations. In explicit matrix form

$$\mathbf{U}_X(\alpha) = \exp(i\alpha\sigma_X) = \begin{pmatrix} \cos\alpha & i\sin\alpha \\ i\sin\alpha & \cos\alpha \end{pmatrix}. \qquad (2.31)$$

In the same manner we find that

$$\mathbf{U}_Y(\alpha) \equiv \exp(i\alpha\sigma_Y) = \begin{pmatrix} \cos\alpha & \sin\alpha \\ -\sin\alpha & \cos\alpha \end{pmatrix}$$

$$\mathbf{U}_Z(\alpha) \equiv \exp(i\alpha\sigma_Z) = \begin{pmatrix} \exp(i\alpha) & 0 \\ 0 & \exp(-i\alpha) \end{pmatrix}. \qquad (2.32)$$

Evaluating the matrix product

$$\mathbf{U}_Z(\frac{\phi+\beta}{2})\mathbf{U}_Y(\theta)\mathbf{U}_Z(\frac{\phi-\beta}{2}) = \begin{pmatrix} \exp(i\phi)\cos\theta & \exp(i\beta)\sin\theta \\ -\exp(-i\beta)\sin\theta & \exp(-i\phi)\cos\theta \end{pmatrix}, \qquad (2.33)$$

we find that unitary operator (2.27) can be expressed as a product of unitary operators whose generators are Pauli matrices and a scalar phase operation $\exp(i\gamma)$. This fact comes in handy in later chapters where we show how quantum gate operations are carried out by unitary operators.

According to Postulate III, measurement devices are associated with Hermitian operators. Operator **n**, which measures a qubit's occupancy number (1 or 0), is such a matrix, but is the converse true? Apparently, there exist an infinite set of Hermitian matrices in this Hilbert space. The Pauli matrices or any linear combination of them are Hermitian, so do they represent possible measurement devices? If so what do they measure? Consider operator σ_X, as it is Hermitian let's assume that it is associated with some measurement device.

Postulate III demands that a measurement with the device results in one of the eigenvalues of σ_X. In order to find the eigenvectors and eigenvalues of σ_X we need to find solutions of

$$\sigma_X|\lambda\rangle = \lambda|\lambda\rangle$$

where $|\lambda\rangle$ is the eigenvector associated with eigenvalue λ. The matrix form of this equation is

$$\begin{pmatrix} 0 & 1 \\ 1 & 0 \end{pmatrix} \begin{pmatrix} c_1 \\ c_2 \end{pmatrix} = \lambda \begin{pmatrix} c_1 \\ c_2 \end{pmatrix}$$

where we expressed vector $|\lambda\rangle$ as a linear combination of the computational basis vectors $|0\rangle, |1\rangle$. Collecting terms we get

$$\lambda c_1 - c_2 = 0$$
$$c_1 - \lambda c_2 = 0. \qquad (2.34)$$

Using one these equations to express c_1 in terms of c_2, and inserting that relation into the other equation one finds that, for non-trivial solutions of (2.34), condition

$$\lambda^2 - 1 = 0 \tag{2.35}$$

must be satisfied. Thus $\lambda = \pm 1$, and inserting these eigenvalues into (2.34) we express c_1 in terms of c_2. For example, with $\lambda = 1$ we get $c_1 - c_2 = 0$ or $c_1 = c_2$. We still have two free parameters because c_2 is a complex number, but the orthonormality condition $\langle \lambda | \lambda \rangle = 1$ fixes $c_1 = c_2 = \exp(i\gamma)/\sqrt{2}$, where γ is an arbitrary number. Thus

$$|\lambda = 1\rangle = \frac{\exp(i\gamma)}{\sqrt{2}} \begin{pmatrix} 1 \\ 1 \end{pmatrix} = \exp(i\gamma)|u\rangle$$

$$|\lambda = -1\rangle = \frac{\exp(i\beta)}{\sqrt{2}} \begin{pmatrix} 1 \\ -1 \end{pmatrix} = \exp(i\beta)|v\rangle \tag{2.36}$$

are eigenvectors of σ_X. γ, β are arbitrary phase factors which by convention we set to zero. For $\lambda = 1$ the state is parameterized by angles $\phi = 0$, $\theta = \pi/2$ on the Bloch sphere, and $\phi = \pi, \theta = \pi/2$ for $\lambda = -1$.

Suppose our qubit system is in state $|\Psi\rangle$. A measurement on this system with device **n** tells us whether the qubit is in the on, or off state but what does a measurement with σ_X tell us? To gain insight and answer that question we first consider a somewhat different physical system, that of a classical electromagnetic wave.

Mathematica Notebook 2.3: Pauli matrices as unitary gate generators. http://www.physics.unlv.edu/%7Ebernard/MATH_book/Chap2/chap2_link.html; See also https://bernardzygelman.github.io

2.3 Polarization of Light: A Classical Qubit

The publication in 1865 of *A Dynamical Theory of the Electromagnetic Field* by the theoretical physicist *James Clerk Maxwell* proved to be a watershed. It sparked both the electric and communication revolutions, and it is inconceivable to imagine the modern world without its transformational insights. Maxwell's synthesis of electricity and magnetism catalyzed the discovery of electromagnetic waves, of which optical phenomena, X-rays, microwaves, etc. are instances. Maxwell's theory predicts that light is a manifestation of electric and magnetic fields that vary in space and time in a specific way.

For a light beam that is coming out of this page, Maxwell's equations predict an electric field that behaves in the following manner

$$\vec{E}(t) = E_0 \exp(i\delta_0) \left(\cos\theta\, \hat{\mathbf{i}} + \sin\theta \exp(i\delta)\, \hat{\mathbf{j}} \right) \exp(i\omega t) \tag{2.37}$$

where the real part of complex function $\vec{E}(t)$ represents the electric field at any point x, y in the plane of the page at time t. $\hat{\mathbf{i}}, \hat{\mathbf{j}}$ are the orthogonal unit vectors along the x, and y directions respectively, δ_0, δ, θ, E_0 are real numbers, and ω is an angular frequency. The magnetic field is perpendicular, at each point in the x, y plane, to the $\vec{E}$ field. Knowledge of $E_0, \delta_0, \delta, \theta, \omega$ provides a complete description of the wave. Instead of using the explicit vector description of (2.37), it is convenient to introduce the *Jones vector*

$$\begin{pmatrix} \cos\theta \\ \exp(i\delta)\sin\theta \end{pmatrix}. \tag{2.38}$$

The first entry in the column matrix is, up to an overall constant and the factor $\exp(i\omega t)$, the x-component of (2.37) and the second entry the y-component. For the value $\theta = 0$ the Jones vector is

$$\begin{pmatrix} 1 \\ 0 \end{pmatrix} = |0\rangle \tag{2.39}$$

and taking the real part of (2.37) with this Jones vector, and setting $\delta = 0$, the electric field in the x, y plane is given by

$$E_0 \hat{\mathbf{i}} \cos(\omega t + \delta_0). \tag{2.40}$$

It describes a vector oscillating along the x-coordinate axis and is called *plane polarized light*. In shorthand, we call it H type light. For $\theta = \pi/2$, $\delta = 0$ the Jones vector is

$$\begin{pmatrix} 0 \\ 1 \end{pmatrix} = |1\rangle \tag{2.41}$$

and the electric field is plane polarized along the y-axis, or V type light. Inserting the values $\theta = \frac{\pi}{4}$ and $\delta = \frac{\pi}{2}$ into (2.38), the Jones vector becomes

$$\frac{1}{\sqrt{2}} \begin{pmatrix} 1 \\ i \end{pmatrix} = \frac{1}{\sqrt{2}} (|0\rangle + i |1\rangle). \tag{2.42}$$

It is a linear combination, containing complex coefficients, of the Joneses vectors that describe linear polarization along the horizontal and vertical directions (x, y axes). With this Jones vector the real part of (2.37), the electric field, is

$$\frac{E_0}{\sqrt{2}} \left(\cos(\omega t + \delta_0)\hat{\mathbf{i}} - \sin(\omega t + \delta_0)\hat{\mathbf{j}} \right). \tag{2.43}$$

Plotting (2.43) on the x, y plane as a function of time, one finds that it describes a vector rotating with angular frequency ω in the clockwise direction about a circle of radius of length $E_0/\sqrt{2}$. The latter is called left circularly, or L, polarized light. The Jones vector

$$\frac{1}{\sqrt{2}} \begin{pmatrix} 1 \\ -i \end{pmatrix} = \frac{1}{\sqrt{2}} (|0\rangle - i |1\rangle) \tag{2.44}$$

2.3 Polarization of Light: A Classical Qubit

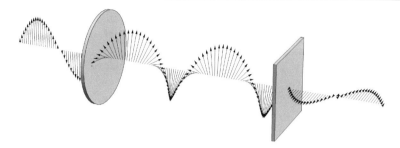

Fig. 2.2 Polarization states of classical light. V-type plane polarized light enters a polariztion filter(disk) which outputs either R, or L-type circular polarized light. The rectangular filter allows only plane H-type of light to pass. The arrows denote the magnitude and direction of the electric field

describes similar time behavior except that $\vec{E}(t)$ rotates in a counter-clockwise manner and is called right, or R, circular polarized light. The Jones vectors (2.39) and (2.41) are eigenstates of **n** whereas the left and right circular polarized Jones vectors (2.42), (2.44), are eigenstates of σ_Y. Monochromatic light can be manipulated by optical instruments as shown in Fig. 2.2. In that figure, plane polarized light along the $\hat{\mathbf{j}}$ axis is incident on a circular polarization filter. If that beam is interupted by a polarization filter that allows only $\hat{\mathbf{i}}$ plane polarized light through, we would find no output as the input was 100% polarized along the $\hat{\mathbf{j}}$ axis. Instead it enters a circular polarization filter (disk object) which outputs (L-type) circular polarized light. Because the output of the left-circular polarizer is a linear combination of the $|0\rangle$, $|1\rangle$ states, a plane horizontal ($\hat{\mathbf{i}}$) polarization filter (rectangular object) does allow passage of that component.

With polarization filters and selectors we possess devices that filter and select a particular polarization state in an incident beam. Suppose we construct a black box, called **N**, that contains two indicator lights labeled H and V. This box filters and detects one of the two components, H or V, polarized light. Another box, called **P**, detects R, L type light.

> Mathematica Notebook 2.4: Visualizing polarization of light. http://www.physics.unlv.edu/%7Ebernard/MATH_book/Chap2/chap2_link.html; See also https://bernardzygelman.github.io

2.3.1 A Qubit Parable

Let me entertain the following imaginary scenario. We provide the aforementioned boxes to researchers in some distant world who have no knowledge of Maxwell's equations and have no other means, except the use of these boxes, to study light phenomena. Passing a (polarized) light beam through two **N** boxes connected in series, the researchers observe either $H\,H$, or $V\,V$ indicator light configurations. They never see the indicator light combinations $H\,V$, or $V\,H$. On a single box, they always find one indicator light on, but never both. A reasonable conclusion from those results is that H, V are intrinsic, independent properties of light. Light appears to consist of V or H type, but never a combination of the two. This binary choice leads the researchers to employ ket notation, i.e., $|H\rangle, |V\rangle$ to formalize a theory of light. So, in this theory, when the H indicator is on, the device detects $|H\rangle$ type light, and $|V\rangle$ light if the other indicator is on. Analogous experiments with the **P** box reveal similar behavior and the researchers define two new states of light, $|R\rangle$, $|L\rangle$ that correspond to those indicators settings. These two pairs of states appear to be mutually exclusive until one researcher performs an experiment in which a beam passes through the box combination **N P N**. In some runs, the researchers observe indicator configurations $H\,R\,V$ in the corresponding boxes as the beam proceeds from left to right. That is, the incident beam leaves the first **N** box in the H state but after passing through the **P** box, the second **N** box detects the presence of V light. This data forces researchers to conclude that the **P** instrument, which measures the R, L properties of light, somehow affects the H, V properties of light. From these results, the alien scientists posit a qubit interpretation and invoke similar hypotheses to those introduced in the previous chapter. The rationale for the theory is that it is consistent with all experiments performed by the pair of measurement devices. Importantly, the scientists conclude that one type of light, e.g. $|R\rangle$, is a combination of $|H\rangle, |V\rangle$ light, a phenomenon they call superposition. They use matrix language to invoke the isomorphism

$$|H\rangle \Rightarrow \begin{pmatrix} 1 \\ 0 \end{pmatrix} \quad |V\rangle \Rightarrow \begin{pmatrix} 0 \\ 1 \end{pmatrix},$$

and realize that these vectors are eigenstates, with eigenvalues 0, 1, of operator

$$\begin{pmatrix} 0 & 0 \\ 0 & 1 \end{pmatrix}$$

which is the matrix representation for their **N** instrument. Its eigenstates are given the special status as the computational basis of Hilbert space. The box **P** should also be represented by a Hermitian matrix. If the eigenvalues of **P** are $1, 0$ corresponding to the eigenstates $|R\rangle, |L, \rangle$ respectively, then the general form for the operator representing box **P** is

$$\mathbf{P} = 1|R\rangle\langle R| + 0|L\rangle\langle L| = |R\rangle\langle R|.$$

Because $|R\rangle, |L\rangle$ are, presumably, linear combinations of the computational basis, $|R\rangle = c_1|0\rangle + c_2|1\rangle$ and so

$$\mathbf{P} = c_1^* c_1 |0\rangle\langle 0| + c_1^* c_2 |1\rangle\langle 0| + c_2^* c_1 |0\rangle\langle 1| + c_2^* c_2 |1\rangle\langle 1| = \begin{pmatrix} c_1^* c_1 & c_2^* c_1 \\ c_1^* c_2 & c_2^* c_2 \end{pmatrix}.$$

2.4 Spin

The complex constants c_1, c_2 are determined by carrying out the series of experiments described above. Suppose experiments reveal that $c_1 = 1/\sqrt{2}$, $c_2 = i/\sqrt{2}$, then

$$\mathbf{P} = \frac{1}{2}\begin{pmatrix} 1 & -i \\ i & 1 \end{pmatrix} = \frac{1}{2}(\mathbb{1} + \sigma_Y).$$

The presence of unit operator $\mathbb{1}$ does not affect eigenstates (but shifts the eigenvalues) since $\mathbb{1}|\Psi\rangle = |\Psi\rangle$ and so we could associate operator **P** with one of the Pauli matrices σ_Y.

This exercise forces us to re-consider our simplistic $q$bulb analogy for a qubit. Though it explains the occupation number measurements of a qubit, as represented by the **N** operator, the Maxwell qubit also possesses the **P** property. It appears that a qubit is more complex than that described by the $q$bulb model. To provide a more realistic model of the qubit we proceed to discuss a purely quantum mechanical phenomenon called *spin*

Before discussing spin, I make a final comment on the qubit interpretation of light. Equation (2.37) offers a complete description of monochromatic electromagnetic waves without the necessity of a qubit interpretation. The alien scientists were forced to employ a Hilbert space interpretation of measurements because we gave them a limited set of tools, the **N** and **P** measurement devices. Those tools provided access only to a course-grained version of Maxwell's theory. If our friends had access to Maxwell's treatise, the qubit interpretation of light would be unnecessary. With the Maxwell theory, the aliens would be able to figure out the underlying mechanism responsible for the **N** and **P** outputs.

That is not the complete story. Maxwell's classical theory has been supplanted by a quantum version, called *Quantum Electrodynamics* or *QED* for short. Its development in the mid-twentieth century is one of the great achievements of quantum field theory. In QED, monochromatic light is described as an excitation of a quantum field, which in many ways has particle properties. This excitation is called a *photon*, and it is a physical realization of a qubit as it has properties similar to those elaborated in our parable. Despite the lesson of our parable, that a classical system can exhibit qubit like behavior, there exists no classical analog for a property shared by a pair of photons, a phenomenon called *entanglement*. The latter plays an important role in quantum information theory and takes center stage in our subsequent discussions.

2.4 Spin

It's time to heed our recommendation and jettison the simple $q$bulb model of the qubit. Instead, we need to identify a physical system that, along with the photon, exhibits all features of the qubit paradigm. The electron serves such a purpose. It was discovered in the mid-late nineteenth century, and to the best of our knowledge, every electron in the universe shares identical values of electric charge and mass.

Unlike protons and neutrons, electrons appear to be fundamental and are point-like. Despite that fact, electrons have a rich internal structure called spin.

Evidence of spin was discovered in experiments in which a beam of neutral atoms, that contains a single valence electron, are guided through a *Stern-Gerlach device (SG)*. In a typical experimental set-up, schematically illustrated in Fig. 2.3, atoms traverse a region where an inhomogeneous magnetic field deflects them. As they exit the device, they impinge on a detection screen that is used to analyze the deviation from their initial trajectory. In classical physics, the deflection force is proportional to the spatial gradient of $\vec{m} \cdot \vec{B}$, where $\vec{B}$ is the magnetic field vector and $\vec{m}$ is a *magnetic moment*. Because of the observed deflection, physicists hypothesized that the electron possesses an intrinsic magnetic moment. Magnetic moments commonly arise when electric charges form current loops. But the electron is a point particle and so attribution of an electron magnetic moment by this mechanism is problematic. In addition to empirical evidence, it was Paul Dirac who provided, in order to reconcile QM with the theory of special relativity, a convincing theoretical argument for the existence of spin (more precisely spin 1/2). Dirac showed how the electron's intrinsic magnetic moment is proportional to its spin property.

In a classical description $\vec{m}$ is distributed over all directions in space, and so in a set-up similar to that shown in Fig. 2.3, the deflection force engenders a continuous spectrum of paths. The SG experiments showed that the atoms are not deflected in this way. Instead, one observes the behavior illustrated in Fig. 2.3. After passing through the SG device, the atoms tend to segregate into two discrete regions on a detection screen. This binary behavior is reminiscent of our *q*bulb analogy in which the bulb is either on or off. Similarly, for the photon, it is either in the $|H\rangle$ or $|V\rangle$ state. We, therefore, postulate that ket $|0\rangle$ is the electron's internal spin state when the SG device detects the atom in the upper region of the screen, and $|1\rangle$ when detected in the lower region. These states are eigenstates of operator **n** which we now associate with the Stern-Gerlach device pointed along the *z*-direction. We perform additional measurements by orienting the SG device along different directions. Obvious choices

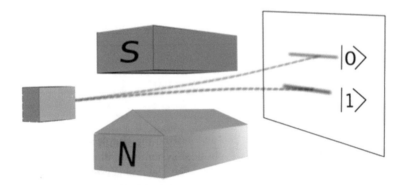

Fig. 2.3 Stern-Gerlach device bifurcates a single stream of atoms, due to their intrinsic spin, into two components. The atoms that segregate into the upper stream are said to be in state $|0\rangle$, while those deflected into the lower path are in the $|1\rangle$ state

2.4 Spin

are the x and y directions, and those measurements also reveal binary segregation of atom trajectories. So for the x-instrument the electron's internal state should also be described by a pair of kets that are eigenstates of a corresponding measurement operator, but which one? Experiments in which a neutral beam is passed through a set of SG devices, with different orientations, led to the quantum mechanical theory of spin discussed below.

2.4.1 Non-commuting Observables and the Uncertainty Principle

Interpreting the results of Stern-Gerlach measurements led physicist to conjecture a new, purely quantum mechanical, property of electrons called spin, or spin-1/2. Spin is represented by three Hermitian operators, corresponding to different orientations of an SG device with respect to axes defined by the beam direction. The operators that correspond to the measurement of electron spin are

$$\mathbf{S}_X \equiv \frac{\hbar}{2} \sigma_X$$
$$\mathbf{S}_Y \equiv \frac{\hbar}{2} \sigma_Y$$
$$\mathbf{S}_Z \equiv \frac{\hbar}{2} \sigma_Z. \qquad (2.45)$$

They are also expressed in vector form $\vec{S} = \mathbf{S}_X \hat{\mathbf{i}} + \mathbf{S}_Y \hat{\mathbf{j}} + \mathbf{S}_Z \hat{\mathbf{k}}$ where $\hat{\mathbf{i}}, \hat{\mathbf{j}}, \hat{\mathbf{k}}$ are the three orthogonal unit vectors of a Cartesian coordinate system. This expression introduces a new dimension-full quantity $\hbar$ into our narrative. It is called the reduced *Planck's constant* and has the physical units of angular momentum so that $\hbar = 6.626176 \times 10^{-34}$ J s. Using the results of the previous discussion, we find that the eigenvalues of each of the components of $\vec{S}$ has the values $\pm\hbar/2$. A measurement with the S_Z SG device produces the binary pattern on the detection screen illustrated in Fig. 2.3. Particles in the top pattern are in the $|0\rangle$ state, whereas atoms in the lower branch are in the $|1\rangle$ state. Because detection is a form of measurement, Postulate IV requires the system to collapse into those corresponding states. Now

$$\mathbf{S}_Z |0\rangle = \hbar/2 \, |0\rangle$$
$$\mathbf{S}_Z |1\rangle = -\hbar/2 \, |1\rangle \qquad (2.46)$$

and so we conclude that state $|0\rangle$ corresponds to a measurement in which $\mathbf{S}_Z$ has the value $\hbar/2$, whereas the value $-\hbar/2$ corresponds to the $|1\rangle$ state. Suppose an $\mathbf{S}_Z$ measurement is performed and we found the system to be in the $|0\rangle$ state. Those filtered atoms are then taken as an incident beam for a new SG device that is oriented along the x-axes, i.e. we measure atoms in the $|0\rangle$ state with SG device $\mathbf{S}_X$. Kets $|u\rangle, |v\rangle$ are eigenstates of $\mathbf{S}_X$ and, according to (2.20), are linear combinations of $|1\rangle$ and $|0\rangle$. Using the Born rule (Postulate III), and the fact that the system state

$$|0\rangle = \frac{1}{\sqrt{2}} \Big(|u\rangle + |v\rangle \Big),$$

we will obtain the eigenvalue $\hbar/2$ 50% of the time, and eigenvalue $-\hbar/2$ 50% of the time following a $\mathbf{S}_X$ measurement. Suppose we find that $\mathbf{S}_X = -\hbar/2$, the collapse postulate (Postulate IV) requires that the system "collapses" into state $|v\rangle$ following that measurement. Finally, we perform yet another measurement with device $\mathbf{S}_Z$. As the system collapsed into state

$$|v\rangle = \frac{1}{\sqrt{2}}\Big(|0\rangle - |1\rangle\Big),$$

there is a 50% probability that measurement with SG device $\mathbf{S}_Z$ finds the system in state $|1\rangle$, despite the fact that we seemed to have filtered this state during the first measurement with $\mathbf{S}_Z$! In summary, we made three consecutive measurements of an initial beam with devices $\mathbf{S}_Z, \mathbf{S}_X$ and then $\mathbf{S}_Z$ again. We found that $|\Psi\rangle$, the qubit Hilbert space amplitude collapses into state $|0\rangle$ following the initial measurement of $\mathbf{S}_Z$. Any subsequent measurement with $\mathbf{S}_Z$ will always find the eigenvalue $S_Z = \hbar/2$, but if we used $\mathbf{S}_X$ instead, followed by an $\mathbf{S}_Z$ measurement there is a 50% chance of obtaining the result $S_Z = -\hbar/2$. This, the propensity of certain measurements to influence the results of subsequent independent measurements on a qubit is a common feature of this theory. We will discuss, in later chapters, how it can be exploited to facilitate secure communications channels. To gain a deeper understanding of the mechanism behind this counterintuitive behavior lets explore the following scenario.

Imagine a qubit in state $|\Psi\rangle$. We perform measurements of it with any of the SG devices $\mathbf{S}_X, \mathbf{S}_Z, \mathbf{S}_Y$. Let's use $\mathbf{S}_Z$ where, according to the Born rule, $|\langle 0|\Psi\rangle|^2$ is the probability that the measurement results in the value $\hbar/2$. If we take many such measurements, always with the same $|\Psi\rangle$, we can calculate the mean value of all possible results obtained. According to probability theory, the mean or expectation, value $\bar{x}$ for a set of quantities x_i, also denoted by $<x>$, is

$$\bar{x} \equiv \sum_i p_i\, x_i \qquad (2.47)$$

where p_i is the probability for event x_i to occur. For the state $|\Psi\rangle = |u\rangle$, $\overline{S}_Z = 0$ since $p_{\hbar/2} = 1/2, x_1 = \hbar/2$ and $p_{-\hbar/2} = 1/2, x_2 = -\hbar/2$. In addition to the mean value, it is also useful to gauge the proclivity for a given measurement result to differ from its mean value. The latter is called the standard deviation, σ, of a measurement. It is defined

$$\sigma = \sqrt{\overline{(x-\bar{x})^2}} = \sqrt{\overline{(x^2 - 2x\bar{x} + \bar{x}^2)}} = \sqrt{\overline{x^2} - \bar{x}^2} \qquad (2.48)$$

where we used the fact that $\overline{x\,\bar{x}} = \bar{x}^2$. σ measures the average deviation of measurement results from the expectation value.

2.4 Spin

Consider now state $|\Psi\rangle$ given by (2.17) and from which we evaluate $\overline{S}_Z$. Using the Born rule and (2.46) we obtain

$$\overline{S}_Z \equiv \frac{\hbar}{2} p_{\hbar/2} - \frac{\hbar}{2} p_{-\hbar/2}$$
$$p_{\hbar/2} = |\langle 0|\Psi\rangle|^2 = \cos^2\theta/2 \quad p_{-\hbar/2} = |\langle 1|\Psi\rangle|^2 = \sin^2\theta/2. \quad (2.49)$$

Or

$$\overline{S}_Z = \frac{\hbar}{2}\cos\theta. \quad (2.50)$$

Since $p_i = |\langle i|\Psi\rangle|^2 = \langle\Psi|i\rangle\langle i|\Psi\rangle$, for $i = 0, 1$ we can re-express

$$\overline{S}_Z = \frac{\hbar}{2}\langle\Psi|0\rangle\langle 0|\Psi\rangle - \frac{\hbar}{2}\langle\Psi|1\rangle\langle 1|\Psi\rangle =$$
$$\langle\Psi|\mathbf{S}_z|0\rangle\langle 0|\Psi\rangle + \langle\Psi|\mathbf{S}_z|1\rangle\langle 1|\Psi\rangle \quad (2.51)$$

where, in deriving the second line, we used (2.46). Now

$$\langle\Psi|\mathbf{S}_z|0\rangle\langle 0|\Psi\rangle + \langle\Psi|\mathbf{S}_z|1\rangle\langle 1|\Psi\rangle = \langle\Psi|\mathbf{S}_z\Big(|0\rangle\langle 0| + |1\rangle\langle 1|\Big)|\Psi\rangle$$

and so, using the closure relation $|0\rangle\langle 0| + |1\rangle\langle 1| = \mathbb{1}$

$$\overline{S}_Z = \langle\Psi|\mathbf{S}_Z\mathbb{1}|\Psi\rangle = \langle\Psi|\mathbf{S}_Z|\Psi\rangle, \quad (2.52)$$

which when evaluated is in harmony with the result given by (2.50). Relation (2.52) informs us that the mean value $\overline{S}_Z$ is equal to the inner product of $|\Psi\rangle$ with state $\mathbf{S}_Z|\Psi\rangle$. It is a general result valid for any Hermitian operator and will come in handy later. Now

$$\mathbf{S}_Z\mathbf{S}_Z = \mathbf{S}_X\mathbf{S}_X = \mathbf{S}_Y\mathbf{S}_Y = \frac{\hbar^2}{4}\mathbb{1}$$

and so the standard deviation for measurement $\mathbf{S}_Z$ is

$$\sigma = \sqrt{\langle\Psi|\frac{\hbar^2}{4}|\Psi\rangle - \frac{\hbar^2}{4}\cos^2\theta} = \frac{\hbar}{2}\sin\theta. \quad (2.53)$$

Equation (2.53) is the average spread in values obtained by measurements with $\mathbf{S}_Z$, provided that the system is in state $|\Psi\rangle$. Because we can also pose this question for measurements with other devices, it is common practice to denote the standard deviation of an operator $\mathbf{A}$ with the symbol ΔA. For the $\mathbf{S}_Z$ instrument

$$\Delta S_Z = \frac{\hbar}{2}\sin\theta \quad 0 \le \theta \le \pi$$

and is also called the uncertainty of a measurement. The larger its value the more uncertain or spread in measured values. The maximum uncertainty for $\mathbf{S}_Z$ is $\frac{\hbar}{2}$. This conclusion makes sense since we know that $\mathbf{S}_Z$ has just two eigenvalues $\pm\frac{\hbar}{2}$, and so the uncertainty cannot be greater than the range of values obtained by a device. However, ΔS_Z can vanish. In that case, there is no uncertainty in the measurement. In other words, if an ensemble of experimenters each had identical copies of $|\Psi\rangle$ and if $\Delta S_Z = 0$, each measurement from each experiment leads to an identical result. According to (2.53) $\Delta S_Z = 0$ when $\theta = 0$ or $\theta = \pi$. Referring to the Bloch sphere in the previous section we recognize ket $|\Psi\rangle$, for $\theta = 0, \pi$ correspond to eigenstates of operator $\mathbf{S}_Z$. Consider the following question; if state $|\Psi\rangle$ leads to zero uncertainty in an SG measurement along the Z direction, i.e., $\Delta S_Z = 0$, does there exist an *SG* measurement along a different orientation axis that also leads to null uncertainty? Suppose the SG device is oriented along the following direction $\hat{\mathbf{n}} = n_x\hat{\mathbf{i}} + n_y\hat{\mathbf{j}} + n_z\hat{\mathbf{k}}$ where $n_x = \sin\Omega\cos\chi$, $n_y = \sin\Omega\sin\chi$, $n_z = \cos\Omega$ and Ω, χ are the polar and azimuthal angles of the point n_x, n_y, n_z. Along this direction

$$\mathbf{S}_n = \hat{\mathbf{n}} \cdot \vec{\mathbf{S}} = \frac{\hbar}{2}\left(\sin\Omega\cos\chi\sigma_X + \sin\Omega\sin\chi\sigma_Y + \cos\Omega\sigma_Z\right)$$

and using (2.17) we find that

$$\overline{S}_n = \langle\Psi|\mathbf{S}_n|\Psi\rangle = \frac{\hbar}{2}\left(\cos\theta\cos\Omega + \cos(\phi - \chi)\sin\theta\sin\Omega\right).$$

Now

$$\langle\Psi|\mathbf{S}_n^2|\Psi\rangle = \frac{\hbar^2}{4}$$

and so

$$\Delta S_n^2 = \frac{\hbar^2}{4} - \overline{S}_n^2 = \frac{\hbar^2}{4}\left(1 - (\cos\theta\cos\Omega + \cos(\phi - \chi)\sin\theta\sin\Omega)^2\right).$$

> Mathematica Notebook 2.5: Experimenting with uncertainty. http://www.physics.unlv.edu/%7Ebernard/MATH_book/Chap2/chap2_link.html; See also https://bernardzygelman.github.io

For the state $\theta = 0, \pi$, so that $\Delta S_Z = 0$,

$$\Delta S_n = \frac{\hbar}{2}\sin\Omega.$$

Thus $\Delta S_n = 0$ only if $\Omega = 0, \pi$, i.e. $\mathbf{S}_n$ must be oriented along the same axis as $\mathbf{S}_Z$. Therefore, it is impossible find an SG device with orientation different from of the $\mathbf{S}_Z$ device, in which the uncertainty ΔS_n also vanishes. This result is a consequence of the fact that $\mathbf{S}_Z$, and $\mathbf{S}_n$, for $\hat{\mathbf{n}}$ not along the Z axis, do not commute, i.e.

$$\mathbf{S}_Z\mathbf{S}_n - \mathbf{S}_n\mathbf{S}_Z \neq 0.$$

2.5 Direct Products

Non-commutivity of operators leads to the *uncertainty principle* which relates the product of uncertainty for two measurement operators with an expectation value of their commutator. Given two operators **A**, **B** that represent measurement devices

$$\Delta A^2 \Delta B^2 \geq \frac{1}{4}|\langle \Psi|[\mathbf{A}, \mathbf{B}]|\Psi\rangle|^2.$$

Proof of this theorem can be found in [1]. If both sides of this equality do not vanish the theorem provides a bound in the degree of uncertainty for each measurement.

2.5 Direct Products

In the previous sections we explored and summarized various properties of the qubit. In Chap. 1, we defined and employed the direct product to construct multi-qubit register states. In this section, we introduce the direct product operation for matrices and use the latter to represent multi-qubit states with matrices.

Given column matrices v_1 and v_2 of dimensions n and m respectively i.e.

$$v_1 = \begin{pmatrix} a_1 \\ a_2 \\ \vdots \\ a_n \end{pmatrix} \quad v_2 = \begin{pmatrix} b_1 \\ b_2 \\ \vdots \\ b_m \end{pmatrix}$$

their tensor, or direct, product $v_1 \otimes v_2$ is a column matrix of dimension $m \times n$ whose elements are arranged in the following manner

$$v_1 \otimes v_2 \equiv \begin{pmatrix} a_1 b_1 \\ a_1 b_2 \\ \vdots \\ a_1 b_m \\ \hline a_2 b_1 \\ \vdots \\ a_2 b_m \\ \hline a_n b_1 \\ \vdots \\ a_n b_m \end{pmatrix}. \tag{2.54}$$

(Note that the horizontal line is a deliminator, not a divisor symbol). This definition allows us to construct matrix representations of multi-qubit registers. For example, consider a register comprised of a qubit pair. In the previous chapter we noted that the basis vectors for this, four dimensional, Hilbert space are $|00\rangle$, $|01\rangle$, $|10\rangle$, and $|11\rangle$. Now since

$$|00\rangle = |0\rangle \otimes |0\rangle$$

we allow the association

$$|0\rangle \otimes |0\rangle \Leftrightarrow \begin{pmatrix} 1 \\ 0 \end{pmatrix} \otimes \begin{pmatrix} 1 \\ 0 \end{pmatrix}$$

and using definition (2.54) we obtain

$$|00\rangle \Rightarrow \begin{pmatrix} 1 \\ 0 \\ 0 \\ 0 \end{pmatrix}. \tag{2.55}$$

In the same manner, we find

$$|01\rangle = \begin{pmatrix} 0 \\ 1 \\ 0 \\ 0 \end{pmatrix}, \ |10\rangle = \begin{pmatrix} 0 \\ 0 \\ 1 \\ 0 \end{pmatrix}, \ |11\rangle = \begin{pmatrix} 0 \\ 0 \\ 0 \\ 1 \end{pmatrix}, \tag{2.56}$$

where we replaced the correspondence symbol with an equality. The four column matrices itemized above, represent the computational basis for the Hilbert space of a two-qubit register. The generalization to any n-qubit register is straightforward.

> Mathematica Notebook 2.6: Constructing matrix Kronecker products. http://www.physics.unlv.edu/%7Ebernard/MATH_book/Chap2/chap2_link.html; See also https://bernardzygelman.github.io

We defined direct products of operators in the previous chapter. Here we extend that definition into matrix language as follows. Given two operators $\mathbf{A}$, $\mathbf{B}$ in a n-dimensional Hilbert space which has as matrix representation the matrices $\underline{A}_{mn}$, $\underline{B}_{pq}$ the direct or Kronecker product $\underline{A} \otimes \underline{B}$ results in a matrix $\underline{C}$

$$\underline{C} = \begin{pmatrix} A_{11}\underline{B} & \cdots & A_{1n}\underline{B} \\ A_{21}\underline{B} & \cdots & A_{2n}\underline{B} \\ \vdots & \ddots & \vdots \\ A_{n1}\underline{B} & \cdots & A_{nn}\underline{B} \end{pmatrix} \tag{2.57}$$

where

$$A_{rs}\underline{B} \equiv A_{rs} \begin{pmatrix} B_{11} & \cdots & B_{1n} \\ B_{21} & \cdots & B_{2n} \\ \vdots & \ddots & \vdots \\ B_{n1} & \cdots & B_{nn} \end{pmatrix}.$$

2.5 Direct Products

In (1.31) we were somewhat nitpickish in making a distinction between a direct product of two kets, and that of operators. Definition (2.57) applies to both cases. For example, consider two qubit kets

$$|a\rangle = \begin{pmatrix} a_1 \\ a_2 \end{pmatrix} \quad |b\rangle = \begin{pmatrix} b_1 \\ b_2 \end{pmatrix}.$$

Since the kets are represented by column matrices we invoke (2.57) and find that the product ket $|a\rangle \otimes |b\rangle$ is identical to the result calculated with (2.54), a special case of expression (2.57). From now on, the symbol $\otimes$ is used to denote the direct product of both operators and kets(bras). For, consider application of (2.57) to operators $\mathbf{A}$, $\mathbf{B}$, and kets $|\psi\rangle$, $|\phi\rangle$ to construct the matrix representation of

$$(\mathbf{A} \otimes \mathbf{B}) \quad \text{and} \quad |\psi\rangle \otimes |\phi\rangle.$$

We find that

$$(\mathbf{A} \otimes \mathbf{B})(|\psi\rangle \otimes |\phi\rangle) = (\mathbf{A}|\psi\rangle) \otimes (\mathbf{B}|\phi\rangle), \tag{2.58}$$

in harmony with definition (1.31).

Problems

2.1 Do the exercises in Mathematica Notebook 2.1

2.2 Give the matrix representations of the states $|\psi\rangle = \frac{1}{\sqrt{2}}(|0\rangle + \exp(i\delta)|1\rangle)$, and $|\phi\rangle = \frac{1}{\sqrt{2}}(|0\rangle - \exp(i\beta)|1\rangle)$, and their dual.

2.3 Using the matrices obtained in Problem 2.2 evaluate $\langle\phi|\psi\rangle$, $\langle\psi|\phi\rangle$. Compare your results with that obtained using the methods discussed in Chap. 1.

2.4 Find the matrix representation for $|\phi\rangle\langle\psi|$ and $|\psi\rangle\langle\phi|$, where $|\psi\rangle$, $|\phi\rangle$ are defined in Problem 2.2.

2.5 Consider the operator

$$\mathbf{O} \equiv |0\rangle\langle 0| + i|1\rangle\langle 0| - i|0\rangle\langle 1| - |1\rangle\langle 1|.$$

(a) Evaluate, using Dirac's method discussed in Chap. 1, $\mathbf{O}|\psi\rangle$, where $|\psi\rangle$ is defined in Problem 2.2. (b) Evaluate by re-expressing $\mathbf{O}$ and $|\psi\rangle$ as matrices. Show that the results obtained in both pictures are isomorphic to each other.

2.6 Identify the following states on the Bloch sphere surface

(a) $|\psi_1\rangle = \dfrac{i}{\sqrt{10}}|0\rangle - \dfrac{3}{\sqrt{10}}|1\rangle$,

(b) $|\psi_2\rangle = \exp(i\pi/4)|0\rangle$,

(c) $|\psi_1\rangle = \dfrac{i}{\sqrt{2}}(|0\rangle - |1\rangle)$.

2.7 Using the matrix representations for the Pauli matrices, verify identities (2.25).

2.8 Given the matrix

$$\begin{pmatrix} 4 & -i\pi \\ 2\exp(i\pi/4) & 3 \end{pmatrix}$$

show that it can be expressed in the form (2.24), by identifying the values of the parameters α, β, b, c.

2.9 Find the conjugate transpose $\mathbf{U}^\dagger$ of expression (2.27). Evaluate the matrix product $\mathbf{U}\mathbf{U}^\dagger$ to confirm that $\mathbf{U}$ is unitary.

2.10 Use Mathematica Notebook 2.3 to exponentiate the operators $\sigma_X, \sigma_Y, \sigma_Z$, as defined in Eqs. (2.31) and (2.32). Using these results to confirm relation (2.33).

2.11 Use Mathematica Notebook 2.3 to construct the operator

$$\mathbf{W} = \mathbf{U}_Z(\phi/2)\mathbf{U}_Y(\theta/2)\mathbf{U}_Z(-\phi/2).$$

Demonstrate that $\mathbf{W}$ is unitary.

2.12 Use the operator that you obtained in Problem 2.11, to evaluate the following, (a) $\mathbf{W}^\dagger \sigma_X \mathbf{W}$, (b) $\mathbf{W}^\dagger \sigma_Y \mathbf{W}$, (c) $\mathbf{W}^\dagger \sigma_Z \mathbf{W}$. Comment on your results.

2.13 Consider the operator $\mathbf{A} = \mathbf{W}\sigma_X\mathbf{W}^\dagger$, where $\mathbf{W}$ is the operator defined in Problem (2.11), find the eigenvalues and eigenstates of $\mathbf{A}$.

2.14 Find the eigenvalues and eigenstates of operator

$$\mathbf{A} = \begin{pmatrix} a & \sqrt{2}+i\sqrt{2} \\ \sqrt{2}-i\sqrt{2} & a \end{pmatrix}$$

where a is a real number.

2.15 Use Mathematica Notebook 2.4 to plot, as a function of time, the electric field given by expression (2.37), for values of the parameters (a) $E_0 = 1, \delta = 0, \delta_0 = \pi, \theta = \pi/2$ (b) $E_0 = 1, \delta = 0, \delta_0 = 0, \theta = 0$, (c) $E_0 = 1, \delta = 0, \delta_0 = 0, \theta = \pi/4$.

2.16 Given the state $|\psi\rangle = \sqrt{\frac{3}{8}}|0\rangle + \sqrt{\frac{5}{8}}\exp(i\pi/4)|1\rangle$. Find the standard deviation of measurement with the operators (a) σ_X, (b) $\sigma_X\sigma_Y$, (c) σ_Z.

2.17 Find the matrix representation for the following multi-qubit kets.

(a) $|1\rangle \otimes |1\rangle \otimes |0\rangle$
(b) $|1\rangle \otimes |0\rangle \otimes |0\rangle$
(c) $|1\rangle \otimes (|1\rangle - |0\rangle) \otimes |0\rangle$
(d) $|1\rangle \otimes (|1\rangle - |0\rangle) \otimes (|1\rangle + |0\rangle)$

2.18 Find the matrix representation of the following operators. (a) $\sigma_X \otimes \mathbb{1}$, (b) $\mathbb{1} \otimes \sigma_X$, (c) $\sigma_X \otimes \sigma_X$, (d) $\sigma_X\sigma_X$.

2.19 Find the matrix representation of the operator,

$$\frac{1}{2}\mathbb{1} \otimes \mathbb{1} + \frac{1}{2}\sigma_Z \otimes \mathbb{1} + \frac{1}{2}\mathbb{1} \otimes \sigma_X - \frac{1}{2}\sigma_Z \otimes \sigma_X.$$

2.20 Using the definition for the Kronecker product of matrices, verify (2.58) for arbitrary one-qubit operators $\mathbf{A}$, $\mathbf{B}$ and states $|\psi\rangle$, $|\phi\rangle$.

Reference

1. K. Gottfried, T.-M. Yan, *Quantum Mechanics: Fundamentals* (Springer, 2003)

Circuit Model of Computation

3

Abstract

We provide a brief review of Boolean logic, the circuit model of computation, and I show how to assemble logic gates from their Boolean components. I present examples of classical circuits, including the half, full and ripple adder, and discuss reversible and irreversible gates. Quantum logic gates, including the Pauli, Hadamard and controlled-not gates are introduced, and we learn how to construct quantum circuits from them. The notions of quantum parallelism and interference enable Deutsch's algorithm, the first proof-of-principle for quantum advantage. We dissect the quantum circuit for the Deutsch-Josza algorithm and demonstrate its ability to perform massively parallel computations; a capability inaccessible to machines based on the bit paradigm. We introduce and discuss the unitary time development of quantum states.

3.1 Boole's Logic Tables

We do not fully understand how the brain intuits that $2 + 3 = 5$, but modern digital computing machines perform similar operations employing the *circuit model of computation*. The latter owes to the work of mathematician *George Boole* whose efforts, in the mid-nineteenth century, led to the logical gate concept, the theoretical underpinning of the electronic digital computer. Boole introduced a systematic procedure using *truth tables* that dissemble complex statements, e.g., $2+3 = 5$, into its smaller logical units. Consider the sum of two bits. Table 3.1 itemizes all possibilities for this computation. The table incorporates two fundamental components, the AND and XOR gates, whose truth tables are tabulated in Table 3.2.

Table 3.1 Truth table for the sum of two bits, including the carry bit value. X,Y are input bit variables, S and C represent the modular sum of two bits, and the carry bit variable respectively. The $\oplus$ symbol represents the base-2 modular addition operator

X	Y	$S(X \oplus Y)$	C
0	0	0	0
0	1	1	0
1	0	1	0
1	1	0	1

Table 3.2 Truth tables for XOR gate (left panel); and the AND gate (right panel). S represents the sum $X \oplus Y$, and C the carry bit aaaa

X	Y	S		X	Y	C
0	0	0		0	0	0
1	0	1		1	0	0
0	1	1		0	1	0
1	1	0		1	1	1

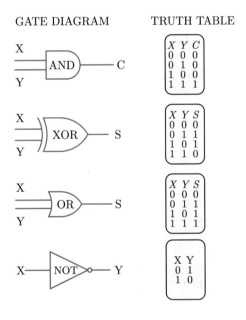

Fig. 3.1 Some important Boolean logic gates and their truth tables

They are examples of *Boolean logic gates*, which output only a single bit value. A pictorial representation of those gates is shown in Fig. 3.1 adjacent to their truth tables. In those diagrams, two horizontal wires (going from left to right) represent the first two columns of the truth tables. They shuttle the input variables X, Y into a "box"-like shape that represents a logical gate. The gate acts as a switch, whose

3.1 Boole's Logic Tables

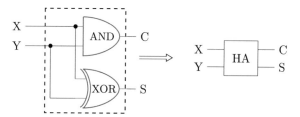

Fig. 3.2 The half-adder logic gate. Circuit components within the dotted line are consolidated to a "black box" labeled **HA**

job is, depending on the values for the X, Y variables, to determine the value of the output bit. Different box shapes represent different truth tables. So, in the AND gate, if $X = 1$ in the top wire and $Y = 0$ in the bottom wire, the output wire bit $C = 0$. Notice that the NOT gate includes one input wire.

In Fig. 3.2 two input leads are connected to a fan-out section that duplicates the input values X, Y and shuttles those values to the XOR and AND gates. Because each has one output, the diagram as a whole contains two output leads. We feed all possible combinations of X, Y into the input leads and find that this gate reproduces truth Table 3.1. It is called the half-adder and requires the two Boolean logic gates shown in Fig. 3.2. A major goal of circuit design is to find the most efficient and economical configuration of Boolean gates for a circuit, or computation.

The half-adder sums two bits, but adding larger numbers requires the accounting of carry bits. Table 3.3 itemizes all possible sums that include an input carry bit. The truth table contains three input columns for variables X, Y and carry bit C_{in}. The 2^3 row entries exhaust all input possibilities; the sum S and output carry C_{out} entries complete the truth table. Figure 3.3, the full adder circuit, is a diagrammatic representation of the logic gate architecture needed to reproduce truth Table 3.3. The gate includes three leads representing X, Y, C_{in} and two output wires that contain the result of the computation. It's often convenient to replace, in a pictorial representation of the full-adder, the three gate components with a "black" box containing three input and two output wires. The series connection of the full-adder "black" boxes in the

Table 3.3 Full adder truth table. C_{in} is the input carry bit value, C_{out} is the output carry bit value

X	Y	C_{in}	S	C_{out}
0	0	0	0	0
0	0	1	1	0
0	1	0	1	0
0	1	1	0	1
1	0	0	1	0
1	0	1	0	1
1	1	0	0	1
1	1	1	1	1

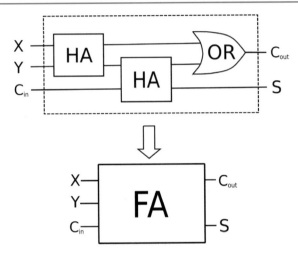

Fig. 3.3 The full-adder logic gate. Circuit components within the dotted line are consolidated to a "black box" labeled **FA**

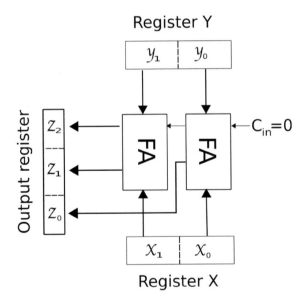

Fig. 3.4 An adding circuit, for a pair of two-bit registers

manner shown in Fig. 3.4 is called a ripple circuit. With it we can compute $2 + 3$. Two registers are needed to store the binary representations of the input and, at least, a three-bit register to store the output.

This circuit not only evaluates $2 + 3$ but correctly adds any pair of two-bit integers. For larger input registers, the generalization of the circuit shown in Fig. 3.4 is straightforward.

3.1 Boole's Logic Tables

> Mathematica Notebook 3.1: Simulating a 4-bit adder circuit. http://www.physics.unlv.edu/%7Ebernard/MATH_book/Chap3/chap3_link.html; See also https://www.bernardzygelman.github.io

3.1.1 Gates as Mappings

It's useful to think of computations, performed by the various gates, as the mapping

$$f : \{0, 1\}^n \to \{0, 1\}^m. \tag{3.1}$$

In this notation, symbol f is a gate identifier, also called a function, and $\{0, 1\}^n$ is shorthand for the set of all configurations of n zeros and ones. For example, $\{0, 1\}^2$ denotes the set $00, 01, 10, 11$.

Table 3.1 is a mapping of two bits, i.e. $\{0, 1\}^2$, into a space of the same dimension, and so the corresponding logic gate is expressed by the mapping

$$f_0 : 00 \to 00$$
$$f_0 : 01 \to 10$$
$$f_0 : 10 \to 10$$
$$f_0 : 11 \to 01. \tag{3.2}$$

That table does not exhaust all possible mappings as entries $00, 01, 10, 11$ can be mapped into a two-bit register in 256 different ways. The adder circuit in Fig. 3.4 defines the mapping

$$f_{ADD} : \{0, 1\}^4 \to \{0, 1\}^3$$

since two input registers, or 4 bits, map into a 3 bit register. Out of the possible 2^{48} unique functions, f_{ADD} is the one that provides the sums for all two-bit integer pairs. Mappings of the form

$$f : \{0, 1\}^n \to \{0, 1\}$$

are called *Boolean functions*. The XOR, AND and OR gates are Boolean functions for $n = 2$. The NOT gate is an example of a Boolean function with $n = 1$. A main insight of the circuit model of computation is that any function $f : \{0, 1\}^n \to \{0, 1\}^m$ can be analyzed into its Boolean components.

Of course, this fact does not tell us how many Boolean functions, or logic gates, are required to perform the calculation. The circuit in Fig. 3.4 includes two full adders, each containing a pair of half adders. Since the half-adders contain four Boolean gates, at least $2 \times 2 \times 3$, elementary Boolean gates are required to perform addition of a pair of two-bit numbers. A sub-discipline of computer science, *computational complexity* [1], concerns itself with questions related to the size and time resources required to perform a computation. There exist $2^{2^n m}$ possible functions for an $\{0, 1\}^n \to \{0, 1\}^m$ mapping, so finding f and expressing it in terms elementary Boolean gates is far from trivial. Indeed, many problems require computational resources that grow in an unbounded manner as n increases. One of the leading discoveries in the past quarter century is that some of these "exponentially hard" problems are tractable with a quantum computer.

It is important to point out that the circuit model of computation is not the only way to frame the question; what is a computation? Before the advent of electronic digital computing machines in the early-mid twentieth century, mathematicians and logicians, notably *Alan Turing* and *Alonzo Church*, developed a theoretical model for computation commonly called the Turing model. For our purposes, the Turing and circuit model of computation are equivalent, in this text we use the latter as our foundational framework.

3.2 Our First Quantum Circuit

In Sect. 3.1 we defined a computation as the implementation, by a suitable configuration of Boolean logic gates, of the mapping $f : \{0, 1\}^n \rightarrow \{0, 1\}^m$. In the circuit model, the n input wires are place-holders for members of the set $\{0, 1\}^n$ and in a typical electronic circuit, the diagrammatic "wire" represents a physical wire whose state, or bit assignment is determined by a current voltage. We assume that the wire "carries" bit value 0 if the current-voltage $V = 0$, and 1 when the current-voltage $V = V_0$. So for set $\{0, 1\}^2$ state 01 consists of two physical wires. The upper wire voltage is set at $V = 0$, and the lower at $V = V_0$. Boolean logic gates are switches that, depending on the possible input voltages $0 V_0$, $V_0 0$, 00, $V_0 V_0$, generate the appropriate output bit, or voltage.

In a quantum circuit, each wire lead represents a qubit ket. For now, we assume that the "wire" in a quantum circuit diagram represents a qubit in a definite quantum state. That state is not altered until the wire enters a quantum gate or a measurement device. By convention, time is assumed to run from left to right in the diagram, and the "wire" lead on the far left-hand side of a diagram denotes the initial qubit state. The qubit state is processed by a quantum gate and the wire lead exiting the gate from its right represents the output state of that qubit. Typically, quantum gates are shown as labeled boxes or solid and empty nodes.

The initial qubit ket could be in the off $|0\rangle$, or on $|1\rangle$, state, or in a linear combination of the two basis kets. Because Hilbert space operators induce maps between vectors (states) in Hilbert space, they serve as quantum analogs of classical gates. One crucial difference between a classical Boolean logic gate and a quantum logic gate is that a quantum gate must be *reversible*. The AND, OR, and other Boolean gates are typically *irreversible*. Consider the AND gate and suppose that its output is shuttled to a gate that we call AND^{-1}. We want AND^{-1} to reproduce the original input for the AND gate. Table 3.4 illustrates a possible truth table for AND^{-1}. However this table describes a one-to-many mapping and so is not a legitimate logic gate. In contrast, the NOT gate does allow an inverse since $NOT \cdot NOT = \mathbb{1}$, and so $NOT^{-1} = NOT$. Gate A that possesses inverse A^{-1}, so that $AA^{-1} = A^{-1}A = \mathbb{1}$

3.2 Our First Quantum Circuit

Table 3.4 Truth table for the illegal AND^{-1} gate

Input	Output
0	00
0	01
0	10
1	11

Table 3.5 Truth tables for the reversible NOT and NOT^{-1} gates

Input	Output
0	1
1	0

is called reversible. Otherwise it is irreversible. Therefore, the classical NOT gate is reversible, but the AND, OR and XOR gate are irreversible. Apparently, if a reversible gate contains n input bits, it must allow at least n output bits. In Chaps. 1 and 2 we showed that $\mathbf{U}$ is a unitary operator if it has the property

$$\mathbf{U}\mathbf{U}^\dagger = \mathbf{U}^\dagger\mathbf{U} = \mathbb{1}.$$

Because $\mathbf{U}^\dagger = \mathbf{U}^{-1}$ unitary operators are quantum logic gate candidates. Notably, the Pauli matrices $\sigma_X, \sigma_Y, \sigma_Z$ are unitary operators and serve as elementary quantum gates (Table 3.5).

Consider the action of the Pauli operator, σ_X, on ket $|0\rangle$. Using the matrix representation,

$$\sigma_X |0\rangle \Rightarrow \begin{pmatrix} 0 & 1 \\ 1 & 0 \end{pmatrix} \begin{pmatrix} 1 \\ 0 \end{pmatrix} = \begin{pmatrix} 0 \\ 1 \end{pmatrix} \Rightarrow |1\rangle. \tag{3.3}$$

In the same manner we find that $\sigma_X|1\rangle = |0\rangle$. Panels (a), (b) of Fig. 3.5 are pictorial representations of those transformations. The box denotes the σ_X gate and the incoming and outgoing wires the initial and final states of the qubit. Note the similarity of the classical NOT gate truth table, with the truth table generated by σ_X acting on the computational basis states. For this reason, the σ_X gate is often referred to as a quantum analog of the NOT gate. Unitary operator σ_Z also is an important quantum gate. It has the feature $\sigma_Z|0\rangle = |0\rangle$, $\sigma_Z|1\rangle = -|1\rangle$, and is an example of a *phase gate*. Because there is no classical analog for a phase, this example hints of an inherent quantum gate versatility not available in its classical counterpart.

Mathematica Notebook 3.2: Some quantum gates. http://www.physics.unlv.edu/%7Ebernard/MATH_book/Chap3/chap3_link.html; See also https://www.bernardzygelman.github.io

Fig. 3.5 The σ_X gate in a quantum logic circuit

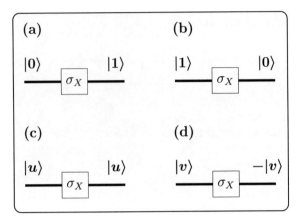

We explored the action of the Pauli gates on the computational basis kets, $|0\rangle$, $|1\rangle$, but superpositions of the latter are also valid qubit states. For example, given that the input qubit state is $|u\rangle = \frac{1}{\sqrt{2}}(|0\rangle + |1\rangle)$,

$$\sigma_X |u\rangle \Rightarrow \begin{pmatrix} 0 & 1 \\ 1 & 0 \end{pmatrix} \frac{1}{\sqrt{2}} \begin{pmatrix} 1 \\ 1 \end{pmatrix} = \frac{1}{\sqrt{2}} \begin{pmatrix} 1 \\ 1 \end{pmatrix} \Rightarrow |u\rangle. \tag{3.4}$$

Likewise, for state $|v\rangle \equiv \frac{1}{\sqrt{2}}(|0\rangle - |1\rangle)$, $\sigma_X|v\rangle = -\sigma_X|v\rangle$. Panels (c), (d) in Fig. 3.5 illustrate the action of the σ_X gate on the basis vectors $|u\rangle$, $|v\rangle$. The figure demonstrates that, for these states, the σ_X, or Pauli-X gate behaves as a phase gate. This chameleon-like, state-dependent, behavior is in stark contrast to that exhibited by classical gates. Nevertheless, convention dictates that the behavior of a quantum gate is characterized by its action on the computation basis vectors and so the σ_X gate is cited as a quantum NOT gate in the literature. Acknowledging the versatility of quantum gates, can we conjure a single qubit quantum gate whose properties are impossible to reproduce in the classical domain? To answer this question consider the following matrix, or operator,

$$\mathbf{U} = \frac{1}{2} \begin{pmatrix} 1+i & 1-i \\ 1-i & 1+i \end{pmatrix}. \tag{3.5}$$

We note that $\mathbf{U}$ is unitary and therefore is a quantum gate candidate. Evaluating the product of the matrix representation of this operator with itself, we find that $\mathbf{UU} = \sigma_X$. Therefore,

$$\mathbf{UU}|0\rangle = \sigma_X|0\rangle = |1\rangle$$
$$\mathbf{UU}|1\rangle = \sigma_X|1\rangle = |0\rangle. \tag{3.6}$$

The quantum circuit for this mapping is given in Fig. 3.6. Because the double action of $\mathbf{U}$ on the computational basis is equivalent to the single action of the σ_X, gate $\mathbf{U}$ is an effective square root operator of the quantum NOT gate, and has no classical

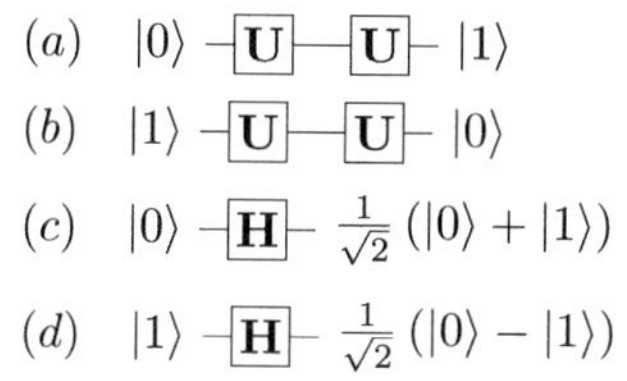

3.2 Our First Quantum Circuit

$$
\begin{aligned}
&(a) \quad |0\rangle -\boxed{U}-\boxed{U}- |1\rangle \\
&(b) \quad |1\rangle -\boxed{U}-\boxed{U}- |0\rangle \\
&(c) \quad |0\rangle -\boxed{H}- \tfrac{1}{\sqrt{2}}(|0\rangle + |1\rangle) \\
&(d) \quad |1\rangle -\boxed{H}- \tfrac{1}{\sqrt{2}}(|0\rangle - |1\rangle)
\end{aligned}
$$

Fig. 3.6 Some single qubit quantum gates. Panels **a**, **b** illustrate the action of the operator $\mathbf{U} \equiv \sqrt{\sigma_X}$, panels **c**, **d** shows the action of the Hadamard gate on the computational basis

analog. The *Hadamard* quantum logic gate is widely used in quantum circuits. A work-horse in applications, its matrix representation is

$$\mathbf{H} \equiv \frac{1}{\sqrt{2}} \begin{pmatrix} 1 & 1 \\ 1 & -1 \end{pmatrix}. \tag{3.7}$$

To investigate its properties we introduce the transformations

$$\begin{aligned}
\mathbf{H}|0\rangle &\Rightarrow \frac{1}{\sqrt{2}} \begin{pmatrix} 1 & 1 \\ 1 & -1 \end{pmatrix} \begin{pmatrix} 1 \\ 0 \end{pmatrix} = \frac{1}{\sqrt{2}} \begin{pmatrix} 1 \\ 1 \end{pmatrix} \Rightarrow |u\rangle \\
\mathbf{H}|1\rangle &\Rightarrow \frac{1}{\sqrt{2}} \begin{pmatrix} 1 & 1 \\ 1 & -1 \end{pmatrix} \begin{pmatrix} 0 \\ 1 \end{pmatrix} = \frac{1}{\sqrt{2}} \begin{pmatrix} 1 \\ -1 \end{pmatrix} \Rightarrow |v\rangle.
\end{aligned} \tag{3.8}$$

With the computational basis $|0\rangle$, $|1\rangle$ as inputs, the Hadamard gate outputs the linear superposition states $|u\rangle$, $|v\rangle$. In the subsequent chapters, we will see how this simple transformation allows for quantum algorithms that leapfrog the capabilities of a classical circuit.

3.2.1 Multi-qubit Gates

We now generalize the quantum gate concept for application in multi-qubit circuits. Let's first focus on a two-qubit system. The computational basis states for the latter are $|00\rangle$, $|01\rangle$, $|10\rangle$, $|11\rangle$. Quantum gates are operators in this Hilbert space and, as is the case for a single qubit, they must be unitary. In Chap. 2 we described how direct products of single qubit operators form multi-qubit operators. Here we demonstrate that the direct product of two single qubit unitary operators, $\mathbf{U}_a \otimes \mathbf{U}_b$, is also unitary. By definition

$$(\mathbf{U}_a \otimes \mathbf{U}_b)^\dagger = \mathbf{U}_a^\dagger \otimes \mathbf{U}_b^\dagger$$

and so

$$(\mathbf{U}_a \otimes \mathbf{U}_b)^\dagger (\mathbf{U}_a \otimes \mathbf{U}_b) = (\mathbf{U}_a^\dagger \mathbf{U}_a) \otimes (\mathbf{U}_b^\dagger \mathbf{U}_b) = \mathbb{1} \otimes \mathbb{1}$$

where we used the unitarity property of $\mathbf{U}_a$ and $\mathbf{U}_b$, and where $\mathbb{1}$ is the identity operator in the single qubit Hilbert space. Therefore,

$$\mathbb{1} \otimes \sigma_X, \sigma_X \otimes \sigma_X, \sigma_X \otimes \mathbb{1}, \sigma_Z \otimes \sigma_Y \tag{3.9}$$

are just a few possible two-qubit quantum gate candidates. In this, and subsequent, discussions it's important to remember and recognize the difference between the two expressions $\sigma_X \otimes \sigma_Y$, and $\sigma_X \sigma_Y$. The latter is a single qubit operator in which the Pauli-Y gate operates on a ket followed by an operation with the Pauli-X gate. The former is a two-qubit operator, each Pauli-X and Pauli-Y operator acts on the qubit in their respective slots. This distinction implies that

$$(\sigma_X \otimes \sigma_Y)^\dagger = \sigma_X^\dagger \otimes \sigma_Y^\dagger = \sigma_X \otimes \sigma_Y$$

$$(\sigma_X \sigma_Y)^\dagger = \sigma_Y^\dagger \sigma_X^\dagger = -\sigma_X \sigma_Y.$$

Some multi-qubit gates cannot be expressed as the direct product of qubits. Of these, one of the most important is the two-qubit controlled-not, or **CNOT**, gate. A pictorial representation of the **CNOT** gate is given in Fig. 3.7. It shows two input and output wires which are place holders for the computational basis $|x_1 x_0\rangle$ vectors, where $x_i \in \{0, 1\}$. According to this diagram coordinate x_1 is not altered under the action of the **CNOT** gate. If $x_1 = 0$, the state in the lower wire is also unaffected, and the gate acts as an identity operator. However if $x_1 = 1$, the lower bit "flips" as in a NOT gate. Symbolically, the gate, denoted by operator $\mathbf{U}_C$, is described by rule

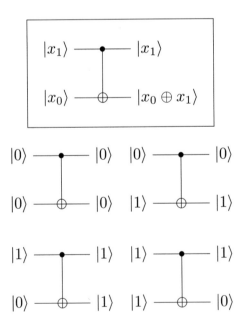

Fig. 3.7 The diagrammatic representation of the two-qubit **CNOT** gate. The lower panels illustrate the action of **CNOT** gate on input the basis kets $|00\rangle$, $|01\rangle$, $|10\rangle$, $|11\rangle$. $|x_1\rangle$ is the control qubit, and $|x_0\rangle$ the target

3.2 Our First Quantum Circuit

Table 3.6 Truth table for the **CNOT** gate

x_1	x_0	y_1	y_0
0	0	0	0
0	1	0	1
1	0	1	1
1	1	1	0

$\mathbf{U}_C |x_1 x_0\rangle \rightarrow |y_1 y_0\rangle$, where $y_1 = x_1$, $y_0 = x_0 \oplus x_1$. The transformation is also encapsulated by Table 3.6.

The table describes how the input ket $|x_1 x_0\rangle$ is mapped into ket $|y_1 y_0\rangle$. Unlike previous diagrams in which the gate proper is represented by a box-like shape, here the input and output wires are connected by a filled node in the top wire, connected to node $\oplus$ in the lower wire. In our convention, the upper wire represents $|x_1\rangle$, and the lower $|x_0\rangle$. In Fig. 3.7, x_1 is the coordinate of the *control bit*, and x_0 that of the *target bit*. In a **CNOT** gate diagram the control bit is always denoted by the wire that contains the solid node symbol.

Lets express $\mathbf{U}_C$ in terms of single q-bit unitary operators. Because $\mathbf{U}_C$ is the identity operator when the control bit coordinate $x_1 = 0$, and "flips" x_0 when $x_1 = 1$, we posit that

$$\mathbf{U}_C = |00\rangle\langle 00| + |01\rangle\langle 10| + |10\rangle\langle 11| + |11\rangle\langle 01|. \tag{3.10}$$

The truth table generated by this expression is in harmony with that of Table 3.6 and so (3.10) is an explicit Hilbert space representation of the **CNOT** gate. We re-express it, using the definition for the direct product, by

$$\mathbf{U}_C = |0\rangle\langle 0| \otimes |0\rangle\langle 0| + |0\rangle\langle 0| \otimes |1\rangle\langle 1| + |1\rangle\langle 1| \otimes |0\rangle\langle 1| + |1\rangle\langle 1| \otimes |1\rangle\langle 0|. \tag{3.11}$$

Now

$$|0\rangle\langle 0| = \frac{1}{2}(\mathbb{1} + \sigma_Z)$$

$$|1\rangle\langle 1| = \frac{1}{2}(\mathbb{1} - \sigma_Z)$$

$$|1\rangle\langle 0| = \frac{1}{2}(\sigma_X - i\sigma_Y)$$

$$|0\rangle\langle 1| = \frac{1}{2}(\sigma_X + i\sigma_Y) \tag{3.12}$$

and so

$$\mathbf{U}_C = \frac{1}{4}(\mathbb{1} + \sigma_Z) \otimes (\mathbb{1} + \sigma_Z) + \frac{1}{4}(\mathbb{1} + \sigma_Z) \otimes (\mathbb{1} - \sigma_Z) + \frac{1}{4}(\mathbb{1} - \sigma_Z) \otimes (\sigma_X + i\sigma_Y) + \frac{1}{4}(\mathbb{1} - \sigma_Z) \otimes (\sigma_X - i\sigma_Y). \tag{3.13}$$

Expanding and collecting terms we obtain

$$\mathbf{U}_C = \frac{1}{2}\mathbb{1} \otimes \mathbb{1} + \frac{1}{2}\sigma_Z \otimes \mathbb{1} + \frac{1}{2}\mathbb{1} \otimes \sigma_X - \frac{1}{2}\sigma_Z \otimes \sigma_X. \qquad (3.14)$$

Therefore, the controlled-not gate is a sum of direct product operators that cannot be factored. We obtain the matrix representation for $\mathbf{U}_C$ by replacing the Pauli, and unit operators, with their matrix representations so that the r.h.s. of (3.14) becomes

$$\frac{1}{2}\begin{pmatrix} 1 & 0 \\ 0 & 1 \end{pmatrix} \otimes \begin{pmatrix} 1 & 0 \\ 0 & 1 \end{pmatrix} + \frac{1}{2}\begin{pmatrix} 1 & 0 \\ 0 & -1 \end{pmatrix} \otimes \begin{pmatrix} 1 & 0 \\ 0 & 1 \end{pmatrix} +$$
$$\frac{1}{2}\begin{pmatrix} 1 & 0 \\ 0 & 1 \end{pmatrix} \otimes \begin{pmatrix} 0 & 1 \\ 1 & 0 \end{pmatrix} - \frac{1}{2}\begin{pmatrix} 1 & 0 \\ 0 & -1 \end{pmatrix} \otimes \begin{pmatrix} 0 & 1 \\ 1 & 0 \end{pmatrix}. \qquad (3.15)$$

Evaluating the matrix direct products we get

$$\mathbf{U}_C = \begin{pmatrix} 1 & 0 & 0 & 0 \\ 0 & 1 & 0 & 0 \\ 0 & 0 & 0 & 1 \\ 0 & 0 & 1 & 0 \end{pmatrix}. \qquad (3.16)$$

Expressing the two-qubit basis as column matrices

$$|00\rangle = \begin{pmatrix} 1 \\ 0 \end{pmatrix} \otimes \begin{pmatrix} 1 \\ 0 \end{pmatrix} = \begin{pmatrix} 1 \\ 0 \\ 0 \\ 0 \end{pmatrix}$$

$$|01\rangle = \begin{pmatrix} 1 \\ 0 \end{pmatrix} \otimes \begin{pmatrix} 0 \\ 1 \end{pmatrix} = \begin{pmatrix} 0 \\ 1 \\ 0 \\ 0 \end{pmatrix}$$

$$|10\rangle = \begin{pmatrix} 0 \\ 1 \end{pmatrix} \otimes \begin{pmatrix} 1 \\ 0 \end{pmatrix} = \begin{pmatrix} 0 \\ 0 \\ 1 \\ 0 \end{pmatrix}$$

$$|11\rangle = \begin{pmatrix} 0 \\ 1 \end{pmatrix} \otimes \begin{pmatrix} 0 \\ 1 \end{pmatrix} = \begin{pmatrix} 0 \\ 0 \\ 0 \\ 1 \end{pmatrix} \qquad (3.17)$$

and using (3.16), we find that $\mathbf{U}_C |00\rangle = |00\rangle$, $\mathbf{U}_C |01\rangle = |01\rangle$, $\mathbf{U}_C |10\rangle = |11\rangle$, $\mathbf{U}_C |11\rangle = |10\rangle$, and is equivalent to truth Table 3.6.

3.2.2 Deutsch's Algorithm

Above, we demonstrated the versatility of quantum gates and showed how the $\sqrt{NOT}$ gate has no classical analog, but have not yet addressed the central question: can one design a quantum circuit that out-performs, in a fundamental way, a classical circuit? The affirmative answer arrived in 1985 from *David Deutsch*, who outlined an algorithm, now called *Deutsch's algorithm*, that hinted at quantum capabilities beyond the reach of the classical circuit model. Although the algorithm is of limited practical value, it introduced the concepts of *quantum parallelism* and *quantum interference*. Both interference and quantum parallelism are invoked in transformational algorithms developed in the 1990s and that we discuss in the next chapter.

Logic gates perform mappings from the set $\{0, 1\}^n$ into $\{0, 1\}^m$ i.e. $f : \{0, 1\}^n \to \{0, 1\}^m$ where f is the name of the gate or function. Consider, for a given n, m, one of $2^{2^n m}$ functions, or gates, and label it with the symbol $\mathbf{U}_f$. In a classical circuit model, $\mathbf{U}_f$ is typically represented by a black box that has n input wire leads and m output wire leads. The black box in a reversible quantum circuit must represent a unitary operator, and so should have an equal number of input and output leads. The following protocol

$$\mathbf{U}_f \, |x\rangle_n \otimes |y\rangle_m = |x\rangle_n \otimes |y \oplus f(x)\rangle_m$$

where

$$|x\rangle_n \equiv |x_{n-1} \ldots x_1 \, x_0\rangle$$
$$|y\rangle_m \equiv |y_{m-1} \ldots y_1 \, y_0\rangle \,. \tag{3.18}$$

insures, as shown below, that quantum gate $\mathbf{U}_f$ is unitary.

Here $f(x)$ is shorthand for the string $f_{m-1}(x) \ldots f_1(x) \, f_0(x)$, where $f_k = \{0, 1\}$ is the k'th component, or binary digit in the kth slot, of map f. $|y \oplus f(x)\rangle_m$ is an m-qubit ket labeled by the set $y'_{m-1} \ldots y'_1 \, y'_0$ where $y'_k = y_k \oplus f_k(x)$. Remember that the modular sum $\oplus$ is defined so that $0 \oplus 0 = 0$, $0 \oplus 1 = 1$, $1 \oplus 0 = 1$, $1 \oplus 1 = 0$. If $n = m = 1$, protocol (3.18) is identical to that of the **CNOT** gate. We can think of (3.18), illustrated in Fig. 3.8, as a multi-qubit generalization of the latter.

For example, consider the mapping for the half-adder

$$f_{ha} : \{0, 1\}^2 \to \{0, 1\}^2 \tag{3.19}$$

defined in Table 3.7. According to protocol (3.18), truth Table 3.7 is reproduced by the four bold face entries in the last column of Table 3.8. Ket $|x\rangle$ contains the domain values x that are mapped into kets $|y'\rangle$. This follows from the fact $|y'\rangle = |y \oplus f(x)\rangle = |f(x)\rangle$ when $y = 00$. Why is it necessary to include the additional mappings, itemized in that table ? To answer this question we use definition (3.18) for $\mathbf{U}_f$ to evaluate

$$\mathbf{U}_f \mathbf{U}_f \, |x\rangle_n \otimes |y\rangle_m = \mathbf{U}_f \, |x\rangle_n \otimes |y \oplus f(x)\rangle_m = |x\rangle_n \otimes |y \oplus f(x) \oplus f(x)\rangle_m \,. \tag{3.20}$$

Now $|y \oplus f(x) \oplus f(x)\rangle = |y\rangle$ and so $\mathbf{U}_f \mathbf{U}_f$ equates to the identity operator. That is, $\mathbf{U}_f = \mathbf{U}_f^{-1}$ is a reversible gate. Let's construct the matrix representation,

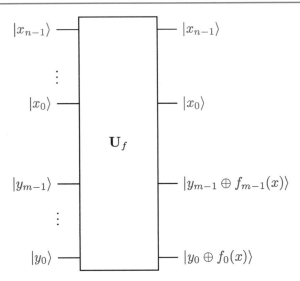

Fig. 3.8 Quantum gate that evaluates function $f(x)$. The function's domain is loaded into the register $|x\rangle \equiv |x_{n-1} \cdots x_1 x_0\rangle$ shown in red. Following processing by $\mathbf{U}_f$ register $|y\rangle \equiv |y_{m-1} \cdots y_1 y_0\rangle$ shown in blue, maps into $|y \oplus f(x)\rangle$. In that register, $f(x)$ is evaluated in entries where $y = 0$

Table 3.7 Explicit representation of function f_{ha}. First column itemizes the domain of the mapping, the second column itemizes the range of f_{ha}

x	$f_{ha}(x)$
0 0	0 0
0 1	0 1
1 0	0 1
1 1	1 0

using Table 3.8, for $\mathbf{U}_{f_{ha}}$. An explicit 16×16 block-diagonal matrix is obtained by connecting the indices in the first column of Table 3.8 with the corresponding indices in the last column of the table. Therefore

$$\mathbf{U}_{f_{ha}} = \begin{pmatrix} A & & \cdots & 0 \\ & B & & \vdots \\ \vdots & & B & \\ 0 & \cdots & & C \end{pmatrix}$$

$$A = \begin{pmatrix} 1&0&0&0 \\ 0&1&0&0 \\ 0&0&1&0 \\ 0&0&0&1 \end{pmatrix} \quad B = \begin{pmatrix} 0&1&0&0 \\ 1&0&0&0 \\ 0&0&0&1 \\ 0&0&1&0 \end{pmatrix} \quad C = \begin{pmatrix} 0&0&1&0 \\ 0&0&0&1 \\ 1&0&0&0 \\ 0&1&0&0 \end{pmatrix} \quad (3.21)$$

Note that $\mathbf{U}_{f_{ha}} \mathbf{U}^{\dagger}_{f_{ha}} = \mathbb{1}$, where $\mathbb{1}$ is the 16×16 identity matrix. Suppose we chose, instead of protocol (3.18),

$$\mathbf{W}_{f_{ha}} |x\rangle_n = |f_{ha}(x)\rangle_n. \quad (3.22)$$

3.2 Our First Quantum Circuit

Table 3.8 Protocol (3.18) for the mapping f_{ha}

Index	$\|x\rangle_n \otimes \|y\rangle_m$	$\|x\rangle_n \otimes \|y \oplus f_{ha}(x)\rangle_m$	$\|x\rangle_n \otimes \|y'\rangle_m$	Index
1	$\|00\rangle \otimes \|00\rangle$	$\|00\rangle \otimes \|00 \oplus f_{ha}(00)\rangle$	$\|00\rangle \otimes \|00\rangle$	1
2	$\|00\rangle \otimes \|01\rangle$	$\|00\rangle \otimes \|01 \oplus f_{ha}(00)\rangle$	$\|00\rangle \otimes \|01\rangle$	2
3	$\|00\rangle \otimes \|10\rangle$	$\|00\rangle \otimes \|10 \oplus f_{ha}(00)\rangle$	$\|00\rangle \otimes \|10\rangle$	3
4	$\|00\rangle \otimes \|11\rangle$	$\|00\rangle \otimes \|11 \oplus f_{ha}(00)\rangle$	$\|00\rangle \otimes \|11\rangle$	4
5	$\|01\rangle \otimes \|00\rangle$	$\|01\rangle \otimes \|00 \oplus f_{ha}(01)\rangle$	$\|01\rangle \otimes \|01\rangle$	6
6	$\|01\rangle \otimes \|01\rangle$	$\|01\rangle \otimes \|01 \oplus f_{ha}(01)\rangle$	$\|01\rangle \otimes \|00\rangle$	5
7	$\|01\rangle \otimes \|10\rangle$	$\|01\rangle \otimes \|10 \oplus f_{ha}(01)\rangle$	$\|01\rangle \otimes \|11\rangle$	8
8	$\|01\rangle \otimes \|11\rangle$	$\|01\rangle \otimes \|11 \oplus f_{ha}(01)\rangle$	$\|01\rangle \otimes \|10\rangle$	7
9	$\|10\rangle \otimes \|00\rangle$	$\|10\rangle \otimes \|00 \oplus f_{ha}(10)\rangle$	$\|10\rangle \otimes \|01\rangle$	10
10	$\|10\rangle \otimes \|01\rangle$	$\|10\rangle \otimes \|01 \oplus f_{ha}(10)\rangle$	$\|10\rangle \otimes \|00\rangle$	9
11	$\|10\rangle \otimes \|10\rangle$	$\|10\rangle \otimes \|10 \oplus f_{ha}(10)\rangle$	$\|10\rangle \otimes \|11\rangle$	12
12	$\|10\rangle \otimes \|11\rangle$	$\|10\rangle \otimes \|11 \oplus f_{ha}(10)\rangle$	$\|10\rangle \otimes \|10\rangle$	11
13	$\|11\rangle \otimes \|00\rangle$	$\|11\rangle \otimes \|00 \oplus f_{ha}(11)\rangle$	$\|11\rangle \otimes \|10\rangle$	15
14	$\|11\rangle \otimes \|01\rangle$	$\|11\rangle \otimes \|01 \oplus f_{ha}(11)\rangle$	$\|11\rangle \otimes \|11\rangle$	16
15	$\|11\rangle \otimes \|10\rangle$	$\|11\rangle \otimes \|10 \oplus f_{ha}(11)\rangle$	$\|11\rangle \otimes \|00\rangle$	13
16	$\|11\rangle \otimes \|11\rangle$	$\|11\rangle \otimes \|11 \oplus f_{ha}(11)\rangle$	$\|11\rangle \otimes \|01\rangle$	14

According to Table 3.7 the matrix representation of gate

$$\mathbf{W}_{f_{ha}} = \begin{pmatrix} 1 & 0 & 0 & 0 \\ 0 & 1 & 1 & 0 \\ 0 & 0 & 0 & 1 \\ 0 & 0 & 0 & 0 \end{pmatrix}. \tag{3.23}$$

As a 4×4 matrix operator it is considerably leaner than the 16×16 gate, $\mathbf{U}_{f_{ha}}$. However, since

$$\mathbf{W}_{f_{ha}} \mathbf{W}_{f_{ha}}^{\dagger} = \begin{pmatrix} 1 & 0 & 0 & 0 \\ 0 & 2 & 0 & 0 \\ 0 & 0 & 1 & 0 \\ 0 & 0 & 0 & 0 \end{pmatrix} \neq \mathbb{1} \tag{3.24}$$

it is not unitary or reversible. Because quantum gates are unitary, relation (3.24) informs us that $\mathbf{W}_{f_{ha}}$ is not a true quantum gate. The price for unitarity is an increase in the number of qubits, as evidenced by our construction of $\mathbf{U}_{f_{ha}}$, in the circuit. The fact that additional resources are needed to assure unitarity seems to counter our claim for quantum supremacy.

Nevertheless, Deutsch's algorithm offers a compelling demonstration for the latter. Let's consider the simplest possible function; one whose domain and range are single qubit registers. In that case we appeal to protocol (3.18) for $n = 1$ and $m = 1$. It supports $2^{2^n m} = 4$ single bit functions that are itemized in Table 3.9. Consider the quantum circuit shown in Fig. 3.9. In addition to the Hadamard gates and gate $\mathbf{U}_f$

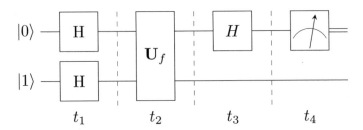

Fig. 3.9 Arrangement of quantum gates that implement Deutsch's algorithm. The dashed lines correspond to various time slices

the figure also contains a measurement instrument denoted by the box with the meter symbol in it. To analyze this circuit we partition the figure into the time slices denoted by t_i. Because time proceeds from left to right we follow the time development of this gate in that order.

Time slice t_0: The two-qubit register is in the direct product state

$$|0\rangle \otimes |1\rangle \equiv |01\rangle$$

Time slice t_1: Each q-bit is processed by a Hadamard gate. According to our previous discussion the bottom wire represents state

$$\mathbf{H}|1\rangle = \frac{1}{\sqrt{2}}(|0\rangle - |1\rangle)$$

whereas the top wire represents state

$$\mathbf{H}|0\rangle = \frac{1}{\sqrt{2}}(|0\rangle + |1\rangle),$$

therefore the two-qubit register is in state

$$|\psi(t_1)\rangle = \frac{1}{2}\Big((|0\rangle + |1\rangle) \otimes (|0\rangle - |1\rangle)\Big) = \frac{1}{2}\Big(|00\rangle + |10\rangle - |01\rangle - |11\rangle\Big)$$

Time slice t_2: State $|\psi(t_1)\rangle$ is processed by $\mathbf{U}_f$, and because it is a linear operator

$$|\psi(t_2)\rangle = \mathbf{U}_f |\psi(t_1)\rangle = \frac{1}{2}\Big(\mathbf{U}_f |00\rangle + \mathbf{U}_f |10\rangle - \mathbf{U}_f |01\rangle - \mathbf{U}_f |11\rangle\Big)$$

Now, according to (3.18),

$$\mathbf{U}_f |00\rangle = |0\rangle \otimes |0 \oplus f(0)\rangle = |0\rangle \otimes |f(0)\rangle$$
$$\mathbf{U}_f |10\rangle = |1\rangle \otimes |0 \oplus f(1)\rangle = |1\rangle \otimes |f(1)\rangle$$
$$\mathbf{U}_f |01\rangle = |0\rangle \otimes |1 \oplus f(0)\rangle = |0\rangle \otimes |f(0) \oplus 1\rangle$$
$$\mathbf{U}_f |11\rangle = |1\rangle \otimes |1 \oplus f(1)\rangle = |1\rangle \otimes |f(1) \oplus 1\rangle .$$

3.2 Our First Quantum Circuit

Therefore

$$|\psi(t_2)\rangle = \frac{1}{2}\Big(|0\rangle \otimes |f(0)\rangle + |1\rangle \otimes |f(1)\rangle - |0\rangle \otimes |f(0) \oplus 1\rangle - |1\rangle \otimes |f(1) \oplus 1\rangle\Big).$$

Time slice t_3: The register is processed by a Hadamard gate in the upper wire, and the identity operator in the lower. Thus

$$|\psi(t_3)\rangle =$$
$$\frac{1}{2} H \otimes \mathbb{1} \Big(|0\rangle \otimes |f(0)\rangle + |1\rangle \otimes |f(1)\rangle - |0\rangle \otimes |f(0) \oplus 1\rangle - |1\rangle \otimes |f(1) \oplus 1\rangle\Big) =$$
$$\frac{1}{2^{3/2}}\Big(|0\rangle + |1\rangle\Big) \otimes |f(0)\rangle + \frac{1}{2^{3/2}}\Big(|0\rangle - |1\rangle\Big) \otimes |f(1)\rangle -$$
$$\frac{1}{2^{3/2}}\Big(|0\rangle + |1\rangle\Big) \otimes |f(0) \oplus 1\rangle - \frac{1}{2^{3/2}}\Big(|0\rangle - |1\rangle\Big) \otimes |f(1) \oplus 1\rangle,$$

Time slice t_4: The measurement device interrogates the state of the qubit represented by the upper wire.

The measurement apparatus points to the value 0 if the upper wire qubit is in state $|0\rangle$, or 1 if it is in state $|1\rangle$. Because $|\psi(t_3)\rangle$ is a linear superposition of those states we need to calculate, using the Born rule, the relative outcome probabilities. The qubit of interest is part of a two-qubit system, so when using the Born rule we need to find the probabilities

$$p(i, j) \equiv |\langle x_i \, y_j | \psi(t_3)\rangle|^2$$

where y_j is the qubit value for the upper wire, and x_i that of the lower. The measuring device probes the value of the upper wire qubit, irrespective of the value x_i, and so

$$p(j) = \sum_{i=0,1} p(i, j).$$

Two outcomes are possible. If $f(0) = f(1)$

$$|\psi(t_3)\rangle = \frac{1}{2^{1/2}} |0\rangle \otimes (|f(0)\rangle - |f(0) \oplus 1\rangle)$$

and so

$$\langle 0\,1 | \psi(t_3)\rangle = 0 \quad \langle 1\,1 | \psi(t_3)\rangle = 0$$
$$\langle 0\,0 | \psi(t_3)\rangle = \frac{1}{\sqrt{2}} \langle 0 | f(0)\rangle - \frac{1}{\sqrt{2}} \langle 0 | f(0) \oplus 1\rangle$$
$$\langle 1\,0 | \psi(t_3)\rangle = \frac{1}{\sqrt{2}} \langle 1 | f(0)\rangle - \frac{1}{\sqrt{2}} \langle 1 | f(0) \oplus 1\rangle.$$

Regardless of the value $f(0)$, we find that

$$|\langle 0\,0 | \psi(t_3)\rangle|^2 = \frac{1}{2} \quad |\langle 1\,0 | \psi(t_3)\rangle|^2 = \frac{1}{2}$$

thus

$$p(j = 0) = 1 \quad p(j = 1) = 0. \tag{3.25}$$

In other words, if $f(0) = f(1)$ the meter is certain to register the value 0, and the value 1 if $f(0) \neq f(1)$, as that event is mutually exclusive from the latter.

Given the map $f : \{0, 1\}^n \rightarrow \{0, 1\}$, f is a *balanced function* if the image of f contains an equal number of zeos and ones. Alternatively, maps that assign $f(x) \rightarrow 0$, or $f(x) \rightarrow 1$, for all x are called *constant functions*. According to Table 3.9 the functions f^1 and f^2 are balanced.

> Mathematica Notebook 3.3: Deutsch's algorithm. http://www.physics.unlv.edu/%7Ebernard/MATH_book/Chap3/chap3_link.html; See also https://www.bernardzygelman.github.io

Imagine a "black box" that computes one of the functions tabulated in Table 3.9. Deutsch's problem asks the question; how many computations are required to infer whether the black box computes a balanced or constant function? To answer this question with a classical circuit we input the value 1 to the black box. If it computes f^1 or f^3 that bit is mapped to output value 1. Otherwise it is mapped to 0, corresponding to the functions f^0 or f^2. But those outcomes do not answer the question; is f balanced or constant? The correct solution is revealed if an additional computation, with input bit value 0, is performed, and so we need at least two computations to answer Deutsch's question definitively. In the quantum circuit that implements Deutsch's algorithm, we are assured of an answer in a single computation. As shown above, if the meter reading takes the value 0 we know with 100% certainty that the function is balanced. It is a constant function otherwise. Because the quantum circuit provides an answer to our query with only a single pass through gate $\mathbf{U}_f$, the latter outperforms the strategy based on classical gates. However, the quantum circuit requires more memory resources than its classical counterpart, and one could argue that shuttling bit values 1 and 0 simultaneously through two parallel classical gates, also delivers the correct solution. However, unlike the quantum gate, a classical gate requires the evaluation of f for all possible inputs. This example just entails the two values 0, 1, but what if the input register contains n bits? Re-phrasing that question; given a Boolean "black box" with a n-bit input register that computes a, single-bit output, function, either balanced or constant, find an algorithm that signals

Table 3.9 Truth table for all single bit gates f^0, f^1, f^2, f^3. First column is the domain $|x\rangle$, remaining columns itemize the target kets $|y\rangle$ for each f^i

x	f^0	f^1	f^2	f^3
1	0	1	0	1
0	0	0	1	1

3.2 Our First Quantum Circuit

which class of functions the black box computes. To interrogate the black box we supply it with some arbitrary bit values x, and so for the pair x_1, x_2, if we find that $f(x_1) \neq f(x_2)$ we know that f must be balanced. But there could be a situation where the first half of the 2^n inputs evaluate to a constant (i.e. 0, or 1), let's say $f = 0$. However, if the $2^n/2 + 1$st input takes the value 1, then the function must be balanced. In that case, we require at least $2^n/2 + 1$ function evaluations, or a massively parallel machine that scales exponentially with n.

3.2.3 Deutsch-Josza Algorithm

A generalization of Deutsch's algorithm, the *Deutsch-Josza algorithm*, answers Deutsch's question for a n-qubit system with only a single query of $\mathbf{U}_f$. It is implemented in a straight-forward generalization of the circuit diagram in Fig. (3.9). With this algorithm the system is initialized to state $|\psi_{in}\rangle = |0\rangle_n \otimes |1\rangle$, where the symbol $|0\rangle_n$ is shorthand for the direct product of n-qubits, all in state $|0\rangle$. At time t_1 the system is found in state

$$|\psi(t_1)\rangle = \mathbf{H}^{\otimes n} |0\rangle_n \otimes \mathbf{H} |1\rangle$$

where the symbol $\mathbf{H}^{\otimes n}$ is a direct product of n Hadamard gates. From our previous discussion we know that

$$\mathbf{H} |0\rangle = \frac{1}{\sqrt{2}} (|0\rangle + |1\rangle)$$

so

$$\mathbf{H} \otimes \mathbf{H} |0\rangle \otimes |0\rangle = \frac{1}{2} (|0\rangle + |1\rangle) \otimes (|0\rangle + |1\rangle) =$$
$$\frac{1}{2} (|00\rangle + |01\rangle + |10\rangle + |11\rangle) = \frac{1}{2} (|0\rangle_2 + |1\rangle_2 + |2\rangle_2 + |3\rangle_2)$$

and

$$\mathbf{H} \otimes \mathbf{H} \otimes \mathbf{H} |0\rangle \otimes |0\rangle \otimes |0\rangle =$$
$$\frac{1}{2^{3/2}} (|0\rangle + |1\rangle) \otimes (|0\rangle + |1\rangle)(|0\rangle + |1\rangle) = \frac{1}{2^{3/2}} \sum_{x=0}^{x=7} |x\rangle_3 .$$

Generalizing to n qubits, we obtain the relation

$$\mathbf{H}^{\otimes n} |0\rangle_n = \frac{1}{2^{n/2}} \sum_{x=0}^{2^n-1} |x\rangle_n . \tag{3.26}$$

Therefore

$$|\psi(t_1)\rangle = \frac{1}{2^{n/2}} \sum_{x=0}^{2^n-1} |x\rangle_n \otimes \frac{1}{\sqrt{2}} (|0\rangle - |1\rangle). \tag{3.27}$$

The set $\mathbf{H}^{\otimes n}$ of Hadamard gates maps the input register, consisting of $n+1$ qubits in state $|0\rangle_n \otimes |1\rangle$ into a linear combination of the 2^n computational basis kets. With a modest number of qubits, let's say $n = 300$, the system amplitude $|\psi(t_3)\rangle$ samples a space of all 2^{300} possible measurement outcomes, a figure larger than the number of electrons in the Universe! The ability of a battery of Hadamard gates to map a given ket, e.g., $|0\rangle_n$ into this linear combination is called quantum parallelism, and it is one of the "secret sauces" that empowers a quantum computer. Because $|\psi(t_2)\rangle$ is a linear combination of all basis kets $|x\rangle_n$, $\mathbf{U}_f |\psi(t_2)\rangle$ evaluates $f(x)$ for all 2^n values of the argument. Contrast this capability with that of a classical computer. The latter can only evaluate $f(x)$ for a single x at a time, or it requires massively parallel capabilities to evaluate f for all x. As n gets larger the resources required to evaluate $f(x)$ by classical computer scale as a power of n. Unfortunately, Postulate IV, discussed in Chap. 1, does not allow access to all information contained in $|\psi(t_3)\rangle$. To gain useful information we take advantage of the other "secret sauce", *interference*. At t_2 the amplitude is processed by $\mathbf{U}_f$ so that

$$|\psi(t_2)\rangle = \mathbf{U}_f |\psi(t_1)\rangle = \frac{1}{2^{n/2}} \sum_{x=0}^{2^n-1} |x\rangle_n \otimes \frac{1}{\sqrt{2}} \Big(|f(x)\rangle - |f(x) \oplus 1\rangle \Big). \quad (3.28)$$

We know that for any x, $f(x)$ is either 0, or 1. If $f(x) = 0$ the term in parentheses in (3.28) assumes the form $(|0\rangle - |1\rangle)$, and $(|1\rangle - |0\rangle)$ if $f(x) = 1$. Or, $(|f(x)\rangle - |f(x) \oplus 1\rangle) = (-1)^{f(x)}(|0\rangle - |1\rangle)$ and

$$|\psi(t_2)\rangle = \frac{1}{2^{n/2}} \sum_{x=0}^{2^n-1} (-1)^{f(x)} |x\rangle_n \otimes \frac{1}{\sqrt{2}} \Big(|0\rangle - |1\rangle \Big). \quad (3.29)$$

We apply gates $\mathbf{H}^{\otimes n} \otimes \mathbb{1}$ on $|\psi(t_2)\rangle$ so that

$$|\psi(t_3)\rangle = \frac{1}{2^{n/2}} \sum_{x=0}^{2^n-1} \mathbf{H}^{\otimes n} (-1)^{f(x)} |x\rangle_n \otimes \frac{1}{\sqrt{2}} \Big(|0\rangle - |1\rangle \Big). \quad (3.30)$$

Remember, the only information we are privy to, is that the "black box" $\mathbf{U}_f$ is either a balanced or constant function and we must find a way to figure out which it is. We measure each qubit in the upper register of state $|\psi(t_3)\rangle$ where, according to the Born rule,

$$p(z) = |\langle z| \otimes \langle 0| |\psi(t_3)\rangle|^2 + |\langle z| \otimes \langle 1| |\psi(t_3)\rangle|^2 =$$
$$\frac{1}{2^n} |\sum_{x=0}^{2^n-1} \langle z_0 z_1 \ldots z_{n-1} | \mathbf{H}^{\otimes n} |x\rangle_n (-1)^{f(x)}|^2$$
$$|z\rangle \equiv |z_{n-1} \ldots z_1 z_0\rangle \quad z_i \in \{0, 1\} \quad (3.31)$$

3.2 Our First Quantum Circuit

is the joint probability that the upper register is found in the n-qubit state $|z\rangle$. Let's calculate the probability that all measurement devices for the top register show the value 0, i.e. $z_0 = z_1 \ldots z_{n-1} = 0$, we then need to evaluate

$$_n\langle 0| \mathbf{H}^{\otimes n} |x\rangle_n = \frac{1}{2^{n/2}} \sum_{z=0}^{2^n-1} \langle z|x\rangle_n = \frac{1}{2^{n/2}} \sum_{z=0}^{2^n-1} \delta_{z,x} = \frac{1}{2^{n/2}}. \quad (3.32)$$

In deriving (3.32) we used (3.26), the fact that $\mathbf{H}^{\otimes n}$ is Hermitian, and the orthogonality property of the register computational basis. Therefore,

$$p(z=0) = \frac{1}{2^n} |\sum_{x=0}^{2^n-1} (-1)^{f(x)} \frac{1}{2^{n/2}}|^2. \quad (3.33)$$

Suppose that $f(x)$ is a constant function, in that case $(-1)^{f(x)}$ is either ± 1 for all x, and the sum (3.33) reduces to

$$p(z=0) = \frac{1}{2^n} |\sum_{x=0}^{2^n-1} \frac{1}{2^{n/2}}|^2 = 1. \quad (3.34)$$

We used interference, as evident in expression (3.33), to enhance the probability for a desired outcome. Indeed, (3.34) tells us that the probability to identify a constant function is 1, or certainty, if all meters point to the 0 position. Likewise, if at least one meter points to the value 1, it is certain that the function is balanced.

A classical computer, confronted with the n-bit Deutsch-Josza problem, requires parallel computing resources, in a worst case scenario, that scale as the n'th power of a constant. In contrast, the "size" of a quantum computer taking advantage of the algorithm outlined above needs to scale only as a fixed power of n, the length of the input register. Using a combination of quantum parallelism and interference, the Deutsch-Josza algorithm provides a compelling demonstration of capabilities not available in the classical circuit model of computation. By the early to mid 1990s additional quantum algorithms addressing Simon's and the Bernstein-Vazirani problems [1] bolstered the emerging consensus that quantum computers could solve certain problems much more efficiently than known classical algorithms. Though they demonstrated the inherent potential of quantum computers, these algorithms had limited utility in applications. Until 1994, quantum computer science was considered a niche discipline, but in that year *Peter Shor* introduced a powerful quantum algorithm, considered by many as the first quantum "killer-app", that has wide-ranging applications spanning many fields of science and technology. We dedicate the next chapter to a discussion of *Shor's algorithm*, but first tie-up a loose end that we have barely addressed in our narrative.

3.3 Hamiltonian Evolution

In Fig. 3.9 we partitioned the circuit into time domains starting from t_0 on the l.h.s. of the diagram to t_4 on the right. We are somewhat cavalier in using the parameter t as a physical time, the notation acknowledges that $t_4 > t_3 > t_2 > t_1 > t_0$ and so in this sense the system amplitude evolves with t. However, for an n-qubit system the true, physical time, development is determined by the following postulate.

> Postulate V: For a closed n-qubit system, the Hilbert space amplitude $|\psi(t)\rangle$ evolves according to
>
> $$|\psi(t)\rangle = \mathbf{U}(t, t_0) |\psi(t_0)\rangle \qquad (3.35)$$
>
> where $\mathbf{U}(t_0, t_0) = \mathbb{1}$,
>
> $$i\hbar \frac{d\mathbf{U}(t, t_0)}{dt} = \mathcal{H}(t)\mathbf{U}(t, t_0) \qquad (3.36)$$
>
> and $\mathcal{H}(t)$ is an Hermitian operator.

The first order differential equation (3.36) is called the *Schroedinger equation* and allows the formal solution, also called the *Volterra-Dyson series*,

$$\mathbf{U}(t, t_0) = \mathbb{1} - i/\hbar \int_{t_0}^{t} \mathcal{H}(t')dt' + (-i/\hbar)^2 \int_{t_0}^{t} \mathcal{H}(t_1)dt_1 \int_{t_0}^{t_1} \mathcal{H}(t_2)dt_2 + \ldots$$
$$t > t_1 > t_2 > \cdots > t_0 \qquad (3.37)$$

The infinite series on the r.h.s. of (3.37) is commonly denoted by the short-hand symbol

$$T \exp(-i \int_{t_0}^{t} dt' \mathcal{H}(t')/\hbar).$$

In special cases, e.g. when $\mathcal{H}$ is time-independent, it simplifies to

$$T \exp(-i \int_{t_0}^{t} dt' \mathcal{H}(t')/\hbar) = \exp(-i \frac{\mathcal{H}\tau}{\hbar}) \quad \tau = t - t_0 \qquad (3.38)$$

where exp denotes exponentiation of operator $\mathcal{H}$.

In general, $\mathbf{U}(t, t_0)$, defined by (3.37), has the property

$$\mathbf{U}(t_n, t_0) = \mathbf{U}(t_n, t_{n-1}) \ldots \mathbf{U}(t_2, t_1)\mathbf{U}(t_1, t_0) \qquad (3.39)$$

3.3 Hamiltonian Evolution

provided that the time ordering $t_n > t_{n-1} \ldots t_2 > t_1 > t_0$, is imposed as a *boundary condition*. Importantly

$$\mathbf{U}(t, t_0)\mathbf{U}^\dagger(t, t_0) = \mathbf{U}^\dagger(t, t_0)\mathbf{U}(t, t_0) = \mathbb{1}, \tag{3.40}$$

and, therefore, $\mathbf{U}(t, t_0)$ describes unitary evolution. Operator $\mathcal{H}(t)$ is called the *Hamiltonian* and its explicit form depends on the details of the physical system. For example, if the system is an electron subjected to a magnetic field, the Hamiltonian is

$$\mathcal{H}(t) = \mu_0(\sigma_X B_X(t) + \sigma_Y B_Y(t) + \sigma_Z B_Z(t)) \tag{3.41}$$

where μ_0, the *Bohr magneton*, is a constant, $B_i(t)$ is the i'th component of a magnetic field, σ_i the Pauli matrices and index i denotes coordinates $\{X, Y, Z\}$. Suppose a constant external magnetic field, of magnitude B_0, is pointed along the z direction. Then

$$\mathcal{H}(t) = \mu_0 \sigma_Z B_0$$

and, because $\mathcal{H}(t)$ is time-independent,

$$\mathbf{U}(t, t_0) = \exp(-i\mu_0 B_0 \sigma_Z \tau/\hbar) = \begin{pmatrix} \exp(-i\omega\tau) & 0 \\ 0 & \exp(i\omega\tau) \end{pmatrix}$$
$$\omega \equiv \mu_0 B_0/\hbar \quad \tau \equiv (t - t_0). \tag{3.42}$$

If the qubit (electron) is in state $|0\rangle$ at $t = t_0$ it evolves, according to the time evolution operator (3.42), into state $\exp(-i\omega(t - t_0))|0\rangle$ at time $t > t_0$. Likewise state $|1\rangle$ evolves into $\exp(i\omega(t - t_0))|1\rangle$. Now

$$\mathbf{U}(t, t_0) = \exp(-i\phi)\mathbf{P}(t, t_0)$$
$$\mathbf{P}(t, t_0) \equiv \begin{pmatrix} 1 & 0 \\ 0 & \exp(2i\phi) \end{pmatrix} \quad \phi \equiv \omega\tau. \tag{3.43}$$

If we neglect the global phase factor $\exp(-i\phi)$ the time development of the qubit is governed by operator $\mathbf{P}$, a *phase shift gate*. Two special cases for this operator are commonly used in circuits, the so-called $\pi/8$, or $\mathbf{T}$ gate, and the $\mathbf{S}$, gate. They correspond to the cases $\phi = \pi/8$ for $\mathbf{T}$ and $\phi = \pi/4$ for $\mathbf{S}$. Explicitly

$$\mathbf{T} = \begin{pmatrix} 1 & 0 \\ 0 & \exp(i\pi/4) \end{pmatrix} \quad \mathbf{S} = \begin{pmatrix} 1 & 0 \\ 0 & i \end{pmatrix}. \tag{3.44}$$

The time development of an (electron) qubit in magnetic field $B_0 \hat{\mathbf{k}}$ is illustrated in Fig. 3.10. In that figure, the bare wire represent the electron qubit. The diagram is labeled with time slices t_0 and t, the interval in which the magnetic field is present. In that interval, the qubit suffers a phase change denoted in the figure by the solid circular shape. The figure, if taken literally, is misleading because the phase change of the qubit is not instantiated in a single instance, as suggested by the illustration. Instead, ϕ changes incrementally in the interval from t_0 to t. So in that figure, the

Fig. 3.10 Phase shift gate, parameters t_0, t_1 give the correct time

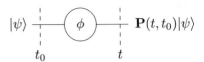

labels t_0 and t_1 are suggestive of the correct time-ordering, if not the actual physical time.

Problems

3.1 Using the four-bit adder circuit given in Notebook 3.1, evaluate the binary sum $0101 + 0001$. Does this circuit predict the correct answer for $1000 + 1000$? If not, generalize it so that the correct result is obtained.

3.2 Itemize all possible functions for the mapping $f : \{0, 1\}^2 \rightarrow \{0, 1\}^1$. Show that the mapping $f : \{0, 1\}^n \rightarrow \{0, 1\}^m$ allows $2^{2^n m}$ unique functions f.

3.3 The state $|\psi\rangle = \alpha |0\rangle + \beta |1\rangle$ is input to a σ_X gate, find the output state. Repeat for the σ_Y and σ_Z gates.

3.4 Find the ouput of the two-qubit gate $\sigma_X \otimes \mathbb{1}$ for the following inputs (a) $|00\rangle$, (b) $|01\rangle$, (c) $|10\rangle$, and (d) $|11\rangle$. Repeat with the gate $\mathbb{1} \otimes \sigma_X$. Are the truth tables for these two gates identical? Comment.

3.5 Below, in Fig. 3.11 panel (a), the symbol σ_i, for $i = X, Y, Z$ refers to the Pauli gates. Evaluate the truth tables for this circuit for each of the Pauli gates. Compare your result for the σ_X gate, with the truth table for the **CNOT** gate shown in Fig. 3.7. Comment.

3.6 Repeat Problem 3.5 for the the gate shown in panel (b) of Fig. 3.11.

3.7 Construct the truth table for the circuit shown in Fig. 3.12. Compare your result with that obtained in Problem 3,6 (b) for the case $\sigma_i = \sigma_X$.

3.8 Construct the matrix representation for the gate shown in Fig. 3.12.

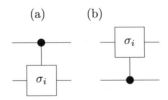

Fig. 3.11 Diagram for Problems 3.5 and 3.6

3.3 Hamiltonian Evolution

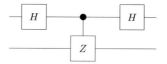

Fig. 3.12 Diagram for Problems 3.7 and 3.8

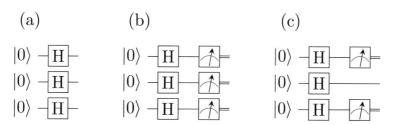

Fig. 3.13 Diagram for Problems 3.9 through 3.11

Table 3.10 Truth table

x	f(x)
0 0	0 1
0 1	1 0
1 0	1 0
1 1	0 1

3.9 In panel (a) of Fig. 3.13 the state $|000\rangle$ is input to the Hadamard gates shown in that figure. Find the output of this three-qubit gate.

3.10 Using Born's rule find the probability that the measurement apparatuses shown in panel (b) of Fig. 3.13 detects the state $|010\rangle$.

3.11 Using Born's rule find the probability that the measurement apparatus at the bottom-most wire of panel (c) in Fig. 3.13 measures the value 1.

3.12 Consider the function f defined by truth Table 3.10. Is this function (a) constant (b) balanced (c) neither of those options ?

3.13 Construct a classical gate, composed of Boolean gates, that evaluates the function defined in Table 3.10.

3.14 Using protocol (3.18), construct a quantum gate that evaluates the function defined in Table 3.10. You should present this gate in the form of a table similar to that given by Table 3.8.

3.15 Write a Mathematica code that expresses the gate obtained in Problem 3.14 as a unitary matrix $\mathbf{U}_f$. Using that matrix, find the output for the following input states $|00\rangle$, $|01\rangle$, $|10\rangle$, $|11\rangle$. Find the output for the input state

$$|\psi\rangle = 1/2\,(|00\rangle - |01\rangle + |10\rangle - |11\rangle).$$

3.16 Repeat Problem 3.15 for the input state given by $|\psi\rangle = \mathbf{H} \otimes \mathbf{H}\,|00\rangle$, where $\mathbf{H}$ is the Hadamard gate.

3.17 Consider the mapping $f : \{0, 1\}^3 \to \{0, 1\}^1$ so that $f(x) = 0$ for all x in $\{0, 1\}^3$, except for $x = 010$ in which $f(010) = 1$. Write a Mathematica code for a quantum gate, $\mathbf{U}_f$, that evaluates this function.

3.18 For the function f defined in Problem 3.17 use $\mathbf{U}_f$ to evaluate $\mathbf{U}_f\,\mathbf{H} \otimes \mathbf{H} \otimes \mathbf{H}\,|000\rangle$. Estimate the probability that a measurement of the last qubit in the output register gives the value 1.

3.19 An electron is subjected to a magnetic field, pointing along the z-axis, of magnitude B_0. At $t = 0$ the electron is in the state $\mathbf{H}\,|0\rangle$. Estimate the probabilities to find the electron in its ground state $|0\rangle$ at the following time intervals

(a) $t = \dfrac{\hbar}{\mu_0 B_0}\dfrac{\pi}{4}$

(b) $t = \dfrac{\hbar}{\mu_0 B_0}\dfrac{\pi}{2}$

(c) $t = \dfrac{\hbar}{\mu_0 B_0}\pi$

3.20 Repeat Problem 3.19 for the input state $\mathbf{H}\,|1\rangle$.

Reference

1. M.E. Nielsen, I.L. Chuang, *Quantum Computation and Quantum Information* (Press, Cambridge U, 2011)

Quantum Killer Apps: Quantum Fourier Transform and Search Algorithms

Abstract

We present a brief overview of the Fourier series, the Fourier and discrete Fourier transforms and their applications. We discuss a quantum algorithm that encodes the Fourier transform of the mapping $f : \{0, 1\}^n \to \{0, 1\}$ in an n-qubit register. It's shown how the quantum Fourier transform (QFT) gate is constructed from single-qubit phase and two-qubit control gates. Due to the collapse postulate, the quantum Fourier transform for f is not available in a register query, but it does allow efficient period estimation. We illustrate how the QFT is exploited in the Shor algorithm for factoring large numbers. On the average, search for an item in an unordered list of size N requires $N/2$ queries. We show how the Grover quantum algorithm improves on this figure of merit as it requires resources that scale as $\sqrt{N}$.

4.1 Introduction

According to its Wikipedia entry [3], a killer application or killer app for short, *"is any computer program that is so necessary or desirable that it proves the core value of some larger technology."* Interestingly, common usage of the catchphrase paralleled the introduction, over twenty years ago, of Shor's algorithm, probably the first quantum killer app. Its entrance immediately demonstrated the core value of a nascent technology, quantum computing and information (QCI).

In this chapter, we focus on the Shor and Grover algorithms. We will learn what they do, why the features of quantum parallelism and quantum interference are central to their transformational capabilities, and why they galvanized a community to build a quantum computer. But first, to set the stage for its quantum analog, we review the classical Fourier transform and its applications.

4.2 Fourier Series

Much of our understanding of heat transfer owes to the seminal contributions of the nineteenth century mathematician *Jean-Baptise Fourier*. He approached the problem of solving partial differential equations, specifically the heat equation, by expressing its solutions as sums of trigonometric functions. In his honor, the representation of a function by an infinite series of trigonometric functions is called a Fourier series. More precisely, a function $f(x)$ that is periodic with period L and square integrable in interval $-L/2 < x < L/2$ allows the representation

$$f(x) = \sum_{n=0}^{\infty} a_n \cos(\frac{2\pi nx}{L}) + \sum_{n=1}^{\infty} b_n \sin(\frac{2\pi nx}{L}), \tag{4.1}$$

where the constants a_n, b_n are given by

$$a_0 = \frac{1}{L} \int_{-L/2}^{L/2} f(x)\, dx$$

$$a_n = \frac{2}{L} \int_{-L/2}^{L/2} f(x) \cos(\frac{2\pi nx}{L})\, dx$$

$$b_n = \frac{2}{L} \int_{-L/2}^{L/2} f(x) \sin(\frac{2\pi nx}{L})\, dx, \tag{4.2}$$

and n is a positive integer. Today, Fourier analysis is found in almost all areas of science, technology, and the arts, from quantum field theory to image processing. Often, it is convenient to express Fourier series in terms of exponential functions. Using the *Euler formula*

$$\exp(\pm 2\pi\, ix/L) = \cos(2\pi x/L) \pm i \sin(2\pi x/L),$$

$$f(x) = \sum_{n=0}^{\infty} \frac{1}{2}\Big((a_n + i b_n) \exp(-2\pi\, i\, nx/L) + (a_n - i b_n) \exp(2\pi\, i\, nx/L)\Big).$$

$$\tag{4.3}$$

Now, $a_n = a_{-n}, b_n = -b_{-n}$ and

$$f(x) = \sum_{n=-\infty}^{\infty} h_n \exp(-2\pi\, i\, nx/L)$$

$$h_n = \frac{1}{L} \int_{-L/2}^{L/2} f(x) \exp(2\pi\, i nx/L)\, dx. \tag{4.4}$$

4.2 Fourier Series

In the limit $L \to \infty$, the discrete parameter n/L is replaced by a continuous index so that [2],

$$f(x) = \int_{-\infty}^{\infty} h(\xi) \exp(-2\pi i \xi x) \, d\xi$$
$$h(\xi) = \int_{-\infty}^{\infty} f(x) \exp(2\pi i \xi x) \, dx \qquad (4.5)$$

where $x, \xi \in \mathscr{R}$. Functions f, h that satisfy (4.5) are called a Fourier transform pair.

> Mathematica Notebook 4.1: The Fourier Series and Transform. http://www.physics.unlv.edu/%7Ebernard/MATH_book/Chap4/chap4_link.html; See also https://www.bernardzygelman.github.io

4.2.1 Nyquist-Shannon Sampling

Often, data is not available in the continuous form of a function. Instead, it is presented as a finite set of discrete values $f(x_0), f(x_1) \ldots f(x_n)$ where $x_0, x_1 \ldots x_n$, $x_n = \Delta n, n = 0, 1, 2 \ldots N-1$. The question arises, is there an analog of the Fourier transform pair (4.5) for discretely sampled data? The Nyquist-Shannon sampling theorem [2] provides the answer.

Consider the data set $f_n = f(\Delta n)$ for $n = 0, 1, 2 \ldots N-1$, where the interval Δ between two consecutive points is the *sampling rate*, We assume that the number of points N sampled is an even integer. The sampling theorem [2] tells us that the Fourier transform $h(\xi)$ should be evaluated at

$$\xi_n = \frac{n}{N\Delta} \quad n = -\frac{N}{2} \ldots \frac{N}{2}.$$

The quantity $1/2\Delta$, the *Nyquist frequency*, sets a limit on the utility of the sampling theorem to functions that are *bandwidth limited*. Using definition (4.5), one finds [2]

$$h(\xi_n) = \int_{-\infty}^{\infty} dx \, f(x) \exp(2\pi i \xi_n x) = \Delta \sum_{k=0}^{N-1} f_k \exp(2\pi i n k/N). \qquad (4.6)$$

In deriving (4.6), we replaced the integral by a Riemann sum.

4.2.2 Discrete Fourier Transform

Motivated by the above discussion, we provide a formal definition for the discrete Fourier transform (*DFT*). Given N complex numbers $f \equiv f_0, f_1, f_2 \ldots f_{N-1}$, the linear combinations

$$h_n = \frac{1}{\sqrt{N}} \sum_{k=0}^{N-1} f_k \exp(2\pi i n k/N) \tag{4.7}$$

where index n ranges from 0 to $N-1$ is called the *DFT* of set f. In this form (4.7) describes a mapping of an N-tuple f, into another N-tuple h, and so it is useful to express f, h as column matrices of dimension N. In matrix form the mapping from f to h is given by the matrix product

$$\underline{h} = \mathbf{U}_{DFT} \underline{f}$$

where

$$\underline{f} = \frac{1}{\sqrt{N}} \begin{pmatrix} f_0 \\ f_1 \\ \vdots \\ f_{N-1} \end{pmatrix}. \tag{4.8}$$

$\mathbf{U}_{DFT}$ is an $N \times N$ square matrix whose jm'th entry

$$(\mathbf{U}_{FT})_{jm} = \exp(2\pi i (j-1)(m-1)/N) = \omega^{(j-1)(m-1)}$$
$$\omega \equiv \exp(2\pi i/N). \tag{4.9}$$

For example if $N = 4$, $\omega = \exp(2\pi i/4) = i$, and

$$\mathbf{U}_{FT} = \begin{pmatrix} 1 & 1 & 1 & 1 \\ 1 & \omega & \omega^2 & \omega^3 \\ 1 & \omega^2 & \omega^4 & \omega^6 \\ 1 & \omega^3 & \omega^6 & \omega^9 \end{pmatrix} = \begin{pmatrix} 1 & 1 & 1 & 1 \\ 1 & i & -1 & -i \\ 1 & -1 & 1 & -1 \\ 1 & -i & -1 & i \end{pmatrix}. \tag{4.10}$$

A list f of N numbers requires N^2 multiplications to generate the *DFT* list h. In 1965, J. Cooley and J. Tukey introduced the *Fast Fourier Transform* or *FFT* algorithm. It scales, in calculations for the *DFT*, as $Nlog(N)$ for large N. If $N \sim 10^6$, *FFT* is a million times more efficient than the N^2 method. Nevertheless, if the size of list f grows exponentially so do the resources required to compute its *DFT*. For example, suppose we want to calculate the *DFT* of all 2^n permutations of ones and zeros in an n-bit register. Since $N = 2^n$ and *FFT* scales as $NLogN$ for large N, a classical computer requires resources that scale as 2^n. With a register of modest length, let's say $n = 300$; one would require 2^{300} parallel processors, more than the number of electrons in the universe, to perform the task in a reasonable amount of time.

4.2 Fourier Series

> Mathematica Notebook 4.2: The Discrete Fourier Transform and *FFT*. http://www.physics.unlv.edu/%7Ebernard/MATH_book/Chap4/chap4_link.html;
> See also https://www.bernardzygelman.github.io

Knowing that the Deutsch-Josza algorithm exploits quantum parallelism, it is natural to ask; does a *DFT* quantum algorithm exist that outperforms a classical computer? We will find that the answer is no. But just as the Deutsch-Josza algorithm reveals global properties of a function profoundly more efficient than a classical computer, so does the quantum analog of the *DFT* allow computation for the period of a function exponentially faster than know classical algorithms.

In 1994 Peter Shor discovered an algorithm that employs the "secret sauces" of quantum parallelism and interference to tackle an immediate and vital real world application in cryptography. Because of its relevance in applications, and its exponential efficiency, the Shor algorithm galvanized the quantum computing community. It is fair to say that its publication proved to be a watershed event. Before we embark describing it, we review properties of the *DFT*, an important component of the Shor algorithm. Foremost, we investigate the *DFT* properties of a periodic function.

A list f_k where k is an index ranging $0, 1, 2, \ldots N - 1$ is periodic, with period r, if

$$f_k = f_{k+nr} \tag{4.11}$$

where $n = \pm 1, \pm 2, \cdots$, and $k + nr \in \{0, 1, \ldots N - 1\}$. Obviously r is an integer, and here, for the sake of simplicity,[1] we assume that N is divisible by the period r. In that case, the *DFT* of f_k can be re-written as a sum of $m = N/r$ partitions,

$$\sqrt{N} h_n =$$

$$\sum_{k=0}^{r-1} f_k \exp(2\pi i n k/N) + \sum_{k=r}^{2r-1} f_k \exp(2\pi i n k/N) + \cdots + \sum_{k=N-r}^{N-1} f_k \exp(2\pi i n k/N) =$$

$$\sum_{k=0}^{r-1} \Big(f_k \exp(2\pi i n k/N) + f_{k+r} \exp(2\pi i n (k+r)/N) + \ldots$$

$$+ f_{k+N-r} \exp(2\pi i n (k+N-r)/N) \Big). \tag{4.12}$$

Because f_k is periodic

$$\sqrt{N} h_n = \sum_{k=0}^{r-1} f_k \exp(2\pi i n k/N) \times$$

$$\Big(1 + \exp(2\pi i r n/N) + \exp(2\pi i 2 r n/N) + \ldots \exp(2\pi i (m-1) r n/N) \Big). \tag{4.13}$$

---

[1] The general case is discussed in Mathematica notebook 4.3.

The geometric series (4.13) sums to

$$1 + z + z^2 + \ldots z^{m-1} = \frac{z^m - 1}{z - 1}, \qquad (4.14)$$

where $z \equiv \exp(2\pi i r n/N)$, and

$$h_n = \frac{1}{\sqrt{N}} \sum_{k=0}^{r-1} f_k \exp(2\pi i n k/N) \left(\frac{-1 + \exp(2 i n \pi)}{-1 + \exp(2 i r n \pi/N)} \right). \qquad (4.15)$$

Because n is an integer, the numerator $-1 + \exp(2\pi i n)$ in (4.15) vanishes for all n as does h_n, provided that $-1 + \exp(2\pi i r n/N) \neq 0$. The denominator vanishes only if

$$p \equiv \frac{rn}{N} \in \mathbb{Z}, \qquad (4.16)$$

and in that case (4.15) is not valid. Instead, for integer p the phase factor z in (4.14) tends to unity, and the geometric sum adds to the value m. In summary, unless p, for a given n, is an integer, $h(n) = 0$. Below, we offer an illustrative example of this behavior.

Panel (a) of Fig. 4.1 illustrates a mapping $f : \{0, 1\}^8 \to \{0, 1\}$ In it, index k on the abscissa labels the first 32 members of the set $\{0, 1\}^8$, and the circles denote the binary-valued function f_k. Visual inspection of that figure reveals a pattern comprised

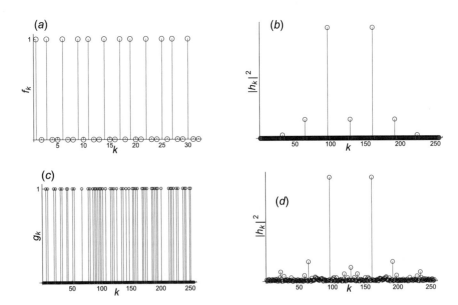

Fig. 4.1 Discrete Fourier transforms. Panel **a**: periodic data set f_k. Panel **b**: *DFT* of f_k. Panel **c**: periodic data set, g_k, contaminated with noise. Panel **d**: *DFT* of g_k

of the repeating set $\{1, 0, 1, 0, 0, 1, 0, 0\}$. Since $r = 8, N = 2^8$, condition (4.16) is satisfied for values of n in set

$$\tilde{n} = \{0, 32, 64, 96, 128, 160, 192, 224\}. \quad (4.17)$$

Panel (b) of Fig. 4.1 is a plot of $|h_n|^2$, the *power spectrum* of this function. It shows that $|h(\tilde{n})|^2 \neq 0$ for values of $\tilde{n}$ (here the case $\tilde{n} = 0$ is not shown) predicted by (4.17). Panel (c) of that figure is a plot of $g_k = f_k \times r_k$, where r_k is a random distribution of 0s and 1s. The set r_k randomly alters the bit values of the unit digits in list f_k and the resulting signal, labeled g_k, does not satisfy periodicity condition (4.11). Its power spectrum is shown in panel (d). Because the signal $f_k \times r_k$ is not periodic, the power spectrum contains non-vanishing contributions at all n. Nevertheless, the amplitudes for the indices predicted in (4.17) is significantly larger than that of the background. The *DFT* power spectrum tells us something about the global properties of the signal that is not readily apparent by visual inspection of the original data.

4.3 Quantum Fourier Transform

We now define the quantum analog of the *DFT*, the quantum Fourier transform or *QFT* for short. *QFT*, being a quantum gate, must have a unitary matrix representation, and so operator **QFT** is defined as follows. Given a register of n qubits where $|j\rangle_n = |j_{n-1} \ldots j_1 j_0\rangle$

$$\mathbf{QFT} |j\rangle_n = \frac{1}{\sqrt{N}} \sum_{k=0}^{N-1} \exp(2\pi i j k/N) |k\rangle_n \quad N = 2^n. \quad (4.18)$$

Unitarity requires that

$$\mathbf{QFT}^\dagger \mathbf{QFT} = \mathbf{QFT}\, \mathbf{QFT}^\dagger = \mathbb{1},$$

and which we demonstrate below. Let's take the inner product of expression (4.18) with ket $|m\rangle_n \equiv |m\rangle$

$$\langle m| \mathbf{QFT} |j\rangle = \frac{1}{\sqrt{N}} \sum_{k=0}^{N-1} \exp(2\pi i j k/N) \langle m|k\rangle \quad (4.19)$$

where, for the sake of economy in notation, we dropped the subscript n attached to the computational basis kets. Because the basis states are orthonormal, i.e $\langle m|k\rangle = \delta_{mk}$

$$\langle m| \mathbf{QFT} |j\rangle = \frac{1}{\sqrt{N}} \sum_{k=0}^{N-1} \exp(2\pi i j k/N) \delta_{mk} = \frac{1}{\sqrt{N}} \exp(2\pi i j m/N). \quad (4.20)$$

The left hand side of this expression is just the m, jth matrix element of operator **QFT**. Using this identity we construct the adjoint conjugate of **QFT**, $(\mathbf{QFT}^\dagger)_{mj} = (\mathbf{QFT}^*)_{jm}$ or

$$(\mathbf{QFT}^\dagger)_{mj} = \frac{1}{\sqrt{N}} \exp(-2\pi i j m/N).$$

Therefore

$$(\mathbf{QFT}^\dagger \mathbf{QFT})_{mj} = \sum_k (\mathbf{QFT}^\dagger)_{mk} (\mathbf{QFT})_{kj} =$$

$$\frac{1}{N} \sum_k^{N-1} \exp(-2\pi i k m/N) \exp(2\pi i k j/N) =$$

$$\frac{1}{N} \sum_k^{N-1} \exp(-2\pi i k (m-j)/N). \tag{4.21}$$

If $m = j$ this expression reduces to

$$\frac{1}{N} \sum_k^{N-1} = 1.$$

If $m \neq j$ the difference is a non-zero integer $q < N$, and

$$\frac{1}{N} \sum_k^{N-1} \exp(-2\pi i k q/N) = \frac{1}{N} \frac{(-1 + \exp(-2\pi i q))}{(-1 + \exp(-2\pi i q/N))} \tag{4.22}$$

where we used (4.14) in summing the l.h.s of this expression. Now, since $0 < q < N$, the denominator in (4.22) does not vanish but, because q is an integer, the numerator does vanish. Therefore,

$$(\mathbf{QFT}^\dagger \mathbf{QFT})_{mj} = \delta_{mj}$$

which is equivalent to the operator relation

$$\mathbf{QFT}^\dagger \mathbf{QFT} = \mathbb{1}. \tag{4.23}$$

4.3.1 QFT Diagrammatics

We established that **QFT** is an acceptable quantum gate, but expressed as a $2^n \times 2^n$ matrix we offered little guidance in its construction. Shor's analysis highlighted a useful fact: the **QFT** gate can be expressed solely in terms of single and two-qubit gates [1].

4.3 Quantum Fourier Transform

Before illustrating the general case, it is worthwhile to explore particular cases. Consider first the single-qubit **QFT** gate. Using definition (4.18) we find

$$\mathbf{QFT}\,|0\rangle = \frac{1}{\sqrt{2}} \sum_{k=0}^{1} \exp(2\pi\,i\,0\,k/2)\,|k\rangle = \frac{1}{\sqrt{2}}\,(|0\rangle + |1\rangle))$$

$$\mathbf{QFT}\,|1\rangle = \frac{1}{\sqrt{2}} \sum_{k=0}^{1} \exp(2\pi\,i\,1\,k/2)\,|k\rangle = \frac{1}{\sqrt{2}}\,(|0\rangle - |1\rangle)) \quad (4.24)$$

where $|0\rangle, |1\rangle$ are computational basis vectors for a single qubit. Note that the single-qubit **QFT** is identical to the Hadamard gate. Now consider the $N = 4$ case, i.e., the two-qubit **QFT**. Following a similar line of attack, we find

$$\mathbf{QFT}\,|0\rangle_2 = \frac{1}{\sqrt{4}} \sum_{k=0}^{3} \exp(2\pi\,i\,0\,k/4)\,|k\rangle_2 =$$
$$\frac{1}{2}(|0\rangle_2 + |1\rangle_2 + |2\rangle_2 + |3\rangle_2) = \frac{1}{2}(|0\rangle + |1\rangle) \otimes (|0\rangle + |1\rangle)$$

$$\mathbf{QFT}\,|1\rangle_2 = \frac{1}{\sqrt{4}} \sum_{k=0}^{3} \exp(2\pi\,i\,1\,k/4)\,|k\rangle_2 =$$
$$\frac{1}{2}(|0\rangle_2 + i\,|1\rangle_2 - |2\rangle_2 - i\,|3\rangle_2) = \frac{1}{2}(|0\rangle - |1\rangle) \otimes (|0\rangle + i\,|1\rangle)$$

$$\mathbf{QFT}\,|2\rangle_2 = \frac{1}{\sqrt{4}} \sum_{k=0}^{3} \exp(2\pi\,i\,2\,k/4)\,|k\rangle_2 =$$
$$\frac{1}{2}(|0\rangle_2 - |1\rangle_2 + |2\rangle_2 - |3\rangle_2) = \frac{1}{2}(|0\rangle + |1\rangle) \otimes (|0\rangle - |1\rangle)$$

$$\mathbf{QFT}\,|3\rangle_2 = \frac{1}{\sqrt{4}} \sum_{k=0}^{3} \exp(2\pi\,i\,3\,k/4)\,|k\rangle_2 =$$
$$\frac{1}{2}(|0\rangle_2 - i\,|1\rangle_2 - |2\rangle_2 + i\,|3\rangle_2) = \frac{1}{2}(|0\rangle - |1\rangle) \otimes (|0\rangle - i\,|1\rangle). \quad (4.25)$$

In (4.25) the action of **QFT** on a two-qubit basis vector is equivalent to the direct product of a pair of single-qubit superposition states. Does this factorization hold for n-qubit basis kets? The answer turns out to be yes. Indeed, we claim that for an n-bit computational basis vector $|j\rangle_n \equiv |j_{n-1} \ldots j_1 j_0\rangle$

$$\mathbf{QFT}\,|j\rangle_n = \frac{1}{\sqrt{N}} \Big(|0\rangle + \exp(2\pi\,i\,[.j_0])\,|1\rangle\Big) \otimes \Big(|0\rangle + \exp(2\pi\,i\,[.j_1 j_0])\,|1\rangle\Big) \otimes \ldots$$
$$\Big(|0\rangle + \exp(2\pi\,i\,[.j_{n-1} \ldots j_2 j_1 j_0])\,|1\rangle\Big) \quad (4.26)$$

where

$$[.j_0] \equiv j_0 \times 2^{-1}$$
$$[.j_1 j_0] \equiv j_1 \times 2^{-1} + j_0 \times 2^{-2}$$
$$[.j_2 j_1 j_0] \equiv j_2 \times 2^{-1} + j_1 \times 2^{-2} + j_0 \times 2^{-3}$$
$$\vdots$$
$$[.j_{n-1} \ldots j_1 j_0] \equiv j_{n-1} \times 2^{-1} + \cdots + j_1 \times 2^{-(n-1)} + j_0 \times 2^{-n}. \quad (4.27)$$

To prove this statement, we first use the fact that

$$|k\rangle_n = |k_{n-1} \ldots k_1 k_0\rangle$$

where the value of index

$$k = k_{n-1} 2^{n-1} + \cdots + k_1 2^1 + k_0 2^0 = \sum_{p=0}^{n-1} k_p 2^p$$

for $k_i \in \{0, 1\}$, and

$$\sum_{k=0}^{N-1} |k\rangle_n = \sum_{k_0=0}^{1} \sum_{k_1=0}^{1} \cdots \sum_{k_{n-1}=0}^{1} |k_{n-1} \ldots k_1 k_0\rangle.$$

Therefore, the summation on the r.h.s. of (4.18) can be re-written as

$$\sum_{k=0}^{N-1} \exp(2\pi i j k/N) |k\rangle = \sum_{k_0=0}^{1} \sum_{k_1=0}^{1} \cdots \sum_{k_{n-1}=0}^{1} \exp\left(2\pi i j \sum_{p=0}^{n-1} k_p 2^p / N\right) |k_{n-1} \ldots k_1 k_0\rangle =$$

$$\sum_{k_0=0}^{1} \sum_{k_1=0}^{1} \cdots \sum_{k_{n-1}=0}^{1} \exp(2\pi i j k_0 2^0/N) \exp(2\pi i j k_1 2^1/N) \ldots \exp(2\pi i j k_{n-1} 2^{n-1}/N) \times$$
$$|k_{n-1} \ldots k_1 k_0\rangle \quad (4.28)$$

or

$$\sum_{k_{n-1}=0}^{1} \exp(2\pi i j k_{n-1} 2^{n-1}/N) |k_{n-1}\rangle \otimes \cdots \otimes \sum_{k_1=0}^{1} \exp(2\pi i j k_1 2^1/N) |k_1\rangle \otimes$$
$$\sum_{k_0=0}^{1} \exp(2\pi i j k_0 2^0/N) |k_0\rangle = \left(|0\rangle + \exp(2\pi i 2^{n-1} j/N) |1\rangle\right) \otimes \cdots$$
$$\otimes \left(|0\rangle + \exp(2\pi i 2^1 j/N) |1\rangle\right) \otimes \left(|0\rangle + \exp(2\pi i 2^0 j/N) |1\rangle\right).$$

4.3 Quantum Fourier Transform

Now $2^{n-1} j/N = \sum_{i=0}^{n-1} j_i 2^i /2$ and so

$$\exp(2\pi i \sum_{i=0}^{n-1} j_i 2^i /2) = \exp(2\pi i [.j_0]),$$

likewise

$$\exp(2\pi i \sum_{i=0}^{n-1} j_i 2^i /4) = \exp(2\pi i [.j_1 j_0]),$$

$$\vdots$$

$$\exp(2\pi i \sum_{i=0}^{n-1} j_i 2^i /2^n) = \exp(2\pi i [.j_{n-1} \ldots j_1 j_0]).$$

Putting this all together we arrive at identity (4.26). According to it, **QFT** is a direct (tensor) product of single qubits states and so allows diagrammatic expression. Instead of finding a diagrammatic representation for **QFT**, it turns out to be easier to construct the diagram for an operator, that I call **TQF**, and which has the property

$$\textbf{TQF} \, |j\rangle_n = \frac{1}{\sqrt{N}} \Big(|0\rangle + \exp(2\pi i [.j_{n-1} \ldots j_2 j_1 j_0]) |1\rangle \Big) \otimes \ldots$$
$$\Big(|0\rangle + \exp(2\pi i [.j_1 j_0]) |1\rangle \Big) \otimes \Big(|0\rangle + \exp(2\pi i [.j_0]) |1\rangle \Big). \quad (4.29)$$

Comparing (4.29) with expression (4.26) we recognize that

$$\textbf{QFT} \, |j\rangle_n = \textbf{S} \, (\textbf{TQF} \, |j\rangle_n)$$

where **S** is a qubit swap operator, i.e., $\textbf{S} \, |j_{n-1} \ldots j_1 j_0\rangle = |j_0 j_1 \ldots j_{n-1}\rangle$. Let's inspect the state $\Big(|0\rangle + \exp(2\pi i [.j_0]) |1\rangle \Big)$ for the first qubit. If $j_0 = 0$, it reduces to $\Big(|0\rangle + |1\rangle \Big)$, and if $j_0 = 1$, to $\Big(|0\rangle - |1\rangle \Big)$. Therefore, this qubit state is equivalent, up to a normalization factor, to $\textbf{H} \, |j_0\rangle$ where **H** is the Hadamard gate. The state of the second qubit is

$$\Big(|0\rangle + \exp(2\pi i [.j_1 j_0]) |1\rangle \Big) = \Big(|0\rangle + \exp(2\pi i j_1 /2) \exp(2\pi i j_0 /4)) |1\rangle \Big).$$

If $j_0 = 0$ that qubit is in a superposition state proportional to $\textbf{H} \, |j_1\rangle$, however if $j_0 = 1$, it becomes

$$\frac{1}{\sqrt{2}} \Big(|0\rangle + \exp(2\pi i [.j_1 j_0]) |1\rangle \Big) = \textbf{R}_2 \, \textbf{H} \, |j_1\rangle$$

where the matrix representation of $\textbf{R}_n$ is

$$\textbf{R}_n = \begin{pmatrix} 1 & 0 \\ 0 & \exp(2\pi i/2^n) \end{pmatrix}. \quad (4.30)$$

(a)

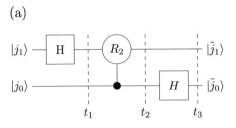

(b)

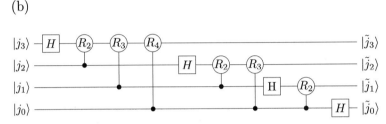

Fig. 4.2 Panel a: A 2-qubit circuit for operator **TQF**. Panel b: 4-qubit circuit. $\left|\tilde{j}_0\right\rangle = |0\rangle + \exp(2\pi i .[j_0]) |1\rangle$, $\left|\tilde{j}_1\right\rangle = |0\rangle + \exp(2\pi i .[j_1 j_0]) |1\rangle$, $\left|\tilde{j}_n\right\rangle = |0\rangle + \exp(2\pi i .[j_n j_{n-1} \ldots j_0]) |1\rangle$. States **QFT** $|j_{n-1} \ldots j_1 j_0\rangle$ are recovered by the action of a qubit swap operation on $\left|\tilde{j}_{n-1} \ldots \tilde{j}_1 \tilde{j}_0\right\rangle$

In the previous chapter, we showed how a target qubit, in a **CNOT** gate, undergoes a bit flip if the control qubit is in state $|1\rangle$. Here, for $|j_1 j_0\rangle$, qubit $|j_0\rangle$ is the control qubit and $|j_1\rangle$ the target. Instead of undergoing a bit flip, the target qubit suffers the phase operation $\mathbf{R}_2$.

A diagrammatic description of the 2-qubit **TQF** gate is given in Fig. 4.2 panel (a). In it, initial state $|\psi\rangle = |j_1 j_0\rangle$ enters at the l.h.s. of the diagram. The bottom wire represents the first qubit in initial state state $|j_0\rangle$. The top wire represents the second qubit in state $|j_1\rangle$. It is processed by a Hadamard gate, after which it is found in the superposition $|0\rangle + \exp(2\pi i .[j_1]) |1\rangle$. The total amplitude at t_1 is

$$|\psi(t_1)\rangle = \frac{1}{\sqrt{2}} \left(|0\rangle + \exp(2\pi i [.j_1]) |1\rangle\right) \otimes |j_0\rangle.$$

Subsequently, it passes through a phase control gate $\mathbf{R}_2$. This gate acts as an identity operator if $j_0 = 0$, but tacks a phase on the top qubit if $j_0 = 1$ so, in that case,

$$\mathbf{R}_2\left(|0\rangle + \exp(2\pi i [.j_1]) |1\rangle\right) \otimes |j_0\rangle =$$
$$\left(|0\rangle + \exp(2\pi i [.j_1]) \exp(2\pi i/4) |1\rangle\right) \otimes |j_0\rangle.$$

Or,

$$|\psi(t_2)\rangle = \frac{1}{\sqrt{2}} \left(|0\rangle + \exp(2\pi i [.j_1 j_0]) |1\rangle\right) \otimes |j_0\rangle.$$

Finally, the bottom qubit passes through a Hadamard gate resulting in

$$|\psi(t_3)\rangle = \frac{1}{2}\Big(|0\rangle + \exp(2\pi i [.j_1 j_0]) |1\rangle\Big) \otimes \Big(|0\rangle + \exp(2\pi i [.j_0]) |1\rangle\Big),$$

in harmony with expression (4.29) for the case $n = 2$. Panel (b) of that figure illustrates the gate diagram for a four-qubit gate. From those figures, it is straightforward to generalize to the n-qubit case.

4.3.2 Period Finding with the QFT Gate

> Mathematica Notebook 4.3: Period Finding with Mathematica's *FFT* Function. http://www.physics.unlv.edu/%7Ebernard/MATH_book/Chap4/chap4_link.html; See also https://www.bernardzygelman.github.io

In the previous section we introduced the function, $f : \{0, 1\}^8 \to \{0, 1\}^1$ tabulated in panel (a) of Fig. 4.1. Because it is periodic with period $r = 8$, the plot of its power spectrum $|h_n|^2$ features discrete peaks at values for n that satisfies (4.16) and is in the set $\tilde{n} \equiv \{0, 32, 64, 96, 128, 160, 192, 224\}$. Since the members of $\tilde{n}$ are multiples of period r, calculation of the latter follows from knowledge of the former. To find $\tilde{n}$, a classical computer needs to calculate the full *DFT* (or its inverse) of f. Though *FFT* is efficient at this task if f_n grows exponentially so do the resources required to compute h_n. Let's explore how a quantum algorithm fares in this effort.

A quantum circuit needs a qubit register of dimension n to accommodate the $N = 2^n$ coordinates. Also, a one-qubit register is required to store $f(n)$ for a given n. Using protocol (3.18) we construct the gate $\mathbf{U}_f$ so that

$$\mathbf{U}_f |k\rangle_n \otimes |0\rangle \to |k\rangle_n \otimes |f(k)\rangle.$$

Quantum parallelism is invoked by a gate sequence in which Hadamard gates operate on each ket $|k_i\rangle$ for $i = 0 \ldots n-1$. Using the symbol $\mathbf{H}^{\otimes n}$ to represent a direct product of n Hadamard gates, we have according to (3.26),

$$\mathbf{H}^{\otimes n} \otimes |0\rangle_n = \frac{1}{\sqrt{N}} \sum_{k=0}^{N-1} |k\rangle. \qquad (4.31)$$

The massively parallel superposition state (4.31) is shuttled through $\mathbf{U}_f$, resulting in

$$\mathbf{U}_f \left(\mathbf{H}^{\otimes n} \otimes \mathbb{1}\right) |0\rangle_n \otimes |0\rangle = \frac{1}{\sqrt{N}} \sum_{k=0}^{N-1} |k\rangle_n \otimes |f(k)\rangle. \qquad (4.32)$$

The r.h.s. of (4.32) is a linear combination of states that includes evaluation of the function $f(k)$ for all values k. In the example introduced in Sect. 4.3, $k = 2^8$, but if the input register is slightly larger, e.g., $n = 100$, then 2^{100} values for $f(k)$ are evaluated by this circuit. Since a classical circuit computes $f(k)$ serially, it would take 2^{100} computational cycles to evaluate $f(k)$ for each k. The quantum gate computes $f(k)$, for each k, in just one pass through the circuit. Unfortunately, the Born rule tells us that a single measurement of the output register reveals only one of all possible $f(k)$. Suppose that measurement leads to the value $f(k = a)$, the collapse postulate tells us that the system immediately collapses into state $|k = a\rangle \otimes |f(a)\rangle$, and all information in the linear superposition state (4.32) is lost.

The quantum circuit does not allow simultaneous access to $f(k)$, for each k, but we can exploit it to uncover global properties of the function. To that end we subject the state on the r.h.s. of (4.32) to a **QFT** gate. That is,

$$\left(\mathbf{QFT} \otimes \mathbb{1}\right) \mathbf{U}_f \left(\mathbf{H}^{\otimes n} \otimes \mathbb{1}\right) |0\rangle_n \otimes |0\rangle = \frac{1}{\sqrt{N}} \sum_{k=0}^{N-1} \left(\mathbf{QFT}\,|k\rangle_n\right) \otimes |f(k)\rangle. \qquad (4.33)$$

Now according to (4.18)

$$\mathbf{QFT}\,|k\rangle_n = \frac{1}{\sqrt{N}} \sum_{j=0}^{N-1} \exp(2\pi\,ijk/N)\,|j\rangle_n,$$

and the action of gate $\left(\mathbf{QFT} \otimes \mathbb{1}\right) \mathbf{U}_f \left(\mathbf{H}^{\otimes n} \otimes \mathbb{1}\right)$ on state $|0\rangle_n \otimes |0\rangle$ results in the state

$$|\psi\rangle = \frac{1}{N} \sum_{k=0}^{N-1} \sum_{j=0}^{N-1} \exp(2\pi\,ijk/N)\,|j\rangle_n \otimes |f(k)\rangle. \qquad (4.34)$$

This sequence of gates is illustrated in Fig. 4.3.

In that diagram, measurement devices are placed on each qubit of the upper register. According to the Born rule, the probability that the measurements yield result $m = m_7 \ldots m_1 m_0$ in binary format is determined by the sum of all joint probabilities, i.e.,

$$p(m) = p_1(m) + p_2(m)$$
$$p_1(m) = |\langle 1\,m\,|\psi\rangle|^2 \quad p_2(m) = |\langle 0\,m\,|\psi\rangle|^2 \qquad (4.35)$$

where $\langle 1\,m| \equiv \langle 1| \otimes \langle m|$, $\langle 0\,m| \equiv \langle 0| \otimes \langle m|$ (for the sake of economy in notation, below we ignore the subscripts for kets $|m\rangle_8$, and simply label them $|m\rangle$). Let's focus on the first term of (4.35)

$$p_1(m) = \frac{1}{N^2} \left| \sum_{k=0}^{N-1} \sum_{j=0}^{N-1} \exp(2\pi\,ijk/N)\,\langle m|j\rangle\,\langle 1|f(k)\rangle \right|^2.$$

4.3 Quantum Fourier Transform

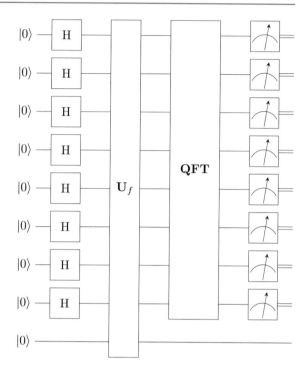

Fig. 4.3 Period finding quantum circuit for an eight-qubit register

Because of the orthonormality condition $\langle m|j\rangle = \delta_{mj}$,

$$p_1(m) = \frac{1}{N^2}\left|\sum_{k=0}^{N-1}\exp(2\pi\,imk/N)f_k\right|^2 \quad f_k \equiv \langle 1|f(k)\rangle. \tag{4.36}$$

The square of the sum on the r.h.s. of (4.36) is proportional to the square of the *DFT* of list f, or its power spectrum. If f_k is periodic with period r, so that $f_k = f_{k+r}$, we apply the analysis framing Eqs. (4.12)–(4.15), and arrive at the expression

$$p_1(m) = \frac{1}{N^2}\left|\sum_{k=0}^{r-1}\exp(2\pi\,imk/N)f_k\left(\frac{-1+\exp(2\,im\pi)}{-1+\exp(2\,irm\pi/N)}\right)\right|^2. \tag{4.37}$$

We already showed how the term inside the parenthesis of (4.37) vanishes unless the ratio rm/N evaluates to an integer. Therefore, a periodic function engenders a probability distribution with well defined, discrete, peaks at values of m where rm/N is an integer. A similar argument applies to the distribution $p_2(m)$. In the example illustrated in Fig. 4.1, those peaks occur at values of m in set $\tilde{n}$. Let's see how this fact affects quantum measurements of the control register. The distribution $p(m)$, calculated using (4.35), is tabulated in Table 4.1. It lists the calculated $p(m)$ for all $m \in \tilde{n}$. The second column in the table itemizes the register occupation configuration for index m, and the third column, $p(m)$. Obviously $\sum_m p(m) = 1$.

Table 4.1 Probability distribution $p(k)$ of outcomes from control register measurements. Configurations not listed have null probability

m	Configuration	$p(m)$
0	00000000	0.531
32	00100000	0.005
64	01000000	0.031
96	01100000	0.182
128	10000000	0.031
160	10100000	0.182
192	11000000	0.031
224	11100000	0.005

Table 4.1 predicts that there is a slightly more than a 50% chance that the outcome of such a measurement is the configuration 00000000, i.e., each meter finds the corresponding qubit in the 0, or "off" position. The configurations corresponding to $m = 96, 160$ each have about an 18% chance of being observed. The remaining configurations listed in Table (4.1) are found with vanishingly smaller, though finite, probabilities. How can we infer the period r from such measurements? Obviously, as a probabilistic computer, we need to run it several times to obtain useful data. Ignoring measurements that result in configuration 00000000, we note non-trivial readings, most likely 01100000, or 96, and 10100000, or 160. Let's take the smaller value $m = 96$ in our sample. According to (4.16) and that $N = 256$,

$$\frac{rm}{N} = r\frac{3}{8} = p \quad p = 1, 2, 3, \ldots$$

from which we can immediately infer that the period r is a multiple of 8. So the period could be $r = 8, 16, \ldots$. By taking a few more measurements (see Mathematica Notebook 4.3) we can deduce the correct answer $r = 8$.

This example involves an eight-qubit register, and so this algorithm does not seem that impressive. But, consider a mapping $f : \{0, 1\}^{200} \to \{0, 1\}$ where $f(x + r) = f(x)$. If r is large and because *FFT* scales on the order of $N = 2^{200}$ in time, or parallel computing resources, a classical computation of it becomes untenable. Consider now feeding f into a modest 200 qubit quantum computer. With the QFT algorithm, quantum parallelism stores the 2^{200} dimensional *DFT* in Hilbert space. It waits for us, within the limits of coherence time, to measure the state of each qubit. Most probably, as was evident in our eight-qubit example, we find a string of 200 zeros in a single run of the machine. Repeating, we eventually measure a non-trivial value for the register content, let's say m. That measurement immediately informs us that the period r is some integer multiple of $m/2^{200}$. Several more runs and measurements are required to converge to an answer, but that is a minor inconvenience compared to the 2^{200} calculations required with a classical algorithm. An alternative method for period finding, uses the Phase Estimation algorithm [1], which is described in Mathematica Notebook 4.3b.

Mathematica Notebook 4.3b: Phase Estimation. https://www.bernardzygelman.github.io

4.3.3 Shor's Algorithm

Because of the sheer size of numerical operations required to execute common algorithms, programmable computers solve problems inaccessible to human capabilities. However, known algorithms for certain tasks need prodigious computing resources that may, even with the most powerful machines, not be available. Factoring a large number falls into that category. By definition, *prime numbers* are divisible only by themselves (or the number 1) and are the atomic building blocks of the number system. Positive integers are either prime numbers or products of prime numbers. I can factor the number $15 = 3 \times 5$ rapidly, but for larger numbers, let's say 78923451, I require a computer to find its prime factors. Using Mathematica, my laptop factors $78923451 = 3 \times 26307817$ within a minuscule fraction of a second. It takes several minutes to factor a 75-digit number, but even the most powerful machine on the planet is unable to factor a sufficiently large number in a reasonable amount of time. For example, the number called RSA-220 is 220 digits long and was first factored, with the help of significant computer resources [4] in 2016. Numbers, such as 78923451 and RSA-220, are called *semi-prime numbers* because they can be factored into only two other prime numbers.

RSA is an acronym for *R. Rivest, A. Shamir, L. Adelman* who developed the *RSA encryption protocol* in the 1970s. Your credit card and online bank transactions probably use RSA encryption. At its heart, the RSA algorithm relies on the fact that it is nearly impossible to factor a very large semi-prime number with present-day technology. The RSA encryption scheme falls in the category of *public key encryption* in which the key, the pair of numbers (n, e), are freely distributed along with a *private key*, d, kept secret by the sender. The sender, let's call her Alice, publishes the public key so that anyone can send an encrypted message to her. Using the public key (n, e), Bob encrypts his message, called a *cipher*, into what appears to be a random string of numbers. Bob publishes that cipher and Alice who possesses the private key, can recover Bob's original message. Because Bob's cipher is public, an eavesdropper could decrypt it provided they can factor a large semi-prime number. So Bob's message is secure because the latter computational task is hard, i.e., it requires time and memory resources that are not available. Or so it was thought until 1994 when Shor published his quantum algorithm. We present a detailed discussion of the RSA algorithm in Mathematica Notebook 4.4 but, in a nutshell, breaking an RSA cipher involves solving the following equations,

$$N = C^d \, Mod \, n \quad C = N^e \, Mod \, n \qquad (4.38)$$

for the exponent d. Here, N is Bob's original message expressed as a string of integers through a *conversion code*. He encrypts it to cipher C, which appears to be a random string of numbers. For example, if Bob's original message is $N = 472653$, he constructs, according to (4.38), the cipher $C = 17837175$ with Alice's public key $n = 78923451$ and $e = 713$. Alice has access to the private key d and evaluates C^d modulus n to reconstruct the original message.

So how does Alice choose the public key values e, n? In RSA, Alice computes $n = pq$ where p, q are two large prime numbers she found in a table of primes. She chooses some value $1 < e < (p-1)(q-1)$ whose greatest common divisor with n is 1, and calculates the private key using the algorithm

$$d\,e = 1\,Mod\,(p-1)(q-1). \tag{4.39}$$

But (e, n) is public and eavesdroppers know that $n = pq$, so could they not also solve (4.39) for d? Fortunately, as argued above, it is extremely difficult to find the prime factors of a large semi-prime number n. It is this feature that makes RSA encryption so useful. The cipher is reasonable safe from prying eyes if n is as large, or larger than RSA220. There are many ways to factor a number, one algorithm called *Miller's algorithm* involves solving the equation

$$f(x, r) = 1 \quad f(x, r) = x^r\,Mod\,n \tag{4.40}$$

for the smallest value r. Here x, called the seed, is an arbitrary number so that $x < r$ and x, n do not share a common factor. The prime factors of n are given (see Mathematica Notebook 4.3, for proofs) by the greatest common divisors (GCD) of $(x^{r/2}+1)$ and $(x^{r/2} - 1)$ with n.[2] Finding the GCD of two numbers is an "easy" problem, but finding r is hard. We now demonstrate that solving Eq. (4.40) is equivalent to finding the period of the function

$$f(r) = x^r\,Mod\,n. \tag{4.41}$$

Proof For a given seed x, in (4.41), the smallest r is called the order. If $f(r+\omega) = f(r)$ then ω is the period of $f(r)$. According to (4.41)

$$f(r + \omega) = x^{r+\omega}\,Mod\,n = (x^r\,Mod\,n)(x^\omega\,Mod\,n)\,Mod\,n \tag{4.42}$$

where we used the multiplication law for modular arithmetic

$$a\,b\,Mod\,n = \Big((a\,Mod\,n)(b\,Mod\,n)\Big)\,Mod\,n.$$

If $(x^\omega\,Mod\,n) = 1$, i.e., a solution to (4.40), relation (4.42) leads to the identity $f(r + \omega) = (x^r\,Mod\,n)Mod\,n = (x^r\,Mod\,n) = f(r)$, and so $r = \omega$, and which completes the proof. □

---

[2] If r is an odd integer, try a different seed x.

In the Miller algorithm, the problem of finding the factors of a semi-prime number n reduces to that of finding the period of a function. In the previous section, we showed how to exploit the power spectrum of a periodic function for that task. We also showed how a quantum computer accomplishes that using a QFT protocol. It turns out that [1] for large n, the latter scales in polynomial time, that is, some fixed power of n. The best classical algorithm, FFT, scales as $N = 2^n$, or exponential time. With a modest quantum computer, containing on the order of several hundred ideal qubits, we could factor RSA-220 in fractions of a second.

> Mathematica Notebook 4.4: An Introduction to Public Key Encryption and RSA. http://www.physics.unlv.edu/%7Ebernard/MATH_book/Chap4/chap4_link.html; See also https://www.bernardzygelman.github.io

> Mathematica Notebook 4.5: Implementing Shor's Algorithm. http://www.physics.unlv.edu/%7Ebernard/MATH_book/Chap4/chap4_link.html; See also https://www.bernardzygelman.github.io

4.4 Grover's Search Algorithm

Ask your robotic assistant, or "the *oracle*" as it insists on being addressed (robots are vain-glorious in this way), to search for an item in a mega-warehouse. The oracle has a scanner that identifies the scanning code on the merchandise. It is limited to a single item scan that takes 1 s per scan. There is a one-in-a-million chance that the oracle locates the item in the first scan of a million unsorted items. At worst, it takes almost 12 days for the search to conclude. On the average the search takes about half that amount, or 6 days.

A random search takes, on the average, about $N/2$ trials, where N is the size of the sample. No known classical algorithm improves on this figure of merit. In 1996, *Lev Grover* introduced a quantum algorithm that improves the above estimate to order $\sqrt{N}$ trials. Instead of an average search time of 6 days, a quantum oracle locates the desired item in 16 min.

Grover's algorithm exploits quantum parallelism in addition to an amplification strategy that we describe below. To implement Grover's algorithm, we construct an oracle i.e., a function that identifies the sought for item. Because we are in the digital domain, the "item" is a particular configuration of an n-qubit register. For example, out of a possible 2^{100} candidates, the sought for article is tagged by a unique string of 1s and 0s in a 100-bit register (e.g., the scanning code). The oracle can distinguish that configuration from all other possibilities but needs to flag that string when found.

Suppose the string corresponds to the computational state $|\xi\rangle_n$ where n is the size of the register. The oracle consists of the mapping $f : \{0, 1\}^n \to \{0, 1\}$

$$f(k) = f(k_{n-1} \ldots k_1 k_0) = 0 \quad k \neq \xi$$
$$f(k) = 1 \quad k = \xi. \quad (4.43)$$

The oracle returns the value 1 for the state in question and 0 otherwise. Consider the direct product of an n-qubit register control register $|k\rangle_n$ and a target qubit $|y\rangle$. We implement the oracle $\mathbf{U}_f$ in the standard way, so that

$$\mathbf{U}_f |k\rangle_n \otimes |y\rangle = |k\rangle_n \otimes |f(k) \oplus y\rangle. \quad (4.44)$$

Let's subject the superposition state $|k\rangle_n \otimes \mathbf{H} |1\rangle$ to the oracle,

$$\mathbf{U}_f |k\rangle_n \otimes \mathbf{H} |1\rangle = \mathbf{U}_f |k\rangle_n \otimes \frac{1}{\sqrt{2}} (|0\rangle - |1\rangle) =$$
$$|k\rangle_n \otimes \frac{1}{\sqrt{2}} (|f(k)\rangle - |f(k) \oplus 1\rangle). \quad (4.45)$$

Now if $f(k) = 0$,

$$\frac{1}{\sqrt{2}} (|f(k)\rangle - |f(k) \oplus 1\rangle) = \frac{1}{\sqrt{2}} (|0\rangle - |1\rangle) = \mathbf{H} |1\rangle$$

likewise, if $f(k) = 1$

$$\frac{1}{\sqrt{2}} (|f(k)\rangle - |f(k) \oplus 1\rangle) = -\mathbf{H} |1\rangle.$$

Therefore,

$$\mathbf{U}_f |k\rangle_n \otimes \mathbf{H} |1\rangle = (-1)^{f(k)} |k\rangle_n \otimes \mathbf{H} |1\rangle. \quad (4.46)$$

The oracle flags the state $|\xi\rangle_n \otimes \mathbf{H} |1\rangle$ by multiplying $\mathbf{H} |1\rangle$ with the phase factor (-1) if $k = \xi$. It is useful to define a new operator, let's call it $\mathbf{V}$, that acts solely on the control register $|k\rangle_n$ so that

$$\mathbf{V} |k\rangle_n = (-1)^{f(k)} |k\rangle_n. \quad (4.47)$$

An explicit form for this operator is

$$\mathbf{V} \equiv \mathbb{1} - 2 |\xi\rangle \langle \xi| \quad (4.48)$$

where $\mathbb{1}$ is the identity in the Hilbert space for the control register. Proof of (4.47) follows from the fact that

$$\left((\mathbb{1} - 2(|\xi\rangle \langle \xi|) \right) |k\rangle = |k\rangle - 2(\langle \xi | k \rangle) |\xi\rangle =$$
$$|k\rangle - 2\delta_{k\xi} |\xi\rangle \quad (4.49)$$

4.4 Grover's Search Algorithm

where we have used Dirac's associative axiom, the orthonormality of the computation basis vectors $|k\rangle$, and we dropped the subscript n in denoting the latter. Now if $k = \xi$ the r.h.s of (4.49) evaluates to $-|\xi\rangle$, and if $k \neq \xi$ to $|k\rangle$, and which proves identity (4.47). Therefore, we can equate

$$\mathbf{U}_f |k\rangle_n \otimes \mathbf{H} |1\rangle = \left(\mathbf{V} |k\rangle_n\right) \otimes \mathbf{H} |1\rangle. \quad (4.50)$$

The parenthesis on the r.h.s. of (4.50) stresses that $\mathbf{V}$ acts only on the control register. Consider the action of $\mathbf{H}^{\otimes n} \otimes \mathbb{1}$ on state $|0\rangle \otimes \mathbf{H} |1\rangle$, i.e.

$$\mathbf{H}^{\otimes n} \otimes \mathbb{1} |0\rangle \otimes \mathbf{H} |1\rangle = \left(\mathbf{H}^{\otimes n} |0\rangle\right) \otimes \mathbf{H} |1\rangle = \frac{1}{\sqrt{N}} \sum_{k=0}^{N-1} |k\rangle \otimes \mathbf{H} |1\rangle \quad (4.51)$$

where we have used the familiar parallelization operation

$$\mathbf{H}^{\otimes n} |0\rangle = \frac{1}{\sqrt{N}} \sum_{k=0}^{N-1} |k\rangle \quad N = 2^n.$$

Using identity (4.50), we find that

$$\mathbf{U}_f \left(\mathbf{H}^{\otimes n} |0\rangle \otimes \mathbf{H} |1\rangle\right) = \mathbf{V} |\phi\rangle \otimes \mathbf{H} |1\rangle$$

$$|\phi\rangle \equiv \frac{1}{\sqrt{N}} \sum_{k=0}^{N-1} |k\rangle. \quad (4.52)$$

We define the operator, also called the *diffusion gate*,

$$\mathbf{W} \equiv 2 |\phi\rangle \langle\phi| - \mathbb{1} \quad (4.53)$$

that acts only on the Hilbert space of the control register. We consider the following sequence of gates

$$(\mathbf{W} \otimes \mathbb{1}) \mathbf{V} |\phi\rangle \otimes \mathbf{H} |1\rangle = \mathbf{W}\mathbf{V} |\phi\rangle \otimes \mathbf{H} |1\rangle. \quad (4.54)$$

This procedure is repeated R times so that the final state is

$$\mathbf{G}^R |\phi\rangle \otimes \mathbf{H} |1\rangle \quad \mathbf{G} \equiv \mathbf{W}\mathbf{V} \quad (4.55)$$

after which a measurement is made on the control register. The probability $p(\xi)$ that a measurement reveals $|\xi\rangle_n$ is, according to the Born rule,

$$|\langle\xi| \mathbf{G}^R |\phi\rangle|^2 |\langle 0| \mathbf{H} |1\rangle|^2 + |\langle\xi| \mathbf{G}^R |\phi\rangle|^2 |\langle 1| \mathbf{H} |1\rangle|^2 = |\langle\xi| \mathbf{G}^R |\phi\rangle|^2. \quad (4.56)$$

We call operator $\mathbf{G}$ the *Grover operator*. Its effect on the massively parallel superposition $|\phi\rangle = \frac{1}{\sqrt{N}} \sum_{k=0}^{N-1} |k\rangle$ is to move it ever closer to $|\xi\rangle$ as R approaches the optimal value $\sqrt{N}$. That is, after $\sqrt{N}$ iterations by the G gate the probability function

$p(\xi)$ approaches its optimal value, preferably close to 1. The sequence of gates is outlined and illustrated in Mathematica Notebook 4.6.

To understand how Grover iterations work, lets first consider the case $R = 0$ for which $\mathbf{G}^0 = \mathbb{1}$. In that case

$$p(\xi) = |\langle \xi | \mathbf{G}^0 | \phi \rangle |^2 = |\langle \xi | \phi \rangle|^2 = \frac{1}{N}.$$

If $n = 100, N = 2^{100}$ the probability $p(\xi)$ of success is vanishingly small. To evaluate the matrix element for $R = 1$, we make use of the identities

$$\mathbf{V} |\phi\rangle = |\phi\rangle - \frac{2}{\sqrt{N}} |\xi\rangle \quad \mathbf{V} |\xi\rangle = -|\xi\rangle$$

$$\mathbf{W} |\xi\rangle = \frac{2}{\sqrt{N}} |\phi\rangle - |\xi\rangle \quad \mathbf{W} |\phi\rangle = |\phi\rangle \quad (4.57)$$

to derive the relations,

$$\mathbf{G} |\phi\rangle = (1 - 4/N) |\phi\rangle + \frac{2}{\sqrt{N}} |\xi\rangle$$

$$\mathbf{G} |\xi\rangle = |\xi\rangle - \frac{2}{\sqrt{N}} |\phi\rangle. \quad (4.58)$$

Therefore

$$p(\xi) = |\langle \xi | \mathbf{G} | \phi \rangle |^2 = \left| (1 - 4/N) \langle \xi | \phi \rangle + \frac{2}{\sqrt{N}} \right|^2 \approx \frac{9}{N} \quad (4.59)$$

since $\langle \xi | \phi \rangle = 1/\sqrt{N}$ and we assumed $N \gg 1$. Thus, with one pass through the Grover gate $\mathbf{G}$ the probability of finding $|\xi\rangle$ improves by a factor 9 from that by a random pick from the list. This fact is already significant since it improves by almost an order of magnitude the chance of success. However, N is large and so $p(\xi)$ is still very small. We can do much better than this by finding an optimum value for p, the number of Grover gates, in the circuit. We could proceed as above and evaluate matrix elements $\langle \xi | \mathbf{G}^p | \phi \rangle$ for various p until the optimum value is reached. Instead, it is instructive to follow a slightly different approach. To that end, we define state $|\eta\rangle$

$$|\eta\rangle \equiv \sqrt{\frac{1}{N-1}} \sum_{k \neq \xi} |k\rangle = \sqrt{\frac{N}{N-1}} |\phi\rangle - \frac{1}{\sqrt{N-1}} |\xi\rangle \quad (4.60)$$

where the sum of index k ranges over all $2^n - 1$ values that exclude state $|\xi\rangle$. Obviously,

$$\langle \xi | \eta \rangle = 0, \quad \langle \eta | \eta \rangle = 1.$$

4.4 Grover's Search Algorithm

We express $|\phi\rangle$ in terms of $|\xi\rangle$ and $|\eta\rangle$, so that

$$|\phi\rangle = \sqrt{\frac{N-1}{N}} |\eta\rangle + \frac{1}{\sqrt{N}} |\xi\rangle. \qquad (4.61)$$

Let's evaluate, using (4.58),

$$\mathbf{G} |\eta\rangle = \sqrt{\frac{N}{N-1}} \mathbf{G} |\phi\rangle - \frac{1}{\sqrt{N-1}} \mathbf{G} |\xi\rangle =$$
$$\frac{N-2}{\sqrt{N(N-1)}} |\phi\rangle + \frac{1}{\sqrt{N-1}} |\xi\rangle. \qquad (4.62)$$

Inserting (4.61) into this expression and in the second line of (4.58), we obtain

$$\mathbf{G} |\eta\rangle = \frac{N-2}{N} |\eta\rangle + 2\frac{\sqrt{N-1}}{N} |\xi\rangle$$
$$\mathbf{G} |\xi\rangle = -2\frac{\sqrt{N-1}}{N} |\eta\rangle + \frac{N-2}{N} |\xi\rangle. \qquad (4.63)$$

Let's define $\sin\theta \equiv \frac{2\sqrt{N-1}}{N}$, and since $\sqrt{1 - (\frac{2\sqrt{N-1}}{N})^2} = \frac{(N-2)}{N}$, (4.63) is equivalent to

$$\mathbf{G} |\eta\rangle = \cos\theta |\eta\rangle + \sin\theta |\xi\rangle$$
$$\mathbf{G} |\xi\rangle = -\sin\theta |\eta\rangle + \cos\theta |\xi\rangle. \qquad (4.64)$$

Equations (4.64) can be re-cast as a matrix equation or

$$\begin{pmatrix} \mathbf{G} |\eta\rangle \\ \mathbf{G} |\xi\rangle \end{pmatrix} = \begin{pmatrix} \cos\theta & \sin\theta \\ -\sin\theta & \cos\theta \end{pmatrix} \begin{pmatrix} |\eta\rangle \\ |\xi\rangle \end{pmatrix} \qquad (4.65)$$

and,

$$\begin{pmatrix} \mathbf{G}^R |\eta\rangle \\ \mathbf{G}^R |\xi\rangle \end{pmatrix} = \begin{pmatrix} \cos\theta & \sin\theta \\ -\sin\theta & \cos\theta \end{pmatrix}^R \begin{pmatrix} |\eta\rangle \\ |\xi\rangle \end{pmatrix} =$$
$$\begin{pmatrix} \cos R\theta & \sin R\theta \\ -\sin R\theta & \cos R\theta \end{pmatrix} \begin{pmatrix} |\eta\rangle \\ |\xi\rangle \end{pmatrix}. \qquad (4.66)$$

We now evaluate $p(\xi)$ after R iterations of the Grover operator (see Fig. 4.4), that is $|\langle\xi| \mathbf{G}^R |\phi\rangle|^2$. According to (4.66) operator $\mathbf{G}^R$ induces a unitary transformation in a 2d-vector space spanned by kets $|\eta\rangle$, $|\xi\rangle$ and $|\xi\rangle\langle\xi| + |\eta\rangle\langle\eta| = \mathbb{1}$, the identity operator in this space. It follows that

$$p(\xi) = |\langle\xi| \mathbf{G}^R \mathbb{1} |\phi\rangle|^2 = |\langle\xi| \mathbf{G}^R |\xi\rangle \langle\xi| \phi\rangle + \langle\xi| \mathbf{G}^R |\eta\rangle \langle\eta| \phi\rangle|^2,$$

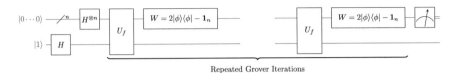

Fig. 4.4 Schematic illustration of Grover iterations

or

$$p(\xi) = \left| \langle \xi | \mathbf{G}^R | \xi \rangle \frac{1}{\sqrt{N}} + \langle \xi | \mathbf{G}^R | \eta \rangle \sqrt{\frac{N-1}{N}} \right|^2, \quad (4.67)$$

where we used the fact $\langle \xi | \phi \rangle = \frac{1}{\sqrt{N}}$, $\langle \eta | \phi \rangle = \sqrt{\frac{N-1}{N}}$. From (4.66) we find that $\langle \xi | \mathbf{G}^R | \xi \rangle = \cos R\theta$, $\langle \xi | \mathbf{G}^R | \eta \rangle = \sin R\theta$, and so

$$p(\xi) = \frac{1}{N} \left| \cos R\theta + \sin R\theta \sqrt{N-1} \right|^2. \quad (4.68)$$

In the limit $N \to \infty$,

$$p(\xi) \approx \sin^2(R\theta) \quad (4.69)$$

and has its maximum value when $R\theta = \pi/2$, or

$$R = \left[\frac{\pi \sqrt{N}}{4} \right] \quad (4.70)$$

where, for large N, $\sin \theta \approx \theta = 2/\sqrt{N}$.

> Mathematica Notebook 4.6: Implementing Grover's Algorithm. http://www.physics.unlv.edu/%7Ebernard/MATH_book/Chap4/chap4_link.html; See also https://www.bernardzygelman.github.io

Problems

4.1 In the interval $-L/2 < x < L/2$, the set of real functions $u_n(x)$, parameterized by integer index n, are said to be orthonormal if

$$\int_{-L/2}^{L/2} dx\, u_n(x)\, u_m(x) = \delta_{nm}$$

4.4 Grover's Search Algorithm

for all values of n, m. Prove that the functions

$$v_n(x) \equiv \sqrt{\frac{2}{L}} \cos(\frac{2\pi n x}{L}) \quad w_m(x) \equiv \sqrt{\frac{2}{L}} \sin(\frac{2\pi m x}{L})$$

are orthonormal for positive integer n, m.

4.2 Using the orthonormality conditions proved in Problem (4.1), verify identities (4.2).

4.3 Derive relations (4.4). Provide a detailed description of each step in the outline of the proof.

4.4 Evaluate the Fourier sums in the exercises given in Mathematica Notebook 4.1.

4.5 Evaluate the discrete Fourier transforms (DFT) in the exercises given in Mathematica Notebook 4.2.

4.6 Using Mathematica, construct the matrix representation of operator **QFT** for a register containing eight qubits. Using that matrix, demonstrate explicitly, that

$$\mathbf{QFT}^\dagger \, \mathbf{QFT} = \mathbf{QFT} \, \mathbf{QFT}^\dagger = \mathbb{1}.$$

4.7 Using Mathematica, explicitly verify identities (4.26), for a four-qubit gate.

4.8 Write a Mathematica script that implements the code, for an arbitrary input state $|j\rangle_2$, illustrated in panel (a) of Fig. 4.2. Construct a two-qubit swap operator, and calculate the action of the swap operator on the output of this gate. Compare your result with that of $\mathbf{QFT} \, |j\rangle_2$.

4.9 Do the exercises in Mathematica Notebook 4.3.

4.10 Write a Mathematica script that implements the oracle function $\mathbf{U}_f$, defined in (4.44), for a five-qubit control register. The oracle function is defined so that

$$f(21) = 1$$

and is zero otherwise. Construct a matrix representation of the oracle function and show that it is unitary.

4.11 With operator $\mathbf{U}_f$ obtained in Problem (4.10), show, explicitly, that identity (4.46) is satisfied.

4.12 For the system defined in Problem (4.10), construct the matrix representation of operator **V** defined in (4.47). Verify (4.48).

4.13 Construct the matrix representation of operator $\mathbf{H}^{\otimes n}$, where $n = 5$.

4.14 For the system defined in Problem (4.10), construct the ket $|\phi\rangle$ defined in (4.52). Using your previous results for $\mathbf{U}_f$ and $\mathbf{V}$ verify the equality

$$\mathbf{U}_f \left(\mathbf{H}^{\otimes n} |0\rangle \otimes \mathbf{H} \right) |1\rangle = \mathbf{V} |\phi\rangle \otimes \mathbf{H} |1\rangle$$

4.15 For the system defined in Problem (4.10), construct the diffusion gate $\mathbf{W}$, and the Grover operator $\mathbf{WV}$.

4.16 With the Grover operator constructed in Problem (4.15), verify identities (4.58).

4.17 Using the Grover operator obtained in Problem (4.16), construct a table of the probabilities $p(\xi) = |\langle \xi | \mathbf{G} | \phi \rangle|^2$, and make a plot of $p(\xi)$ for all integer $0 \leq \xi < 2^5$.

4.18 Repeat Problem (4.17) for operator $\mathbf{G}^p$ where $p = 1, 2, 3 \ldots$. Comment on the trend observed with this data. Is there an optimum choice for p?

4.19 Write a Mathematica script that simulates Grover's algorithm for a ten-qubit register.

References

1. M.E. Nielsen, I.L. Chuang, *Quantum Computation and Quantum Information* (Cambridge University Press, Cambridge, 2011)
2. W.H. Press, S.A. Teukolsky, W.T. Vetterling, B.P. Flannery, *Numerical Recipes: The Art of Scientific Computing*, 3rd edn. (Cambridge University Press, 2007)
3. Wikipedia entry for Killer application. https://en.wikipedia.org/wiki/Killer_application
4. P. Zimmermann, Factorisation of RSA-220 with CADO-NFS, Cado-nfs-discuss. https://lists.gforge.inria.fr/pipermail/cado-nfs-discuss/2016-May/000626.html

5. Quantum Mechanics According to Martians: Density Matrix Theory

Abstract

We discuss the density matrix approach introduced by John von Neumann to quantum mechanics. We illustrate how it is applied in the calculation of expectation values and Born probabilities. The postulates of quantum mechanics, according to the Copenhagen interpretation, are summarized from the vantage point of the density operator framework. We learn how density operators are used to discern coherent, or pure, state ensembles from statistical mixtures of pure states. It is shown that, for entangled states, the traced density operator to a lower dimensional Hilbert space, results in reduced density operators that describe a mixed state. The Schmidt decomposition theorem and the Schmidt number, which measures the degree of entanglement of quantum states, is introduced and discussed. We define and illustrate the concept of von Neumann entropy.

5.1 Introduction

John von Neumann and *Eugene Wigner*, who made seminal contributions to the quantum theory, were members of a group [1] of èmigrès that possessed oversized intellects, a thick accent, and shared a common geographic lineage. An inside joke, among colleagues, posited that they were descendants of a Martian scout force that landed near Budapest around the turn of the last century.

More likely, the quality of their schooling was a determining factor in nurturing their otherworldly talents. Among fellow students who became leading figures in science and mathematics, von Neumann and Wigner were mentored by *Làszlò Ràtz*, a mathematics teacher at the Budapest-Fasori Evangèlikus Gimnàzium.

My favorite "Martian", John von Neumann, made unique contributions to the foundations of mathematics and logic, quantum theory, game theory and pioneered the development of one of the first digital electronic computers. His highly influential treatise *The Mathematical Foundations of Quantum Mechanics*, published in 1932, codified the Copenhagen interpretation. In it, von Neumann introduced an alternative formalism, called the density operator or matrix formulation, for describing quantum systems. Though equivalent to the bra-ket formalism, it is more powerful than the latter in describing statistical mixtures of quantum states. Today, the density operator formulation is indispensable in QCI applications. In this chapter, we review that theory and translate the foundational postulates introduced in previous chapters, to the language of density operators.

5.2 Density Operators and Matrices

Consider scenario (A); each member of a group of physicists possesses a qubit. Half of the qubits are in the $|0\rangle$ state, and the other half in state $|1\rangle$. The qubits are electrons, and the scientists employ a Stern-Gerlach device to measure $\mathbf{S}_Z = \hbar/2\,\sigma_Z$, the component of spin along the z-axis of a laboratory reference frame. Because the kets $|0\rangle$, $|1\rangle$ are eigenstates of $\mathbf{S}_Z$ with eigenvalues $\pm\hbar/2$ respectively, the measurements yield the expectation value

$$<\mathbf{S}_Z> = 0.$$

Alternatively, in scenario (B), we assume that each scientist's qubit is in the superposition state

$$|u\rangle = \frac{1}{\sqrt{2}}\,(\,|0\rangle + |1\rangle\,) \tag{5.1}$$

and measurements are made with device $\mathbf{S}_Z$. Because $|u\rangle$ is a superposition state, the Born rule predicts outcomes $\pm\hbar/2$. You should convince yourself that about 1/2 of the scientists observe $\hbar/2$, whereas the remainder observes $-\hbar/2$. On the average they, again, find

$$<\mathbf{S}_Z> = 0.$$

Because both scenarios lead to the same expectation value for $<\mathbf{S}_Z>$, we might, at first, conclude that the two descriptions are equivalent. Below we show why that inference is incorrect.

In scenario (B); each scientist possesses a qubit in state $|u\rangle$ but now takes measurements with instrument $\mathbf{S}_X = \hbar/2\,\sigma_X$, the spin along the x-axis. In Chap. 2 we learned that $|u\rangle$, $|v\rangle$, defined in (2.20), are eigenstates of $\mathbf{S}_X$ with eigenvalues $\pm\hbar/2$ respectively. Therefore, every measurement with $\mathbf{S}_X$ yields the result $\hbar/2$, and so the ensemble expectation value is

$$<\mathbf{S}_X> = \hbar/2. \tag{5.2}$$

5.2 Density Operators and Matrices

In case (A) half the scientists possess state $|0\rangle = \frac{1}{\sqrt{2}}(|u\rangle + |v\rangle)$, and the other half $|1\rangle = \frac{1}{\sqrt{2}}(|u\rangle - |v\rangle)$. According to the Born rule, the probability that a measurement with $\mathbf{S}_X$ yields $\hbar/2$ is

$$p_X(\hbar/2) = |\langle u|0\rangle|^2 = 1/2,$$

likewise

$$p_X(-\hbar/2) = |\langle v|0\rangle|^2 = 1/2.$$

In summary, if half of the qubits are in state $|0\rangle$, and the other half $|1\rangle$, measurements with $\mathbf{S}_X$ again lead to

$$<\mathbf{S}_X> = 0. \tag{5.3}$$

The fact that the expectation values (5.2) and (5.3) differ demonstrates that the two scenarios are not equivalent. In Hilbert space, any state, whether computational basis $|0\rangle$, $|1\rangle$ or superposition $|u\rangle$, $|v\rangle$, is called a coherent or pure state. In fact, a coherent state is just a synonym for a Hilbert space state. So is there such a thing as an incoherent state? In scenario (B) each scientist possesses a qubit in coherent state $|u\rangle$, but case (A) describes an ensemble of measurement values obtained from a statistical mixture of two different coherent states. The postulates itemized in the previous chapters taught us how to compile, via the Born rule, statistics for systems described by a coherent, or pure, state. But we seem to lack a vocabulary that allows a convenient framework for statistical mixtures of coherent states. The von-Neumann *density matrix* or density operator, provides that framework.

To incorporate both pure and mixed ensembles, von Neumann introduced the concept of a density operator. The operator

$$\rho \equiv \sum_i p_i |\psi_i\rangle\langle\psi_i|$$
$$\sum_i p_i = 1 \tag{5.4}$$

where $|\psi_i\rangle$ is a unit vector in Hilbert space and $p_i \geq 0$, is called a density operator. The set $|\psi_i\rangle$ may, or may not be orthogonal or independent, and the constraint $\sum_i p_i = 1$ suggests that set p_i is a probability measure. We can think of a density operator as describing a collection, or ensemble,

$$\{p_1, |\psi_1\rangle; p_2, |\psi_2\rangle; \cdots p_n, |\psi_n\rangle\}. \tag{5.5}$$

It is interpreted as a mixture of states, where $p_1 \times 100$ percent are in state $|\psi_1\rangle$, $p_2 \times 100$ percent are in state $|\psi_2\rangle$, and so on. With this notation scenario (A) is described by

$$\{p_1 = 1/2, |0\rangle; p_2 = 1/2, |1\rangle\}$$

and (B) by,

$$\{p_1 = 1, |u\rangle; p_2 = 0, |v\rangle\}.$$

The density operator formalism avoids this cumbersome notation and describes each scenario by the corresponding density operators

$$\rho_A = \frac{1}{2} |0\rangle \langle 0| + \frac{1}{2} |1\rangle \langle 1|$$
$$\rho_B = |u\rangle \langle u|, \quad (5.6)$$

where the subscripts denote case (A) and (B) respectively.

In Chap. 2 we showed that the expectation value $<\mathbf{A}>$ of an ensemble of outcomes obtained by measurements on state $|\Psi\rangle$ with operator $\mathbf{A}$ is given by the expression

$$<\mathbf{A}> = \langle \Psi | \mathbf{A} | \Psi \rangle.$$

We claim that

$$Tr\, \rho \mathbf{A} = \langle \Psi | \mathbf{A} | \Psi \rangle \quad (5.7)$$

where the symbol Tr refers to the trace of an operator. Typically, trace is an operation on a matrix. Given matrix elements O_{nm}, of operator $\mathbf{O}$, its trace is defined as the sum of its diagonal elements $\sum_n O_{nn}$. The matrix trace is invariant with respect to the basis vectors used to construct the matrix representation of an operator. Therefore expression (5.7) is independent of the representation used in calculating it. Below we offer a proof of identity (5.7). The proof for the general case is left as an exercise.

Proof Using the closure property for a complete set of eigenstates $|a\rangle$, the r.h.s of (5.7) is

$$\langle \Psi | \mathbb{1} \mathbf{A} \mathbb{1} | \Psi \rangle = \sum_a \sum_{a'} \langle \Psi | a \rangle \langle a | \mathbf{A} | a' \rangle \langle a' | \Psi \rangle =$$

$$\sum_a \sum_{a'} \langle a' | \Psi \rangle \langle \Psi | a \rangle \langle a | \mathbf{A} | a' \rangle =$$

$$\sum_a \sum_{a'} \rho_{a'a} \mathbf{A}_{aa'} = \sum_{a'} [\rho \mathbf{A}]_{a'a'} = Tr[\rho \mathbf{A}] \quad (5.8)$$

where

$$\rho = |\Psi\rangle \langle \Psi|, \quad \rho_{a'a} \equiv \langle a' | \rho | a \rangle, \quad \mathbf{A}_{aa'} \equiv \langle a | \mathbf{A} | a' \rangle.$$

□

Let's apply (5.7) to the density operators corresponding to scenarios (A) and (B). We use the computational basis to express operator ρ_A as a matrix; thus

$$\underline{\rho_A} = \begin{pmatrix} \langle 0 | \rho_A | 0 \rangle & \langle 0 | \rho_A | 1 \rangle \\ \langle 1 | \rho_A | 0 \rangle & \langle 1 | \rho_A | 1 \rangle \end{pmatrix} = \begin{pmatrix} 1/2 & 0 \\ 0 & 1/2 \end{pmatrix}. \quad (5.9)$$

Also, in this basis, the matrix representations for $\mathbf{S}_Z$ and $\mathbf{S}_X$ are

$$\underline{S}_Z = \frac{\hbar}{2} \begin{pmatrix} 1 & 0 \\ 0 & -1 \end{pmatrix} \quad \underline{S}_X = \frac{\hbar}{2} \begin{pmatrix} 0 & 1 \\ 1 & 0 \end{pmatrix}. \quad (5.10)$$

5.2 Density Operators and Matrices

Using (5.9) and (5.10) we obtain

$$Tr\,[\rho_A\,S_Z] = \frac{\hbar}{2}Tr\begin{pmatrix}1/2 & 0\\ 0 & 1/2\end{pmatrix}\begin{pmatrix}1 & 0\\ 0 & -1\end{pmatrix} = \frac{\hbar}{4}Tr\begin{pmatrix}1 & 0\\ 0 & -1\end{pmatrix} = 0 \quad (5.11)$$

$$Tr\,[\rho_A\,S_X] = \frac{\hbar}{2}Tr\begin{pmatrix}1/2 & 0\\ 0 & 1/2\end{pmatrix}\begin{pmatrix}0 & 1\\ 1 & 0\end{pmatrix} = \frac{\hbar}{4}Tr\begin{pmatrix}0 & 1\\ 1 & 0\end{pmatrix} = 0. \quad (5.12)$$

Similarly, for

$$\rho_B = |u\rangle\langle u| = \frac{1}{2}(|0\rangle + |1\rangle)(\langle 0| + \langle 1|) =$$
$$\frac{1}{2}(|0\rangle\langle 0| + |0\rangle\langle 1| + |1\rangle\langle 0| + |1\rangle\langle 1|) \quad (5.13)$$

whose matrix representation

$$\underline{\rho}_B = \begin{pmatrix}\langle 0|\rho_B|0\rangle & \langle 0|\rho_B|1\rangle\\ \langle 1|\rho_B|0\rangle & \langle 1|\rho_B|1\rangle\end{pmatrix} = \begin{pmatrix}1/2 & 1/2\\ 1/2 & 1/2\end{pmatrix}, \quad (5.14)$$

we find

$$Tr\,[\rho_B\,S_Z] = 0 \quad Tr\,[\rho_B\,S_X] = \hbar/2.$$

In summary,

$$\langle S_Z\rangle_A = Tr\,[\rho_A\,S_Z] = 0 \quad \langle S_X\rangle_A = Tr\,[\rho_A\,S_X] = 0$$
$$\langle S_Z\rangle_B = Tr\,[\rho_B\,S_Z] = 0 \quad \langle S_X\rangle_B = Tr\,[\rho_B\,S_X] = \hbar/2, \quad (5.15)$$

in harmony with the expectation values obtained previously.

The density operator formalism facilitates an economical and convenient method for estimating expectation values. We now show how density operators can also be used to implement the Born rule. In quantum mechanics, a measurement is associated with a Hilbert space Hermitian operator **M**. The eigenvalues, m_i, of **M** represent possible results of a measurement, and the Born rule determines the probability for observation of m_i. Measurement operators can always be expressed in the form

$$\mathbf{M} = \sum_i m_i\,|m_i\rangle\langle m_i|$$

where $|m_i\rangle$ is an eigenstate of **M** with eigenvalue m_i, and obey the closure relation

$$\sum_i |m_i\rangle\langle m_i| = \mathbb{1}.$$

According to the Born rule, if the system is in quantum state $|\psi\rangle$, there exists a probability measure

$$p(m_i) = |\langle m_i|\psi\rangle|^2 = \langle m_i|\psi\rangle\langle \psi|m_i\rangle. \quad (5.16)$$

According to Dirac's axioms, we are allowed to express the r.h.s of (5.16) as an expectation value of operator

$$\mathbf{M}_i \equiv |m_i\rangle \langle m_i|$$

with respect to state $|\psi\rangle$. That is,

$$\langle\psi||m_i\rangle \langle m_i||\psi\rangle = \langle\psi|\mathbf{M}_i|\psi\rangle = Tr[\rho \mathbf{M}_i] \tag{5.17}$$

and so Born's rule is encapsulated in the relation

$$p(m_i) = Tr[\rho \mathbf{M}_i]. \tag{5.18}$$

Our proof of (5.18) relied on the fact that ρ represents a *coherent state*, but it also turns out to be true for any valid density operator [3], and whose proof is left as an exercise.

With relation (5.18) we re-frame our foundational postulates, according to von-Neumann, using the language of density operators or matrices.

Postulate I	An n-qubit register defines a Hilbert space spanned by basis vectors that are direct products of n qubits $|a\rangle \otimes |b\rangle \otimes |c\rangle \cdots \otimes |n\rangle$, where $a, b, c \cdots n \in 0, 1$.
Postulate II	A full description of the system is encapsulated by a density operator $\rho = \sum_i p_i |\psi_i\rangle \langle\psi_i|$ where $|\psi_i\rangle$ is a normalized state in Hilbert space and $\sum_i p_i = 1$.
Postulate III (Born's rule)	The act of measurement associated with Hermitian operator $M = \sum_i m_i |m_i\rangle \langle m_i|$ results in one of its eigenvalues m_i. The probability of obtaining a nondegenerate eigenvalue m_i, is given by the expression $$p(m_i) = Tr[\rho \mathbf{M}_i]$$ where $\mathbf{M}_i \equiv |m_i\rangle \langle m_i|$ is a projection operator. If the eigenvalues are degenerate, with value m, the probability to find that value is $\sum_i Tr[\rho \mathbf{M}_i]$ where the sum is over all i in which $m_i = m$.
Postulate IV	For a closed system, the density operator evolves in time so that $$\rho(t) = \mathbf{U}(t, t_0)\, \rho(t_0)\, \mathbf{U}^\dagger(t, t_0)$$ where $\rho(t)$ is the density operator at time $t > t_0$. $\mathbf{U}(t, t_0)$ is the unitary time evolution operator.
Postulate V (collapse hypothesis)	In a closed system let $\mathbf{M}_m$ represent a measuring device whose eigenvalue is m. If a measurement on the system in state ρ yields m, the system is then characterized by a density operator [3]

$$\rho' = \frac{\mathbf{M}_m \rho \mathbf{M}_m^\dagger}{Tr(\mathbf{M}_m^\dagger \mathbf{M}_m \rho)}.$$

These postulates are equivalent to the list itemized in previous chapters. However, the von Neumann density matrix formulation provides a unified description that includes both coherent and mixed state systems. We have already explored the density matrix approach for the pure and mixed state scenarios (A) and (B), but now focus in detail, on some important properties of the density matrix.

5.3 Pure and Mixed States

Previously, we noted that ρ_A describes an ensemble generated by a statistical distribution of quantum states, whereas ρ_B describes a pure, or coherent, state. Given a density matrix, how can we determine whether it describes a coherent state or not? We know that, for a coherent state,

$$\rho = |\psi\rangle\langle\psi|, \tag{5.19}$$

and for any normalized state $|\psi\rangle$

$$\rho^2 \equiv \rho\rho = |\psi\rangle\langle\psi|\psi\rangle\langle\psi| = |\psi\rangle\langle\psi| = \rho \tag{5.20}$$

where we have used Dirac's axiom to equate $\langle\psi|\,|\psi\rangle = \langle\psi|\psi\rangle = 1$.

Defining the m, nth matrix element of ρ,

$$[\rho]_{mn} = \langle m|\rho|n\rangle$$

we find

$$Tr\,\rho = \sum_m \rho_{mm} = \sum_m \langle m|\rho|m\rangle = \sum_m \langle m|\psi\rangle\langle\psi|m\rangle =$$
$$= \sum_m \langle\psi|m\rangle\langle m|\psi\rangle = \langle\psi|\psi\rangle = 1 \tag{5.21}$$

where we used the closure property of the basis $|m\rangle$. Therefore,

$$Tr\,\rho^2 = Tr\,\rho = 1. \tag{5.22}$$

Property (5.22) is the defining feature of a density operator for a pure, or coherent, state. Now consider the density matrix describing a mixed ensemble

$$\rho = \sum_i p_i |\psi_i\rangle\langle\psi_i|$$

and assume that the states $|\psi_i\rangle$ are orthonormal. Thus,

$$\begin{aligned}\rho^2 &= \sum_k \sum_j p_k\, p_j\, |\psi_k\rangle \langle\psi_k| |\psi_j\rangle\langle\psi_j| \\ &= \sum_k \sum_j p_k\, p_j\, |\psi_k\rangle\langle\psi_j|\, \delta_{jk} = \sum_j p_j^2\, |\psi_j\rangle\langle\psi_j|\end{aligned} \quad (5.23)$$

where we have made use of Dirac's bra-ket axioms and the orthogonality relation $\langle\psi_j|\psi_k\rangle = \delta_{kj}$. Therefore,

$$Tr\,\rho^2 = \sum_i p_i^2\, Tr\, |\psi_i\rangle\langle\psi_i| = \sum_i p_i^2 \quad (5.24)$$

where we have used the fact $Tr\, |\psi_i\rangle\langle\psi_i| = 1$. Since $\sum_i p_i = 1$ it follows

$$\sum_i p_i^2 < 1$$

if more than one $p_i \neq 0$. Thus,

$$Tr\,\rho^2 \leq 1. \quad (5.25)$$

The equality is satisfied if ρ describes a *pure ensemble* and the inequality follows if ρ describes a *mixed ensemble*. Inequality (5.25) holds even if states $|\psi_i\rangle$ are not orthogonal [3].

5.4 Reduced Density Operators

Consider a direct product state of two qubits,

$$|\psi_{AB}\rangle = |\psi_A\rangle \otimes |\psi_B\rangle. \quad (5.26)$$

The subscripts identify the qubit subspace, so in this notation $|\psi_B\rangle$ is a normalized vector in Hilbert sub-space B, as is $|\psi_A\rangle$ in sub-space A. $|\psi_{AB}\rangle$ "lives" in the direct product Hilbert space. The density operator corresponding to this state is

$$\rho_{AB} = |\psi_{AB}\rangle\langle\psi_{AB}| = |\psi_A\rangle\langle\psi_A| \otimes |\psi_B\rangle\langle\psi_B| \quad (5.27)$$

and satisfies the equality

$$Tr\,\rho_{AB}^2 = 1 \quad (5.28)$$

as it represents a coherent state (5.26). We define the partial trace by the operation

$$\begin{aligned}Tr_B\,\rho_{AB} &\equiv |\psi_A\rangle\langle\psi_A|\, Tr(|\psi_B\rangle\langle\psi_B|) \\ Tr_A\,\rho_{AB} &\equiv Tr(|\psi_A\rangle\langle\psi_A|)\, |\psi_B\rangle\langle\psi_B|.\end{aligned} \quad (5.29)$$

5.4 Reduced Density Operators

From our previous discussion we know that $Tr(|\psi_B\rangle\langle\psi_B|) = 1$, $Tr(|\psi_A\rangle\langle\psi_A|) = 1$, and so the reduced density matrices

$$Tr_B \, \rho_{AB} = \rho_A \quad Tr_A \, \rho_{AB} = \rho_B$$
$$\rho_A = |\psi_A\rangle\langle\psi_A| \quad \rho_B = |\psi_B\rangle\langle\psi_B| \tag{5.30}$$

are single qubit density operators. Since $\rho_A = |\psi_A\rangle\langle\psi_A|$, $Tr\rho_A^2 = 1$, and so ρ_A also describes a coherent state. The same is true for ρ_B. To appreciate the significance of this result, imagine probing the system in state (5.26) with instrument $\mathbf{M} \otimes \mathbb{1}$, where $\mathbf{M} = m_1 |m_1\rangle\langle m_1| + m_2 |m_2\rangle\langle m_2|$, to measure qubit A. Device $\mathbf{M} \otimes \mathbb{1}$ reveals eigenvalues m_1, m_2 corresponding to eigenstates $|m_1\rangle$, $|m_2\rangle$ respectively. In a measurement that reveals eigenvalue m_1, the collapse hypothesis requires that

$$|\psi_A\rangle \otimes |\psi_B\rangle \rightarrow |m_1\rangle \otimes |\psi_B\rangle. \tag{5.31}$$

or, in the von-Neumann formulation

$$\rho_{AB} \rightarrow |m_1\rangle\langle m_1| \otimes |\psi_B\rangle\langle\psi_B|, \tag{5.32}$$

In this example, a measurement of sub-space A or B, does not disturb the partner state, but as we show below, this is not true for entangled states.

5.4.1 Entangled States

Not every multi-qubit state can be factored. For example, consider the following two-qubit state

$$|\psi_{AB}\rangle = \frac{1}{\sqrt{2}}\Big(|0\rangle \otimes |1\rangle + |1\rangle \otimes |0\rangle\Big) = \frac{1}{\sqrt{2}}\Big(|01\rangle + |10\rangle\Big) \tag{5.33}$$

the density operator associated with it is

$$\rho_{AB} = |\psi_{AB}\rangle\langle\psi_{AB}| = \frac{1}{2}\Big(|01\rangle\langle 10| + |01\rangle\langle 01| + |10\rangle\langle 10| + |10\rangle\langle 01|\Big). \tag{5.34}$$

Now,

$$\rho_{AB}^2 = |\psi_{AB}\rangle\langle\psi_{AB}|\psi_{AB}\rangle\langle\psi_{AB}| = |\psi_{AB}\rangle\langle\psi_{AB}|$$
$$Tr \, \rho_{AB}^2 = Tr \, |\psi_{AB}\rangle\langle\psi_{AB}| = 1 \tag{5.35}$$

and so ρ_{AB} describes a pure, or coherent, two-qubit state. Using linearity of the partial trace operation,

$$Tr_B \, \rho_{AB} =$$
$$\frac{1}{2}\Big(|0\rangle\langle 0| \, (Tr \, |1\rangle\langle 1|) + |0\rangle\langle 1| \, (Tr \, |1\rangle\langle 0|) +$$
$$|1\rangle\langle 0| \, (Tr \, |0\rangle\langle 1|) + |1\rangle\langle 1| \, (Tr \, |0\rangle\langle 0|)\Big), \tag{5.36}$$

and the fact

$$Tr \, |1\rangle \langle 1| = Tr \, |0\rangle \langle 0| = 1$$
$$Tr \, |0\rangle \langle 1| = Tr \, |1\rangle \langle 0| = 0,$$

we get

$$\rho_A \equiv Tr_B \, \rho_{AB} = \frac{1}{2}\Big(|0\rangle \langle 0| + |1\rangle \langle 1|\Big). \tag{5.37}$$

Now

$$\rho_A^2 = \frac{1}{4}\Big(|0\rangle \langle 0| |0\rangle \langle 0| + |0\rangle \langle 0| |1\rangle \langle 1| + |1\rangle \langle 1| |0\rangle \langle 0| + |1\rangle \langle 1| |1\rangle \langle 1|\Big) =$$
$$\frac{1}{4}\Big(|0\rangle \langle 0| + |1\rangle \langle 1|\Big), \tag{5.38}$$

thus,

$$Tr \, \rho_A^2 = \frac{1}{4}\Big(Tr \, |0\rangle \langle 0| + Tr \, |1\rangle \langle 1|\Big) = \frac{1}{2}. \tag{5.39}$$

Because $Tr \, \rho_A^2 < 1$ we recognize that the reduced single-qubit density matrix describes a mixed state, despite the fact that we started with a pure two-qubit state. To gain insight let's again measure the two-qubit system, in state (5.33), with instrument $\mathbf{M}_{AB} \equiv \mathbf{M} \otimes \mathbb{1}$. To use the Born rule, we need to itemize the eigenstates of $\mathbf{M}_{AB}$. They are

$$|m_1\rangle \otimes |0\rangle \quad |m_1\rangle \otimes |1\rangle \quad |m_2\rangle \otimes |0\rangle \quad |m_2\rangle \otimes |1\rangle$$

where $|m_1\rangle \, |m_2\rangle$ are eigenstates of $\mathbf{M}$. Because the eigenvalues m_1, m_2 are degenerate, e.g.,

$$\mathbf{M}_{AB} \, |m_1\rangle \otimes |0\rangle = m_1 \, |m_1\rangle \otimes |0\rangle$$
$$\mathbf{M}_{AB} \, |m_1\rangle \otimes |1\rangle = m_1 \, |m_1\rangle \otimes |1\rangle, \tag{5.40}$$

the Born rule states that the probability for the experimenter to measure m_1 is given by the sum

$$p(m_1) = |\langle 0| \otimes \langle m_1| \, |\psi_{AB}\rangle|^2 + |\langle 1| \otimes \langle m_1| \, |\psi_{AB}\rangle|^2 \tag{5.41}$$

Using (5.33), we have,

$$\langle 0| \otimes \langle m_1| \, |\psi_{AB}\rangle = \frac{1}{\sqrt{2}} \langle m_1| 1\rangle$$
$$\langle 1| \otimes \langle m_1| \, |\psi_{AB}\rangle = \frac{1}{\sqrt{2}} \langle m_1| 0\rangle \tag{5.42}$$

5.4 Reduced Density Operators

Or

$$p(m_1) = \frac{1}{2}\left(|\langle m_1|1\rangle|^2 + |\langle m_1|0\rangle|^2\right) = 1/2 \tag{5.43}$$

likewise, we find that $p(m_2) = 1/2$. In deriving (5.43) we used the fact that $|0\rangle\langle 0| + |1\rangle\langle 1| = \mathbb{1}$. We now ask the following; can we find a unique single-qubit state $|\psi\rangle$ that reproduces the probability distribution (5.43)? if we choose

$$|\psi\rangle = \frac{1}{\sqrt{2}}(|0\rangle + |1\rangle)$$
$$\mathbf{M} = \sigma_Z$$

then $p(m_1 = 1) = 1/2$, $p(m_2 = -1) = 1/2$. However, if we choose $\mathbf{M} = \sigma_X$ for the measurement device then $p(m_1 = 1) = 1$, $p(m_2 = 0) = 0$. According to (5.43) $p(m_1) = 1/2$, $p(m_2) = 1/2$ regardless of the choice for $\mathbf{M}$. Therefore, that probability distribution cannot be produced from this pure single-qubit state. However, it can result from a mixed state, defined by density operator (5.37).

A qubit measurement device is represented by a Hermitian matrix, whose most general form is

$$\mathbf{M} = \begin{pmatrix} z_1 & x+iy \\ x-iy & z_2 \end{pmatrix} \tag{5.44}$$

where z_1, z_2, x, y are real numbers. The eigenvalues of $\mathbf{M}$ are

$$m_1 = \frac{z_1 + z_2}{2} + \frac{1}{2}\sqrt{(z_1 - z_2)^2 + 4(x^2 + y^2)}$$
$$m_2 = \frac{z_1 + z_2}{2} - \frac{1}{2}\sqrt{(z_1 - z_2)^2 + 4(x^2 + y^2)}$$

corresponding to eigenstate

$$|m_1\rangle = \begin{pmatrix} \exp(i\phi)\cos\theta/2 \\ \sin\theta/2 \end{pmatrix} \quad |m_2\rangle = \begin{pmatrix} -\exp(i\phi)\sin\theta/2 \\ \cos\theta/2 \end{pmatrix}. \tag{5.45}$$

where $\tan\phi = y/x$, $\cos\theta = (z_1 - z_2)/\sqrt{4(x^2 + y^2) + (z_1 - z_2)^2}$. Using this relation we construct the *projection operator*, corresponding to measurement of eigenvalue m_1,

$$\mathbf{M}_1 \equiv |m_1\rangle\langle m_1| = \begin{pmatrix} \cos^2\left(\frac{\theta}{2}\right) & \frac{1}{2}\exp(-i\phi)\sin(\theta) \\ \frac{1}{2}\exp(i\phi)\sin(\theta) & \sin^2\left(\frac{\theta}{2}\right) \end{pmatrix} \tag{5.46}$$

The matrix representation, with respect to the computational basis, for ρ_A given in (5.37) is

$$\rho_A = \frac{1}{2}\begin{pmatrix} 1 & 0 \\ 0 & 1 \end{pmatrix} \tag{5.47}$$

and so

$$p(m_1) = Tr\,[\rho_A \mathbf{M}_1] =$$
$$\frac{1}{2} Tr\left[\begin{pmatrix} 1 & 0 \\ 0 & 1 \end{pmatrix} \begin{pmatrix} \cos^2\left(\frac{\theta}{2}\right) & \frac{1}{2}\exp(-i\phi)\sin(\theta) \\ \frac{1}{2}\exp(i\phi)\sin(\theta) & \sin^2\left(\frac{\theta}{2}\right) \end{pmatrix}\right] = \frac{1}{2}. \quad (5.48)$$

Using the reduced density matrix formalism, we find that $p(m_1) = 1/2$ is independent of the parameters z_1, z_2, x, y that define the measurement operator **M**, and in agreement with the analysis leading to (5.43). To summarize: for a two-qubit state that can be factored, i.e., $|\psi_{AB}\rangle = |\psi_A\rangle \otimes |\psi_B\rangle$, the reduced density operator describes a single-qubit pure state. In contrast, the reduced density operator for an entangled two-qubit state, describes a mixed ensemble of single-qubit states.

Mathematica Notebook 5.1: Entanglement as a route to de-coherence. http://www.physics.unlv.edu/%7Ebernard/MATH_book/Chap5/chap5_link.html; See also https://bernardzygelman.github.io

5.5 Schmidt Decomposition

So far, we focused on two types of states; the kind given by (5.26), a pure two-qubit state that can be factored into a product of single qubit states, and entangled states that cannot be factored. The Schmidt decomposition theorem allows us to quantify the degree of entanglement by introducing the concept of a *Schmidt number*.

Theorem 5.1 *In a direct product Hilbert space (AB) where $|\phi_i\rangle$ are a complete orthonormal basis in space A of dimension N, and $|\lambda_i\rangle$ are a complete orthonormal basis in space B of dimension N, any vector $|\psi_{AB}\rangle$ can be expressed in the following manner,*

$$|\psi_{AB}\rangle = \sum_{i=1}^{N} \sqrt{p_i}\,|\phi_i\rangle \otimes |\lambda_i\rangle \quad (5.49)$$

where $p_i \geq 0$, $\sum_i p_i = 1$.

Proof Assume that $|\gamma_i\rangle, |\chi_j\rangle$ are complete basis vectors for space A and B respectively, so that

$$|\psi_{AB}\rangle = \sum_\alpha \sum_\beta c_{\alpha\beta}\,|\gamma_\alpha\rangle\,|\chi_\beta\rangle, \quad \sum_\alpha \sum_\beta |c_{\alpha\beta}|^2 = 1. \quad (5.50)$$

5.5 Schmidt Decomposition

The coefficients $c_{\alpha\beta}$ in this expansion are taken as the α, β components of matrix $\underline{c}$. Now according to the *singular-value decomposition* [2], any well-defined square matrix can be written in the form

$$\underline{c} = \underline{u}\,\underline{d}\,\underline{v}$$

where $\underline{u}$, $\underline{v}$ are square unitary matrices and $\underline{d}$ a diagonal matrix whose elements are non-negative. Therefore

$$c_{\alpha\beta} = \sum_i u_{\alpha i}\, d_i\, v_{i\beta}$$

and

$$|\psi_{AB}\rangle = \sum_\alpha \sum_\beta \sum_i u_{\alpha i}\, d_i\, v_{i\beta}\, |\gamma_\alpha\rangle \otimes |\chi_\beta\rangle.$$

We define

$$|\phi_i\rangle \equiv \sum_\alpha u_{\alpha i}\, |\gamma_\alpha\rangle$$
$$|\lambda_i\rangle \equiv \sum_\beta v_{i\beta}\, |\chi_\beta\rangle \tag{5.51}$$

and so

$$|\psi_{AB}\rangle = \sum_i d_i\, |\phi_i\rangle \otimes |\lambda_i\rangle. \tag{5.52}$$

Now both $|\lambda_i\rangle$ and $|\phi_i\rangle$ are related to basis $|\gamma_i\rangle$, $|\chi_j\rangle$ via the unitary transformations (5.51) and are orthonormal. Therefore

$$\langle \psi_{AB}|\psi_{AB}\rangle = 1 = \sum_i d_i^2$$
$$d_i = \sqrt{p_i} \tag{5.53}$$

□

Our proof of the Schmidt decomposition theorem is limited to sub-spaces of equal dimensions N. In fact, the theorem is valid [3] for any two direct product spaces of arbitrary dimension. For example, suppose we partition a five-qubit Hilbert space into two product spaces, one with dimension 2 (a two-qubit space A) and the other (space B) with dimension 3. In that case, space A is spanned by four basis states $|\gamma_i\rangle$, for $i = 1, \ldots 4$. In space B the there are 2^3 basis vectors $|\chi_j\rangle$, for $j = 1, 2, \ldots 8$. The Schmidt decomposition is now written

$$|\psi_{AB}\rangle = \sum_{i=1}^{4} \sqrt{p_i}\, |\varphi_i\rangle \otimes |\lambda_i\rangle \tag{5.54}$$

where the summation index in the sum ranges only over the dimension of the smaller Hilbert space.

Given a state expressed in bi-partite form $|\psi_{AB}\rangle$ as above, the density operator

$$\rho_{AB} = |\psi_{AB}\rangle\langle\psi_{AB}| = \sum_{i=1}\sum_{j=1} \sqrt{p_i}\sqrt{p_j}(|\varphi_i\rangle\langle\varphi_j|) \otimes (|\lambda_i\rangle\langle\lambda_j|) \quad (5.55)$$

taking the partial trace over sub-space B

$$\rho_A = Tr_B\, \rho_{AB} = \sum_{i=1}\sum_{j=1} \sqrt{p_i}\sqrt{p_j}(|\varphi_i\rangle\langle\varphi_j|)Tr(|\lambda_i\rangle\langle\lambda_j|). \quad (5.56)$$

Because $Tr(|\lambda_i\rangle\langle\lambda_j|) = \delta_{ij}$, we find that

$$\rho_A = \sum_{i=1}^{n} p_i\, |\varphi_i\rangle\langle\varphi_i|. \quad (5.57)$$

Because $\sum_i p_i = 1$ and $|\varphi_i\rangle$ are basis kets for space A, ρ_A is indeed a density operator for space A. The number of non-zero values for p_i in (5.57) is called the Schmidt number and flags the degree of entanglement in state $|\psi_{AB}\rangle$. For Schmidt number 1, ρ_A describes a pure state, and a basis exists in which $|\psi_{AB}\rangle$ can be factored into a direct product states. In our examples, the state given by (5.26) has Schmidt number 1, whereas state (5.33) has Schmidt number 2.

Suppose a quantum system is described by the density operator

$$\rho = \sum_i p_i\, |\psi_i\rangle\langle\psi_i| \quad (5.58)$$

in a Hilbert space with basis $|\psi_i\rangle$. If the Schmidt number is greater than one, then (5.58) describes a mixed state. We know that mixed states can be produced by taking the partial trace of a pure state in a higher dimensional Hilbert space, so it is natural to ask is the converse possible? That is, given mixed state (5.58) can we find a higher dimensional Hilbert space pure state whose partial trace leads to (5.58). This operation is called *purification* and is described below. Let's define a Hilbert space that is a direct product of space R with A, because of the Schmidt theorem we can always construct a pure state $|\psi_{RA}\rangle$ by an appropriate choice of Schmidt coefficients p_i, i.e.

$$|\psi_{RA}\rangle = \sum_i \sqrt{p_i}\, |\psi_i\rangle \otimes |\phi_i\rangle \quad (5.59)$$

where $|\psi_i\rangle$, $|\phi_i\rangle$ are basis vectors in the R and A spaces respectively. Following the partial trace over space A, we arrive at the desired result (5.58).

5.6 von Neumann Entropy

We already met Claude Shannon, widely heralded as the father of *information theory*. At the formative ages of the modern communications era, Shannon was concerned with the question, how does one quantify the information content of a message ? Intuitively, we suspect that the string of letters *" Strive not to be a success, but rather to be of value"* contains more information than let's say *"ghyer domd uiol hujik hyret ooopt sderg."* But how does one know that for sure? Perhaps the second message is an encrypted version of the first. Like the proverbial monkey who, by chance, typed out a Shakespeare sonnet, I might have inadvertently put together a string of letters that has a profound meaning in a language unknown to me. So we should be careful to distinguish the common notion of information, as conveying a meaning, with a more objective measure. In the words of Shannon [4], *"The fundamental problem of communication is that of reproducing at one point either exactly or approximately a message selected at another point. Frequently the messages have meaning; that is they refer to or are correlated according to some system with certain physical or conceptual entities. These semantic aspects of communication are irrelevant to the engineering problem. The significant aspect is that the actual message is one selected from a set of possible messages."*

With that proviso, Shannon introduced a measure of information content of an event (message) i given by the expression

$$I_i = -Log_2(p_i)$$

where p_i is the probability of that event, out of a set of all possible messages, and Log_2 is the base-2 logarithm. Note that the smaller the probability for event i, the larger the information content I_i. Given a distribution of events with probabilities $p_1 \ldots p_n$, the mean value of a stochastic variable X_i is

$$<X> = \sum_{i=1}^{n} p_i X_i$$

and if we take I_i to be a stochastic variable, its mean value is given by the expression

$$H \equiv -\sum_{i=1}^{n} p_i \, Log_2(p_i). \tag{5.60}$$

The quantity H is called *Shannon information entropy*. In a quantum system characterized by a density matrix ρ, the von Neumann entropy, S, is defined by

$$S = -Tr\,[\rho \, ln(\rho)] \tag{5.61}$$

where the natural logarithm, ln, of the density matrix is

$$ln(\rho) \equiv (\rho - \mathbb{1}) - \frac{1}{2}(\rho - \mathbb{1})^2 + \frac{1}{3}(\rho - \mathbb{1})^3 + \ldots \tag{5.62}$$

Suppose the density operator is that of a pure state so that $\rho = |\psi\rangle\langle\psi|$ then

$$\rho \ln(\rho) = \rho\left((\rho - 1) - \frac{1}{2}(\rho - 1)^2 + \frac{1}{3}(\rho - 1)^3\right) =$$
$$\rho(\rho - 1)\left[1 - \frac{1}{2}(\rho - 1) + \frac{1}{3}(\rho - 1)^2 + \ldots\right]$$

but since, for a pure state $\rho^2 = \rho$, the factor in the parenthesis vanishes and so

$$S(|\psi\rangle\langle\psi|) = 0. \tag{5.63}$$

The von Neumann entropy for a pure state is zero. Consider a representation where the density matrix is diagonal, i.e

$$\rho = \begin{pmatrix} p_1 & 0 & 0 & \cdots \\ 0 & p_2 & 0 & \cdots \\ 0 & 0 & p_3 & \cdots \\ \vdots & \vdots & \vdots & \ddots \end{pmatrix}$$

then

$$\ln(\rho) = \begin{pmatrix} \ln(p_1) & 0 & 0 & \cdots \\ 0 & \ln(p_2) & 0 & \cdots \\ 0 & 0 & \ln(p_3) & \cdots \\ \vdots & \vdots & \vdots & \ddots \end{pmatrix}$$

and so

$$S = -Tr[\rho \ln(\rho)] = -\sum_i p_i \ln(p_i) \tag{5.64}$$

and which is proportional to the Shannon entropy. Interestingly, von-Neumann entropy pre-dates Shannon entropy by about 15 years. Earlier, in the nineteenth century, *Ludwig Boltzmann* and *Willard Gibbs* defined similar measures to gauge the amount of disorder in a physical system.

Problems

5.1 For scenario (A) introduced in Sect. 5.2, evaluate the following expectation values
$$<\sigma_X>, <\sigma_Y>, <\sigma_Z>, <\sigma_X\sigma_Z>.$$

5.2 For scenario (B) introduced in Sect. 5.2 evaluate the following expectation values
$$<\sigma_X>, <\sigma_Y>, <\sigma_Z>, <\sigma_X\sigma_Z>.$$

Problems

5.3 Consider the state

$$|\psi\rangle = \frac{1}{\sqrt{2}} (|0\rangle - \exp(-i\pi/2) |1\rangle).$$

(a) Construct the density operator expressed in bra-ket notation.

(b) Using the computation basis, construct the density matrix for this state.

5.4 Consider the state

$$|\psi\rangle = \frac{1}{\sqrt{2}} (|0\rangle - \exp(-i\pi/2) |1\rangle).$$

(a) Construct the density operator ρ expressed in bra-ket notation and evaluate $\rho\rho$.

(b) Using the computation basis construct the density matrix $\underline{\rho}$ and evaluate the matrix product $\underline{\rho}\,\underline{\rho}$. Evaluate the trace of ρ^2.

5.5 Consider the state

$$|\psi\rangle = \frac{1}{\sqrt{2}} (|0\rangle - \exp(-i\pi/2) |1\rangle).$$

Using basis $|u\rangle$, $|v\rangle$ vectors, defined in (2.20), construct the density matrix $\underline{\rho}$ and evaluate the matrix product $\underline{\rho}\,\underline{\rho}$. Evaluate the trace ρ^2.

5.6 Given a qubit in state $|\psi(t_0)\rangle = |0\rangle$ at time $t_0 = 0$, and the Hamiltonian operator $\mathbf{H} = \hbar\sigma_X$, find $|\psi(t)\rangle$ for $t > 0$. Construct the the time dependent density operator.

$$\rho(t) = |\psi(t)\rangle\langle\psi(t)|$$

Find the density matrix, with respect basis $|0\rangle$, $|1\rangle$.

5.7 Given Hamiltonian $\mathbf{H} = \hbar\sigma_Z$, and state $|\psi(t_0)\rangle = \frac{1}{\sqrt{2}}(|0\rangle + |1\rangle)$, evaluate, in the computational basis, density matrix

$$\rho(t) = \mathbf{U}(t, t_0)\rho(t_0)\mathbf{U}^\dagger(t, t_0)$$

where $\mathbf{U}(t, t_0)$ is the time development operator. You can assume that $t_0 = 0$.

5.8 A system's time evolution is governed by the time-independent Hamiltonian $\mathbf{H}$. Show that the density operator $\rho(t)$ obeys the following first order differential equation

$$\frac{d\rho(t)}{dt} = \frac{1}{i\hbar}[\mathbf{H}, \rho(t)].$$

5.9 Generalize the time evolution equation given in Problem (5.8), if Hamiltonian $\mathbf{H}(t)$ is a function of time.

5.10 Using $\rho(t)$ obtained in Problem (5.7), find the time dependent expectation value
$$\sigma_Z(t) \equiv \langle \psi(t)|\sigma_Z|\psi(t)\rangle.$$
Do the same for $\sigma_X(t)$ and $\sigma_Y(t)$.

5.11 For each single-qubit density matrix itemized below, determine whether it represents a pure or mixed state.

(a) $\begin{pmatrix} 1 & 0 \\ 0 & 0 \end{pmatrix}$ (b) $\begin{pmatrix} \frac{1}{2} & 0 \\ 0 & \frac{1}{2} \end{pmatrix}$ (c) $\begin{pmatrix} \frac{1}{4} & 0 \\ 0 & \frac{3}{4} \end{pmatrix}$ (d) $\begin{pmatrix} \frac{1}{2} & \frac{1}{2} \\ \frac{1}{2} & \frac{1}{2} \end{pmatrix}$

5.12 Which of the following is not a valid density matrix.

(a) $\begin{pmatrix} 1 & 0 \\ 0 & 1 \end{pmatrix}$ (b) $\begin{pmatrix} 0 & 0 \\ 0 & 1 \end{pmatrix}$ (c) $\frac{1}{4}\begin{pmatrix} 3 & \sqrt{3} \\ \sqrt{3} & 1 \end{pmatrix}$ (d) $\begin{pmatrix} \frac{1}{2} & -\frac{1}{2} \\ -\frac{1}{2} & \frac{1}{2} \end{pmatrix}$

5.13 Construct the density matrices for the following 2-qubit states in the computational basis $|00\rangle, |01\rangle, |10\rangle, |11\rangle$.

(a) $\frac{1}{4}(|00\rangle - |01\rangle + |10\rangle - |11\rangle)$

(b) $\frac{1}{4}(|00\rangle + |01\rangle + |10\rangle + |11\rangle)$

(c) $\frac{1}{4}(|00\rangle + i\,|10\rangle + i\,|01\rangle - |11\rangle)$

(d) $\frac{1}{\sqrt{2}}(|01\rangle - i\,|10\rangle)$

5.14 For each two-qubit density matrix ρ_{AB} obtained in Problem (5.13), evaluate the partial traces
$$Tr_A \rho_{AB}, \quad Tr_B \rho_{AB}.$$
Which of the states in Problem (5.13) are entangled states?

5.15 Consider the measurement operator $\mathbf{M} = \sigma_X \otimes \sigma_Y$. (a) Find the expectation value of $\mathbf{M}$ for states (a), (b), (c), (d) in Problem (5.13). (b) Define $\rho_A = Tr_B \rho_{AB}$, obtained in Problem (5.14), and use it to evaluate
$$Tr(\rho_A\, \sigma_X).$$
Compare the results obtained in parts (a) and (b). Comment.

5.16 Consider the state
$$\frac{1}{4}(|00\rangle - |01\rangle + |10\rangle - |11\rangle).$$

Express it in terms of a Schmidt decomposition. What is its Schmidt number ?

5.17 Repeat Problem (5.16) for state

$$\frac{1}{2}(|01\rangle - |10\rangle).$$

5.18 Give a general proof of relations (5.7) and (5.18).

5.19 Give a proof of theorem (5.1), for a bipartite Hilbert space of dimension n, m respectively.

5.20 Find the von Neumann entropy for

$$\rho = \begin{pmatrix} \frac{1}{4} & 0 \\ 0 & \frac{3}{4} \end{pmatrix}, \quad \rho = \begin{pmatrix} \frac{1}{2} & \frac{1}{2} \\ \frac{1}{2} & \frac{1}{2} \end{pmatrix}.$$

References

1. I. Hargittai, *Martians of Science: Five Physicists Who Changed the Twentieth Century* (Oxford University Press, Oxford, 2006)
2. W.H. Press, S.A. Teukolsky, W.T. Vetterling, B.P. Flannery, *Numerical Recipes: The Art of Scientific Computing*, 3rd edn. (Cambridge University Press, Cambridge, 2007)
3. M.E. Nielsen, I.L. Chuang, *Quantum Computation and Quantum Information* (Cambridge University Press, Cambridge, 2011)
4. C. Shannon, The bell system. Tech. J. **27**(379–423), 623–656 (1948)

6. No-Cloning Theorem, Quantum Teleportation and Spooky Correlations

> **Abstract**
>
> The no-cloning theorem states that an arbitrary quantum state cannot be copied from one qubit and duplicated on another qubit. We offer a proof of this theorem and illustrate how quantum states can be teleported between two qubits. The Bell and Clauser-Horne-Shimony-Holt inequalities are introduced and shown to be demonstrable features of entangled quantum systems. We discuss how the private key distribution problem is dealt with using quantum key distribution (QKD). The BB84 and Ekert protocols are examples of the latter, and we review and illustrate their implementation. We show how entangled states enable dense coding and offer a brief synopsis of Greenberger-Horne-Zeilinger (GHZ) states and their application.

6.1 Introduction

In addition to giving us the theories of relativity, *Albert Einstein* was a quantum pioneer. He introduced the photon quantum in 1905; in a work that influenced a young generation of physicists, including *Niels Bohr*, *Louis de-Broglie* and others. Nevertheless, Einstein was deeply disturbed by the final product, the Copenhagen interpretation of a theory, which he was instrumental in its conception and development. So in 1935, along with his colleagues, *Boris Podolsky* and *Nathan Rosen*, Einstein published a paper that sought to illustrate, through a gedanken experiment, deficiencies in the accepted interpretation of the quantum theory. At first largely ignored, that effort now commonly referred to as the EPR paper or argument, holds the distinction as one the most highly cited articles in modern physics literature. Though the thesis introduced therein has since been neutered, questions posed in EPR have occupied and vexed generations of physicists and philosophers.

In the early 1950s, *David Bohm* resurrected the EPR scenario under a different, more accessible, guise that made a strong impression on a physicist working at the CERN laboratory, *John Bell*. Bell published a paper that sought to address some of the issues raised in EPR. He developed a set of criteria, now called the *Bell inequalities*, that could be tested in the laboratory.[1] A central theme in the EPR and Bell scenarios concerns correlations, highlighted by the famous Einstein epigram, "*...spooky action at a distance..*", of measurements made on a quantum system by devices that are not causally connected. In this chapter, we focus on these questions and review Bell's and other's arguments. We show how issues that for a long time were considered to lie in the domain of philosophical discourse are enabling transformative technologies and applications.

6.2 On Quantum Measurements

It's now a good time to review and apply our knowledge of states and measurements. Suppose Alice is in possession of a qubit in state

$$|\psi\rangle = a|0\rangle + b|1\rangle \quad a^2 + b^2 = 1,$$

on which she performs measurements with occupation number operator $\mathbf{n} = |1\rangle\langle 1|$.

- Question: What are the possible results of Alice's measurements?

  – Answer: She can only observe the values 0 or 1, the eigenvalues of $\mathbf{n}$.

- Question: Suppose $a = 1/2$. If a measurement reveals the value 0, what result does a measurement, immediately after this one, give?

  – Answer: According to the collapse postulate, that measurement forces quantum state $|\psi\rangle$ into state $|0\rangle$. Therefore, the probability of finding 0 is 100%.

- Question: Alice has access to a thousand identical copies of a system, each described by state $|\psi\rangle$. She measures each copy, how many measurement lead to the result 0? How many result in the value 1?

  – Answer: She finds, approximately, $a^2(1000) = 250$ instances of 0. Approximately 750 measurements reveal the value 1.

---

[1] Correlation experiments where anticipated by Chien-Shiung Wu as early as 1950 [12].

- Question: From this set of measurements, can Alice guess the value of b?

    - Answer: Sort of. She knows, according to the Born rule, that $b^2 = 3/4$ and guesses that $b = \sqrt{3}/2$. But would not $b = \sqrt{3}/2\,i$, also work?

- Question: (i) How many copies of $|\psi\rangle$ does Alice need to recover full information of that state? (ii) Is there a copy machine that allows Alice to make any number (e.g., 1000) of copies from a single version of $|\psi\rangle$?

    - Answer: (i) Because Alice has access to only one measurement device, **n**, all she will ever learn by an indefinite number of measurements are the values $|b|^2$. (ii) No. The answer to item (ii) is a consequence of the no-cloning theorem, which we describe and prove below.

6.3 The No-Cloning Theorem

Alice wants a facsimile of a qubit in state $|\psi\rangle$. To make a copy, she needs to find another qubit (e.g., an electron) that can store the information contained in her original qubit. Diagrammatically, Alice needs to do something like this

$$|\psi\rangle \otimes |\phi\rangle \rightarrow |\psi\rangle \otimes |\psi\rangle. \tag{6.1}$$

After the copy operation the two qubits in (6.1) each encode the same information. With additional electrons, she should be able to make an unlimited number of copies of $|\psi\rangle$. Alice would like to take the state $|\psi\rangle \otimes |\phi\rangle \otimes |\beta\rangle \cdots$ and turn it into $|\psi\rangle \otimes |\psi\rangle \otimes |\psi\rangle \cdots$. In this way, she could perform the type of measurements described in Sect. 6.2. She would measure the state of the first electron, and record the result. So if she obtained 0 from that measurement, the new, collapsed, state is

$$|\psi\rangle \otimes |\psi\rangle \cdots \otimes |\psi\rangle \otimes |0\rangle.$$

With repeated measurements on each electron, Alice would have data that looks something like this ...01100101010. Each of the zeros and ones is the result of a measurement on an identical system, and Alice could perform the necessary statistics to glean information contained in $|\psi\rangle$. Ideally, she would like an unlimited supply of states $|\psi\rangle$ at her disposal, since, with the clever use of non-commuting measurement devices, she could deduce the region of the Bloch sphere where $|\psi\rangle$ is located. Unfortunately, the no-cloning theorem states that (6.1) is impossible to implement.

Proof Equation (6.1) is a gate operation, and we know that gates are represented by unitary operators. This means there should exist a gate $\mathbf{U}$ so that $\mathbf{U}\mathbf{U}^\dagger = \mathbf{U}^\dagger\mathbf{U} = \mathbb{1}$, and

$$\mathbf{U}|\psi\rangle \otimes |\phi\rangle = |\psi\rangle \otimes |\psi\rangle. \tag{6.2}$$

Now **U** should be universal, i.e., it should implement the transformation

$$\mathbf{U}|\Omega\rangle \otimes |\phi\rangle = |\Omega\rangle \otimes |\Omega\rangle \tag{6.3}$$

for all states $|\Omega\rangle$ and $|\phi\rangle$ in Hilbert space, not just the state of interest $|\psi\rangle$. If (6.3) is satisfied so must

$$\langle\phi| \otimes \langle\Omega|\mathbf{U}^\dagger = \langle\Omega| \otimes \langle\Omega|. \tag{6.4}$$

Taking the inner product of (6.2) with (6.4) we find

$$\langle\phi| \otimes \langle\Omega|\mathbf{U}^\dagger\mathbf{U}|\psi\rangle \otimes |\phi\rangle = \langle\Omega| \otimes \langle\Omega||\psi\rangle \otimes |\psi\rangle,$$

but $\mathbf{U}^\dagger\mathbf{U} = \mathbb{1}$ and so

$$\langle\Omega|\psi\rangle = \langle\Omega|\psi\rangle^2 \tag{6.5}$$

which is satisfied only if $\langle\Omega|\psi\rangle = 0$, or $\langle\Omega|\psi\rangle = 1$. Because $|\Omega\rangle$ is arbitrary, (6.5) is too restrictive to be satisfied for all possible $|\Omega\rangle$. In other words, there exists no universal copying machine **U** that has the property

$$\mathbf{U}|\psi\rangle \otimes |\phi\rangle = |\psi\rangle \otimes |\psi\rangle. \tag{6.6}$$

□

Consider the gate $\mathbf{U}_X = \mathbb{1} \otimes \sigma_X$ operating on state $|1\rangle \otimes |0\rangle$, so that

$$\mathbf{U}_X |1\rangle \otimes |0\rangle = |1\rangle \otimes |1\rangle. \tag{6.7}$$

Does (6.7) violate the no-cloning theorem? It does not. Though (6.7) is true; **U** acting on a different state, let's say $|0\rangle \otimes |0\rangle$, does not copy the contents of the second qubit into the first. As Alice does not know the contents of the second qubit, $\mathbf{U}_X$ would be useless to her unless, by chance, that qubit is in state $|1\rangle$.

The no-cloning theorem implies that a quantum version of a "photocopy" machine does not exist. This is a good thing if you wish to protect copyright information, e.g., pictures, songs etc. It is a bad thing if you are Alice and want to make enough copies to carry out independent repeated measurements of a single state. Whereas state $|\psi\rangle$ of a qubit cannot be copied, it can be teleported.

6.4 Quantum Teleportation

In previous chapters, we introduced the concept of an entangled two-qubit state. A hallmark of these states is that they cannot be factored into a direct product of single qubit states. Entangled states have many quantum information applications,

6.4 Quantum Teleportation

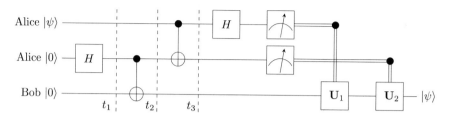

Fig. 6.1 Teleportation of Alice's qubit state $|\psi\rangle$, to Bob's qubit. Bob chooses, after consulting Alice on a classical channel (double lines), to process his qubit with gates $\mathbf{U}_1$ and $\mathbf{U}_2$

quantum teleportation of a qubit state is one of those. Using the shorthand notation, $|a\rangle \otimes |b\rangle \equiv |a\,b\rangle$, we itemize the following two-qubit states

$$|\beta_{00}\rangle \equiv \frac{1}{\sqrt{2}}\left(|00\rangle + |11\rangle\right)$$

$$|\beta_{01}\rangle \equiv \frac{1}{\sqrt{2}}\left(|01\rangle + |10\rangle\right)$$

$$|\beta_{10}\rangle \equiv \frac{1}{\sqrt{2}}\left(|00\rangle - |11\rangle\right)$$

$$|\beta_{11}\rangle \equiv \frac{1}{\sqrt{2}}\left(|01\rangle - |10\rangle\right), \tag{6.8}$$

also called *Bell states*, after the twentieth century physicist John Bell who demonstrated far-reaching consequences implied by their existence. Alice has a qubit, let's say an electron, in state

$$|\psi\rangle = \alpha|0\rangle + \beta|1\rangle$$
$$\alpha^2 + \beta^2 = 1. \tag{6.9}$$

She wants to teleport $|\psi\rangle$ to Bob, who also possesses a qubit. Now Alice does not need to transfer the material body (e.g., the electron) to Bob, all she needs to do is make certain that Bob's qubit finds itself in state $|\psi\rangle$. Unlike the ill-fated scientist in the 1958 horror flick *The Fly*, who transmits each dissembled atom of his body to be re-constituted in a teleportation booth,[2] Alice realizes that it is not the medium but the information that needs to be teleported.

Consider the diagram in Fig. 6.1 which describes the evolution of three qubits. The electron qubit at the uppermost wire in that figure belongs to Alice, and it is in state $|\psi\rangle$ given by (6.9). She is also in possession of a second qubit, a photon, that is denoted by the middle wire in the diagram and labeled with $|0\rangle$. Bob, located hundreds of kilometers from Alice, possesses an electron qubit in state $|0\rangle$. At time t_1 Alice's photon qubit passes through a Hadamard gate, after which the system is described by

---

[2] As you might guess from the film's title, things did not work out well for our anti-hero.

$$|\Psi(t_1)\rangle = \left(\mathbb{1} \otimes \mathbf{H} \otimes \mathbb{1}\right)|\psi\rangle \otimes |0\rangle \otimes |0\rangle =$$
$$|\psi\rangle \otimes \frac{1}{\sqrt{2}}(|0\rangle + |1\rangle) \otimes |0\rangle. \tag{6.10}$$

Subsequently, Alice's photon, acts as a control qubit in a **CNOT** gate to Bob's target qubit. The **CNOT** gate is represented by operator

$$\mathbf{U}_{C_1} = \mathbb{1} \otimes (|00\rangle\langle 00| + |01\rangle\langle 10| + |10\rangle\langle 11| + |11\rangle\langle 01|),$$

and acting on $|\Psi(t_1)\rangle$ we find

$$|\Psi(t_2)\rangle = \mathbf{U}_{C_1}|\Psi(t_1)\rangle = |\psi\rangle \otimes \frac{1}{\sqrt{2}}\Big(|0\rangle \otimes |0\rangle + |1\rangle \otimes |1\rangle\Big) =$$
$$|\psi\rangle \otimes \frac{1}{\sqrt{2}}(|00\rangle + |11\rangle) = |\psi\rangle \otimes |\beta_{00}\rangle. \tag{6.11}$$

In state $|\Psi(t_2)\rangle$, Alice's photon and Bob's electron are entangled. Expanding and re-arranging (6.11), we get

$$\sqrt{2}|\Psi(t_2)\rangle = \alpha|00\rangle \otimes |0\rangle + \alpha|01\rangle \otimes |1\rangle + \beta|10\rangle \otimes |0\rangle + \beta|11\rangle \otimes |1\rangle. \tag{6.12}$$

In the next gate progression $\mathbf{U}_{C_2}$, Alice's qubit is the control and her photon is the target qubit. Since

$$\mathbf{U}_{C_2} = (|00\rangle\langle 00| + |01\rangle\langle 10| + |10\rangle\langle 11| + |11\rangle\langle 01|) \otimes \mathbb{1}, \tag{6.13}$$

$$\sqrt{2}\mathbf{U}_{C_2}|\Psi(t_2)\rangle =$$
$$\alpha|00\rangle \otimes |0\rangle + \beta|10\rangle \otimes |1\rangle + \alpha|01\rangle \otimes |1\rangle + \beta|11\rangle \otimes |0\rangle. \tag{6.14}$$

Finally, Alice's electron qubit passes through a Hadamard gate, after which

$$2|\Psi(t_3)\rangle = \alpha\,(|0\rangle + |1\rangle) \otimes |0\rangle \otimes |0\rangle +$$
$$\beta\,(|0\rangle - |1\rangle) \otimes |0\rangle \otimes |1\rangle +$$
$$\alpha\,(|0\rangle + |1\rangle) \otimes |1\rangle \otimes |1\rangle +$$
$$\beta\,(|0\rangle - |1\rangle) \otimes |1\rangle \otimes |0\rangle \tag{6.15}$$

or,

$$|\Psi(t_3)\rangle = |00\rangle \otimes \frac{1}{2}(\alpha|0\rangle + \beta|1\rangle) +$$
$$|01\rangle \otimes \frac{1}{2}(\alpha|1\rangle + \beta|0\rangle) +$$
$$|10\rangle \otimes \frac{1}{2}(\alpha|0\rangle - \beta|1\rangle) +$$
$$|11\rangle \otimes \frac{1}{2}(\alpha|1\rangle - \beta|0\rangle). \tag{6.16}$$

Now Alice performs measurements on her qubits. They reveal one of these choices, {00}, {01}, {10}, or {11}, each with a probability of 25%. Suppose Alice's measurement finds the value {00}. The collapse hypothesis demands that the system finds itself in state

$$|00\rangle \otimes (\alpha |0\rangle + \beta |1\rangle). \tag{6.17}$$

- Alice calls Bob on a classical channel (e.g., a telephone) denoted in Fig. 6.1 by the double wires, to inform him that she observed {00}. With this (classical) information, Bob knows, without needing to measure his qubit, that it is in the desired state $(\alpha |0\rangle + \beta |1\rangle)$. Teleportation of the original qubit state has been achieved. But Alice observes {00} only a quarter of the time. What about the other values that she might measure?
- If Alice measures {01}, the system collapses into

$$|01\rangle \otimes (\alpha |1\rangle + \beta |0\rangle). \tag{6.18}$$

Alice now tells Bob she observed {0 1}, and so Bob knows he is in possession of a qubit in the state (6.18). But it is not the desired state. Bob processes that state with a Pauli-X gate, so that

$$\sigma_X (\alpha |1\rangle + \beta |0\rangle) \Rightarrow (\alpha |0\rangle + \beta |1\rangle),$$

and his electron qubit is placed into the teleported state $|\psi\rangle$.
- If Alice measures {10}, the system collapse into

$$|10\rangle \otimes (\alpha |0\rangle - \beta |1\rangle). \tag{6.19}$$

Bob is informed by Alice of this result and processes that state with a Pauli-Z gate.
- Finally, if Alice measures {11}, Bob will know he is in possession of state

$$\alpha |1\rangle - \beta |0\rangle.$$

The post-processing of this state to obtain the desired state is left as an exercise.

In summary, all scenarios itemized above allow Alice to teleport state $|\psi\rangle$ to Bob's qubit.

6.5 EPR and Bell Inequalities

Consider the entangled, or Bell, state

$$|\psi\rangle = \frac{1}{\sqrt{2}} (|0\rangle \otimes |0\rangle + |1\rangle \otimes |1\rangle) = \frac{1}{\sqrt{2}} (|00\rangle + |11\rangle) = |\beta_{00}\rangle. \tag{6.20}$$

Bell states are generic and occur in a wide variety of quantum systems. For example, the stable form of atomic Helium, an atom of two electrons bound by an equal and opposite nuclear charge, is found in a spin singlet state, a type of Bell state. Another example in which Bell states manifest is in the decay of an excited atom into its ground state. In this event, energy and momentum are conserved, and the decay is accompanied by the release of one or more photons. In two-photon decay, conservation of momentum requires that they have equal and opposite momenta. Suppose that $|0\rangle$ denotes the state of definite, let's call it H-type, polarization, and $|1\rangle$ represents a photon in a V-type polarization state. For the sake of argument, we assume that the following Bell state describes the emitted photons

$$|\psi\rangle = \frac{1}{\sqrt{2}} \left(\left|\vec{k}\,0\right\rangle \otimes \left|-\vec{k}\,0\right\rangle - \left|\vec{k}\,1\right\rangle \otimes \left|-\vec{k}\,1\right\rangle \right). \tag{6.21}$$

In this notation $\left|\vec{k}\,0\right\rangle$ represents an H-polarized photon with momentum $\vec{k}$ and so $\left|\vec{k}\,0\right\rangle \otimes \left|-\vec{k}\,0\right\rangle$ is a state where the first and second photons are found to have H-type polarizations, but travel in opposing directions. Vector $\vec{k}$ points from the atom decay site to Bob's laboratory, whereas $-\vec{k}$ to Alice's lab (See Fig. 6.2). Alice's device measures the momentum of a single photon. Furthermore, it registers a photon only if it is in state $\left|-\vec{k}\right\rangle$, likewise, Bob's device detects photons in state $\left|\vec{k}\right\rangle$. Therefore we re-express (6.21) in the form

$$|\psi_{AB}\rangle = \frac{1}{\sqrt{2}} (|0\rangle_B \otimes |0\rangle_A + |1\rangle_B \otimes |1\rangle_A) \tag{6.22}$$

where the subscripts A refers to Alice's qubit and B to Bob's. In other words, Alice measures the photon qubit with subscript A, and Bob the photon with subscript B. Remember that Bob and Alice are separated by the twice the distance a light beam travels, in the time interval it takes to register a photon in the detection device. Though Alice and Bob possess qubits in a common state $|\psi_{AB}\rangle$, events associated with measurement of their respective qubits are not causally connected.

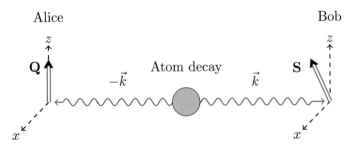

Fig. 6.2 EPR correlations experiment

6.5 EPR and Bell Inequalities

Before proceeding further, we need to acknowledge that we are cheating a bit. We are implicitly assuming localization in space, of state $|\vec{k}\rangle$. A more accurate description requires the use of photon wave-packets. Our picture is justified by the fact that its predictions are in harmony with that of a more nuanced analysis.

Although the original experiment measured the polarization properties of photons, in the following discussion we analyze an updated version of this experiment using electrons [6]. Instead of measuring photon polarizations, we consider the measurement of a component of an operator that is proportional to the electron spin $\vec{S} = \hbar/2\, \vec{\sigma}$. Alice's instrument measures the quantity

$$\vec{P}_A \cdot \vec{n} \quad \vec{P}_A = \sigma_Z(A)\,\hat{k} + \sigma_X(A)\,\hat{i}$$
$$\vec{n} = \cos\theta_1\,\hat{k} + \sin\theta_1\,\hat{i} \tag{6.23}$$

where θ_1 is the angle between the z axis, shown in Fig. 6.2, and $\vec{n}$. $\sigma_X(A), \sigma_Z(A)$ are Pauli operators for qubit A. Similarly, Bob measures his qubit with instrument $\vec{P}_B = \sigma_Z(B)\,\hat{k} + \sigma_X(B)\,\hat{i}$, pointed along direction $\vec{m} = \cos\theta_2\,\hat{k} + \sin\theta_2\,\hat{i}$ where θ_2 is the angle with respect to the z-axis in the xz plane of Bob's coordinate system. Together, Alice's and Bob's simultaneous measurements of their respective qubits is represented by operator

$$\vec{P}_A \cdot \vec{n} \otimes \vec{P}_B \cdot \vec{m}. \tag{6.24}$$

We interpret device $\vec{P}_A \cdot \vec{n}$, whose eigenvalues are ± 1, as a spin analyzer. If a measurement responds with eigenvalue $+1$, we say an electron's spin is directed along the $\vec{n}$ direction; if the measurement reveals -1, it's spin is directed along $-\vec{n}$. Suppose Alice sets her instrument to the setting $\theta_1 = 0$, and Bob to the setting $\theta_2 = \pi/4$. We label them as

$$\mathbf{Q} \equiv \vec{P}_A|_{\theta_1=0} = \sigma_Z(A)$$
$$\mathbf{S} \equiv \vec{P}_B|_{\theta_2=\pi/4} = \frac{1}{\sqrt{2}}(\sigma_X(B) + \sigma_Z(B)). \tag{6.25}$$

Repeated measurements of $\mathbf{Q}$ and $\mathbf{S}$, for an ensemble of N measurements, leads to data that might look like that given in Table 6.1.

For a sufficiently large set of measurements, Alice and Bob can calculate various statistical moments such as the expectation values $<\mathbf{Q}>$, $<\mathbf{S}>$, or the correlation expectation value $<\mathbf{Q}\mathbf{S}>$. The latter is especially interesting and, according to the postulates of the quantum theory, it should approach the value

$$<\mathbf{QS}> = \langle \psi_{AB}|\mathbf{Q}\,\mathbf{S}|\psi_{AB}\rangle. \tag{6.26}$$

We learned several different ways of evaluating this quantity, let's use the density matrix approach in which

$$<\mathbf{QS}> = Tr[\rho_{AB}\,\mathbf{Q} \otimes \mathbf{S}] \tag{6.27}$$

Table 6.1 A possible ensemble of Alice's and Bob's measurement results with devices **Q**, **S** for the two-qubit state (6.22)

Trial #	Alice	Bob	QS
1	1	−1	−1
2	1	1	1
3	−1	−1	1
4	−1	1	−1
⋮	⋮	⋮	⋮
N	−1	−1	1

where ρ_{AB} is the density matrix for state $|\psi_{AB}\rangle$. Now

$$\rho_{AB} = \frac{1}{2}(|00\rangle\langle 00| + |00\rangle\langle 11| + |11\rangle\langle 00| + |11\rangle\langle 11|) = \frac{1}{2}\begin{pmatrix} 1 & 0 & 0 & 1 \\ 0 & 0 & 0 & 0 \\ 0 & 0 & 0 & 0 \\ 1 & 0 & 0 & 1 \end{pmatrix} \quad (6.28)$$

where we used the matrix representations of $|00\rangle$ and $|11\rangle$. Likewise,

$$\vec{P}_A \cdot \vec{n} \otimes \vec{P}_B \cdot \vec{m} =$$
$$(\cos\theta_1 \sigma_Z(A) + \sin\theta_1 \sigma_X(A)) \otimes (\cos\theta_2 \sigma_Z(B) + \sin\theta_2 \sigma_X(B)) =$$
$$\begin{pmatrix} \cos\theta_1 & \sin\theta_1 \\ \sin\theta_1 & -\cos\theta_1 \end{pmatrix} \otimes \begin{pmatrix} \cos\theta_2 & \sin\theta_2 \\ \sin\theta_2 & -\cos\theta_2 \end{pmatrix} =$$
$$\begin{pmatrix} \cos\theta_1\cos\theta_2 & \cos\theta_1\sin\theta_2 & \cos\theta_2\sin\theta_1 & \sin\theta_1\sin\theta_2 \\ \cos\theta_1\sin\theta_2) & -\cos\theta_1\cos\theta_2 & \sin\theta_1\sin\theta_2 & -\cos\theta_2\sin\theta_1 \\ \cos\theta_2\sin\theta_1 & \sin\theta_1\sin\theta_2 & -\cos\theta_1\cos\theta_2 & -\cos\theta_1\sin\theta_2 \\ \sin\theta_1\sin\theta_2 & -\cos\theta_2\sin\theta_1 & -\cos\theta_1\sin\theta_2 & \cos\theta_1\cos\theta_2 \end{pmatrix} \quad (6.29)$$

and so

$$Tr\left[\rho_{AB}\left(\vec{P}_A \cdot \vec{n} \otimes \vec{P}_B \cdot \vec{m}\right)\right] = \cos(\theta_1 - \theta_2). \quad (6.30)$$

For measurement **Q**, $\theta_1 = 0$, and for measurement **S**, $\theta_2 = \pi/4$, we get

$$<QS> = \cos\pi/4 = \frac{1}{\sqrt{2}}. \quad (6.31)$$

Result (6.31) represents a positive correlation between Alice's measurement outcomes with that of Bob's. For example, if Alice measures the value +1, then it is more likely that Bob also finds the value +1. If the correlation came out to have the value 0, Bob would just as likely obtain −1 as +1, that is there is no correlation between Alice's and Bob's measurements. A negative correlation implies that Bob is more likely to measure −1 if Alice measures +1.

6.5 EPR and Bell Inequalities

There is nothing special about the **Q** or **S** measurements. Alice could rotate her instrument to a different angle, as could Bob. We define

$$\mathbf{R} \equiv \mathbf{P}_A|_{\theta_1=\pi/2} = \sigma_X(A)$$
$$\mathbf{T} \equiv \mathbf{P}_B|_{\theta_2=3\pi/4} = \frac{1}{\sqrt{2}}(\sigma_X(B) - \sigma_Z(B)) \quad (6.32)$$

so Alice has two measurement devices **Q**, **R**. For every trial she uses **Q** or **R**, the order of measurement is immaterial. She records her observed data and tabulates the results of her measurement values in one column and the instrument used in the other column. Bob does the same with data obtained using instruments **S**, **T**. After a large number of runs, Alice and Bob tabulate, similar to that shown in Table 6.1, and compare their data. They calculate the correlation expectation values $<\mathbf{QS}>, <\mathbf{QT}>, <\mathbf{RS}>, <\mathbf{RT}>$, and find the approaching values

$$<\mathbf{QS}> \to \frac{1}{\sqrt{2}} \quad <\mathbf{QT}> \to -\frac{1}{\sqrt{2}}$$
$$<\mathbf{RS}> \to \frac{1}{\sqrt{2}} \quad <\mathbf{RT}> \to \frac{1}{\sqrt{2}}. \quad (6.33)$$

The results are in perfect agreement with the predictions of (6.30), a result that should not be surprising if you believe in the formalism and postulates of the quantum theory. Why are we then expending all this effort in plodding through yet another quantum exercise? Before answering this question, let's introduce a thought experiment below.

6.5.1 Bertlmann's Socks

Reinhold Bertlmann, a friend and colleague of John Bell, liked to wear socks of different colors [2]. If on a particular day one of the socks was red, you could be certain that the other sock was some other color. John Bell invoked this observation to illustrate how correlations manifest in everyday life and as a foil to advance his thesis concerning correlations predicted by the quantum theory.

To illustrate that thesis, we use an argument, put forth by *J. Clauser, M. Horn, A. Shimony* and *R. Holt* and know as the CHSH inequalities. Inspired by the Bertlmann sock example, we invoke an imaginary Bertlmann, Bart, who also wears socks of different colors, but on a given day, may or not be wearing a tie, may or may not be sporting a watch, and may or may not shave his beard. For each of these characteristics, we assign a stochastic variable that assigns the values ± 1 to the outcomes tabulated in Table 6.2.

To get to know Bart, we calculate various correlations of Bart's behavior. If, one a given day, he wears socks of a different color does he also shave his beard, etc.? We assign the stochastic variable Q for the color of his socks, the value of $Q, q = 1$ if Bart's socks are of the same color, $q = -1$ if they are different. In the same way, the variables S, R, T characterize the three other traits. Tabulating each trait assignment, per day, we construct a picture of Bart's sartorial habits. With enough data, we get a

Table 6.2 Stochastic variable assignments for Bart

Sock color (Q)	Tie (T)	Watch (S)	Beard shaved (R)
Same $q = 1$	Yes $t = 1$	Yes $s = 1$	Yes $r = 1$
Different $q = -1$	No $t = -1$	No $s = -1$	No $r = -1$

reliable estimate for $<QS>$. If that value is zero, there is no correlation with Bart wearing socks of the same color and wearing a watch. Positive and negative values imply positive and negative correlations. There are constraints on the values $<QS>$. Because both Q and S are bounded by ± 1, it is evident that $|<QS>| \leq 1$. This bound also applies to correlations of the type $<XY>$ for $X,Y \in Q, R, S, T$. With a minimum number of assumptions and knowledge of Bart's behavior, we can impose additional bounds on the correlations. We have no idea what Bart is going wear on a given day, but we assume that, for a large ensemble of events, it can be described by a probability distribution

$$p(q, s, t, r).$$

It is the probability that Bart, on a given day, is characterized by a given q, s, r and t value. So if $q = 1, s = -1, r = -1, t = -1$, $p(q, r, s, t)$ is the probability that Bart is wearing socks of different colors, is not wearing a tie, is not wearing a watch, and has not shaved his beard. We don't know what that probability is, but we assume that it does exist. This is a very general assumption; implicitly it acknowledges that there is a sample space or an underlying objective reality. The only constraint on it is

$$\sum_{qrst} p(q, r, s, t) = 1 \qquad (6.34)$$

where the sum is over $q = \pm 1, s = \pm 1, r = \pm 1, t = \pm 1$. We also assume that two events, q_1, r_1, s_1, t_1 and q_2, r_2, s_2, t_2 cannot both be true at the same time, that is, they are mutually exclusive. In that case the laws of probability demand that,

$$p(q, s) = \sum_{r,t} p(q, r, s, t) \quad p(q, t) = \sum_{r,s} p(q, r, s, t)$$
$$p(r, s) = \sum_{q,t} p(q, r, s, t) \quad p(r, t) = \sum_{s,q} p(q, r, s, t). \qquad (6.35)$$

By definition, the expectation value,

$$<QS> = \sum_{q,s} q\, s\, p(q, s) = \sum_{q,s,r,t} q\, s\, p(q, r, s, t). \qquad (6.36)$$

6.5 EPR and Bell Inequalities

Similarly

$$<QT> = \sum_{q,s,r,t} q\,t\, p(q,r,s,t)$$

$$<RS> = \sum_{q,s,r,t} r\,s\, p(q,r,s,t)$$

$$<RT> = \sum_{q,s,r,t} r\,t\, p(q,r,s,t)$$

and

$$<QS> + <RS> + <RT> - <QT> =$$
$$\sum_{q,r,s,t} (q s + r s + r t - q t)\, p(q,r,s,t) =$$
$$\sum_{q,r,s,t} ((q+r)s + (r-q)t)\, p(q,r,s,t). \tag{6.37}$$

Now the term $q + r$ ranges in values from $+2$ (if both $r = q = 1$) to -2, but if $q + r = 2$, then $(r - q) = 0$. Likewise, if $(q + r) = -2$, then $(r - q) = 0$. Also, because $s = \pm 1, t = \pm 1$, it should be evident that the maximum absolute value of the term $((q+r)s + (r-q)t)$ is 2. Therefore we obtain the CHSH inequality

$$|<QS> + <RS> + <RT> - <QT>| \leq 2. \tag{6.38}$$

As pointed out this is a very general result, valid for any $p(q,r,s,t)$ distribution of your liking. As an example, suppose $p(q = 1, r = -1, s = -1, t = -1) = 1$ and all other $p(q,r,s,t)$ vanish. Then

$$<QS> = -1,\ <RS> = 1,\ <RT> = 1,\ <QT> = -1$$

and the l.h.s. of (6.38) sums to 2, in harmony with the CHSH inequality.

Having established this identity, we use predictions (6.33) to evaluate $<QS> + <RS> + <RT> - <QT>$ for Alice's and Bob's measurements. The sum

$$<\mathbf{QS}> + <\mathbf{RS}> + <\mathbf{RT}> - <\mathbf{QT}> = 2\sqrt{2}, \tag{6.39}$$

contradicts inequality (6.38). Now Alice and Bob appear to be measuring or uncovering the traits of an object, in this case, the polarization properties of electron spins. If these traits are imprinted on the object, in the same way that variables Q, R, S, T describe Bart, then inequality (6.38) should be satisfied. The fact that it does not has profound implications. Inequality (6.38) is based on very basic and general assumptions. One of which is that the space of events allows a probability measure $p(q,r,s,t)$. Another assumption is that a given assignment q,r,s,t is mutually exclusive of other points in sample space. The latter assumption could be invalidated in the following scenario.

Suppose Bart possesses superpowers that allow him to control the outcome of trait measurements on the fly. For example, if on a given day Bart had trait assignments $q = 1, r = 1, s = 1, t = 1$ an honest measurement would reveal those values. Let's assume that Alice measures Q and R, and Bob measures S and T. But if Bart has the power to change the value of s, t if a measurement for Q reveals the value $q = 1$, he may be able to manipulate the outcome so that, $<QS>=<RS>=<RT>= 1$, and $<QT>= 0$, thereby invalidating the CHSH inequality.

Can we appeal to this scenario to explain Alice's and Bob's data? John Bell summarized this state of affairs by introducing two interpretations. In the first, we accept the results, which have now been validated by numerous experiments, of quantum mechanical predictions, but give up the notion of an objective reality. That is, the measured properties of an object do not completely describe it, or we cannot ascribe the propensity of an object to display traits $Q = q, R = r, S = s, T = t$ by assignment of $p(q, r, s, t)$. In the second interpretation, similar to the way Bart uses his superpowers to alter measurements, Alice's polarization measurements somehow influence Bob's measurements in such a way that violates the CHSH inequality. But Bob and Alice's instruments are separated by a non-causal interval, and such a conspiracy implies faster-than-light communication, a heresy.

If we accept the predictions of quantum mechanics, we forgo the possibility of an objective reality, and/or allow for faster-than-light communication. Together these features constitute what is called *local realism* [11]. Experiments [1,3,4,6,10] have since demonstrated the breaking of the Bell inequalities, and therefore, call into question the nature of what we perceive as reality. Today, the physics community is divided along this fault-line. The widely accepted orthodox dogma, the Copenhagen interpretation, is best summed up by a quote [9] from the celebrated theoretical physicist *Stephen Hawking*, *"I don't demand that a theory corresponds to reality because I don't know what it is. Reality is not a quality you can test with litmus paper. All I am concerned with is that the theory should predict the results of measurements."*

6.5.2 Bell's Theorem

John Bell asked the question, how does strictly enforced local realism constrain measurement correlations? Local realism requires that the random variable $A = \pm 1$ associated with Alice's device depends only on the orientation of Alice's device, i.e., $A = A(\vec{n}_A)$. It should not depend on $\vec{n}_B$ the orientation of Bob's device. However, both may depend on other local, *hidden variables*, that we collectively call λ but are inaccessible to us. These parameters determine the properties of Alice's and Bob's qubits. Therefore we let $A = A(\vec{n}_A, \lambda)$. In the same manner, for the random variable measured by Bob, we assume $B = B(\vec{n}_B, \lambda)$ so the expectation value

$$<AB> = \int d\lambda \rho(\lambda) A(\vec{n}_A, \lambda) B(\vec{n}_B, \lambda) \tag{6.40}$$

6.5 EPR and Bell Inequalities

is the central assumption of Bell's local realism requirement. Here $\rho(\lambda)$ is a probability for the system to be in a state for a given value λ of the hidden variables. Now if $\rho(\lambda)$ is a probability distribution

$$\int d\lambda \, \rho(\lambda) = 1.$$

If $\vec{n} = \vec{m}$, i.e. $\theta_1 = \theta_2$ and, (6.30) demands that $<AB> = 1$. Therefore, for $\vec{n} = \vec{m}$

$$<AB> = \int d\lambda \rho(\lambda) \, A(\vec{n}, \lambda) \, B(\vec{n}, \lambda) = 1 \tag{6.41}$$

which implies that $B(\vec{n}, \lambda) = A(\vec{n}, \lambda)$ since $A^2(\vec{n}, \lambda) = 1$. Therefore,

$$<AB> = \int d\lambda \rho(\lambda) \, A(\vec{n}_A, \lambda) \, A(\vec{n}_B, \lambda) \tag{6.42}$$

and since there is nothing special about directions $\vec{n}_A, \vec{n}_B$ we also have

$$<AC> = \int d\lambda \rho(\lambda) \, A(\vec{n}_A, \lambda) \, A(\vec{n}_C, \lambda)$$

$$<BC> = \int d\lambda \rho(\lambda) \, A(\vec{n}_B, \lambda) \, A(\vec{n}_C, \lambda). \tag{6.43}$$

Now,

$$<AB> - <AC> = \int d\lambda \, \rho(\lambda) \, A(\vec{n}_A, \lambda) \, (A(\vec{n}_B, \lambda) - A(\vec{n}_C, \lambda)) =$$

$$\int d\lambda \, \rho(\lambda) \, A(\vec{n}_A, \lambda) \, A(\vec{n}_B, \lambda) \left[1 - A(\vec{n}_B, \lambda) \, A(\vec{n}_C, \lambda) \right] \tag{6.44}$$

where we used the fact that $A(\vec{n}_B, \lambda)^2 = 1$. Since

$$1 - A(\vec{n}_B, \lambda) \, A(\vec{n}_C, \lambda) \geq 0$$

we are led to the Bell inequality,

$$|<AB> - <AC>| \leq \int d\lambda \, \rho(\lambda) \left[1 - A(\vec{n}_B, \lambda) \, A(\vec{n}_C, \lambda) \right] =$$
$$1 - <BC>. \tag{6.45}$$

Mathematica Notebook 6.1: The Bell Inequalities. http://www.physics.unlv.edu/%7Ebernard/MATH_book/Chap6/chap6_link.html; See also https://bernardzygelman.github.io

6.6 Applications

In Chap. 4 we introduced the Shor algorithm and discussed its application in the RSA public key encryption protocol. In this chapter, we explore how the quantum theory can be exploited in applications to *private key encryption*. Suppose Alice wants to send a secure message to Bob. First, she encodes a message as a set of bits. Bob, as well as eavesdroppers, knows how to convert those bits back into a message. Alice wants Bob, and only Bob, to get those bits. So Alice chooses to encrypt her message with a private key. Suppose her message is the string

$$Alice\_message = \{0, 1, 1, 0, 1, 0, 1, 1, 1, 1, 0, 0\}. \tag{6.46}$$

How does Alice encrypt this? Because the message contains 12 bits, she generates a set of 12 random bits and adds them, using modular arithmetic, to her message. Her private key is a string of random binary entries, e.g.,

$$key = \{1, 1, 1, 1, 1, 0, 1, 1, 0, 1, 0, 1\}. \tag{6.47}$$

The encrypted message is formed by taking the modular sum (base 2) of Alice's message and the key so that

$$encrypted\_message = Alice\_message \oplus key =$$
$$\{1, 0, 0, 1, 0, 0, 0, 0, 1, 0, 0, 1\}. \tag{6.48}$$

If Eve intercepts this bit string, it will be very difficult for her to decrypt it, since a modular sum of random bits with a message is also random. However, if Alice reuses this key, Eve can study the intercepted messages for patterns that will eventually allow her to "guess" that key. Therefore, Alice should only use this key once and generate additional random keys for each message. A private key that is used only once is called a *one-time pad*. So Alice feels secure with her one-time pad, but how will Bob be able to decrypt her message? Easy, if Alice sent Bob a one-time pad by secret courier etc., Bob would decode the message by the following prescription,

$$Bob\_message = encrypted\_message \oplus key,$$

where we used the fact that for any binary string a, $a \oplus a = 0$. Maybe Alice's courier is loyal, discreet and does not sell the key to Eve, but you never know. Because it is a one time pad, every time Alice sends a private message, she needs to generate a new key and have the courier deliver it. That strategy can get expensive. Alice is faced with the *private key distribution problem*, but quantum theory comes to the rescue. In 1992, *Arthur Ekart* developed a private key distribution protocol that is based on the availability of EPR pairs or Bell states. Before that, in 1984, *Charles Bennet* and *Gilles Brassard* introduced an implementation, of an idea first put forward by *Stephen Wiesner* in 1970, for quantum key distribution (QKD). Today it is called the BB84 protocol and has already found use in industry and government.

6.6 Applications

6.6.1 BB84 Protocol

In the BB84 protocol, Alice produces a private key based on random bits. She then encodes these bits in terms of the states $|0\rangle$ and $|1\rangle$, i.e., the computational basis. She needs a qubit, i.e., an electron, photon, etc., the medium that is the physical embodiment of that information. So in the computational basis, Alice makes the association

$$|0\rangle \to 0$$
$$|1\rangle \to 1.$$

Her measurements are performed by the gate

$$\mathbf{G}_3 \equiv \frac{1}{2}\mathbb{1} - \frac{1}{2}\sigma(0,0) = \begin{pmatrix} 0 & 0 \\ 0 & 1 \end{pmatrix} \tag{6.49}$$

where

$$\sigma(\theta,\phi) \equiv \begin{pmatrix} \cos\theta & \exp(-i\phi)\sin\theta \\ \exp(i\phi)\sin\theta & -\cos\theta \end{pmatrix}. \tag{6.50}$$

Because

$$\mathbf{H}|0\rangle = \frac{1}{\sqrt{2}}(|0\rangle + |1\rangle)$$
$$\mathbf{H}|1\rangle = \frac{1}{\sqrt{2}}(|0\rangle - |1\rangle) \tag{6.51}$$

are eigenstates of the measuring device $\sigma(\frac{\pi}{2}, 0)$, Alice can also encode her 0, 1 bits in terms of the new basis $\mathbf{H}|0\rangle$, $\mathbf{H}|1\rangle$ by using the gate

$$\mathbf{G}_1 \equiv \frac{1}{2}\mathbb{1} - \frac{1}{2}\sigma(\pi/2, 0) = \mathbf{H}\,\mathbf{G}_3\,\mathbf{H}^\dagger. \tag{6.52}$$

Suppose Alice sends a 12 bit string {011001110101} in terms of her computational basis, she loads her qubits in the following way

$$\{|0\rangle, |1\rangle, |1\rangle, |0\rangle, |0\rangle, |1\rangle, |1\rangle, |1\rangle, |0\rangle, |1\rangle, |0\rangle, |1\rangle\}. \tag{6.53}$$

The right-most entry in the string represents the first qubit, and the left-most entry, the 12th qubit. Note that this qubit string is not a direct product since Alice is not sending a coherent state $|011001110101\rangle$. Instead, she is sending individual qubits one-at-a-time. Bob receives this string and measures each of its qubits. But which device should he use? If he chooses $\mathbf{G}_3$ for each qubit, he will successfully re-construct Alice's message, {011001110101}. Now, Alice chooses to encode her bits in the following way

6 No-Cloning Theorem, Quantum Teleportation and Spooky Correlations

$$\{|0\rangle, \frac{1}{\sqrt{2}}(|0\rangle - |1\rangle), |1\rangle, \frac{1}{\sqrt{2}}(|0\rangle + |1\rangle), |0\rangle, \frac{1}{\sqrt{2}}(|0\rangle - |1\rangle),$$

$$|1\rangle, \frac{1}{\sqrt{2}}(|0\rangle - |1\rangle), |0\rangle, |1\rangle, \frac{1}{\sqrt{2}}(|0\rangle + |1\rangle), \frac{1}{\sqrt{2}}(|0\rangle - |1\rangle)\}. \quad (6.54)$$

Equation (6.54) represents the same bit string as before, but for qubits 1, 2, 5, 7, 9, 11, Alice did not use the computational basis to encode logical bit values. Bob knows that Alice is using a random selection of eigenstates to encode her qubits, and so he arbitrarily chooses the series of measurement devices,

$$\{\mathbf{G}_3, \mathbf{G}_1, \mathbf{G}_1, \mathbf{G}_1, \mathbf{G}_3, \mathbf{G}_3, \mathbf{G}_3, \mathbf{G}_3, \mathbf{G}_1, \mathbf{G}_3, \mathbf{G}_1, \mathbf{G}_3\}. \quad (6.55)$$

Unwittingly, Bob chose the correct devices, used by Alice, for qubits 2,3,6,8,9,11,12. In a measurement, he finds the qubit values

$$\{0\ 1\ 0\ 0\ 0\ 1\ 1\ 0\ 1\ 1\ 0\ 0\}. \quad (6.56)$$

At this point, he informs Alice on a classical channel that he has completed his measurements. After Alice received Bob's confirmation, she publicly announces her measurement gate configuration,

$$\{\mathbf{G}_3, \mathbf{G}_1, \mathbf{G}_3, \mathbf{G}_1, \mathbf{G}_3, \mathbf{G}_1, \mathbf{G}_3, \mathbf{G}_1, \mathbf{G}_3, \mathbf{G}_3, \mathbf{G}_1, \mathbf{G}_1\}. \quad (6.57)$$

Both Bob and Eve are privy to this information. Bob now informs Alice, on a classical channel, which of his qubits where measured with the devices she chose. Alice uses that information to strike out the gates and, bit values, that do not conform to Bob's gate configuration, i.e.,

$$\{\mathbf{G}_3, \mathbf{G}_1, \cancel{\mathbf{G}_3}, \mathbf{G}_1, \mathbf{G}_3, \cancel{\mathbf{G}_1}, \mathbf{G}_3, \cancel{\mathbf{G}_1}, \cancel{\mathbf{G}_3}, \mathbf{G}_3, \mathbf{G}_1, \cancel{\mathbf{G}_1}\}. \quad (6.58)$$

Bob does the same for his gate string so that it looks like

$$\{\mathbf{G}_3, \mathbf{G}_1, \cancel{\mathbf{G}_1}, \mathbf{G}_1, \mathbf{G}_3, \cancel{\mathbf{G}_3}, \mathbf{G}_3, \cancel{\mathbf{G}_3}, \cancel{\mathbf{G}_1}, \mathbf{G}_3, \mathbf{G}_1, \cancel{\mathbf{G}_3}\}. \quad (6.59)$$

Bob and Alice throw away the bit values for the struck gates and, assuming that Eve has not eavesdropped, both possess identical truncated bit strings, whose values are

$$\{0\ 1\ 0\ 0\ 1\ 1\ 0\}. \quad (6.60)$$

This string could be used as a private key, but how can they be sure that no-one has intercepted the *flying qubits*.[3] To find out whether Eve has eavesdropped, Bob and Alice share, on an un-secured classical channel, the bit values of the first three qubits in (6.60), i.e., {1 1 0}. If Eve, by chance, also chose gates $\mathbf{G}_3, \mathbf{G}_3, \mathbf{G}_1$ corresponding to those bit values, then there is no way that Bob and Alice can infer the presence of a snoop. There is, in this example, a one in eight chance for that possibility. However,

---

[3] Flying qubits are physically transported between two locations. Typically they are photons traveling through empty space or some medium.

6.6 Applications

for sufficiently large bit samples, that probability becomes vanishing small. So most likely Eve will have a different measurement gate sequence. Let's suppose she chose G_1, G_3, G_1 for the said qubits. Because Eve's third gate differs from Bob's choice, he now has a 50% chance of finding the bit configuration {010}, instead of {110}. As Bob sent Alice this string, Alice immediately recognizes the mismatch and warns Bob that someone is listening. At this point, Bob and Alice throw out the offending data and repeat the procedure until they share enough common bits to satisfy their privacy concerns. Of course, our demonstration, using only three qubits, is of limited utility. A real application requires a large enough bit string to ascertain Eve's presence.

Once Alice and Bob are satisfied with this test, they throw away the set {110}, (after all, this information was sent on a classical where Charles could be listening) and use the remaining set,

$$\{0\ 1\ 0\ 0\}$$

which they, and only they, share. Alice and Bob are now in possession of a possible joint private key. We have made one important, and unrealistic assumption; that the quantum channel is devoid of noise. In the real world, measurement devices, communication channels, etc., are always subjected to environmental factors that contaminate and compromise data. Those factors, typically called noise, cannot be distinguished from the presence of a snoop but they can be mitigated by increasing the number of qubits and employing error correction codes [7]. The latter topic is reviewed in Chap. 9. Possible loopholes [8] in the *BB*84 protocol is a *man-in-the-middle (MITM) attack*. If Eve fools Alice into believing that she is talking to Bob, all bets are off.

6.6.2 Ekert Protocol

Suppose that Alice and Bob each possess a qubit of a pair in the Bell state

$$|\psi\rangle = \frac{1}{\sqrt{2}} \left(|01\rangle - |10\rangle \right). \tag{6.61}$$

Alice measures the value of the spin projection $\vec{\sigma} \cdot \vec{n}_a$ where $\vec{n}_a$ is the direction of Alice's measuring device. Likewise, Bob measures $\vec{\sigma} \cdot \vec{n}_b$ where $\vec{n}_b$ is the direction of Bob's device. They both agree to choose among two directions along the z and x-axis randomly. So sometimes their choice of axis agree, i.e., both choose either z or x. Other times they disagree, Bob decides on x, and Alice decides z and vice versa. The expectation value for the correlation is

$$\langle \psi | \vec{\sigma}_a \cdot \vec{n}_a\, \vec{\sigma}_b \cdot \vec{n}_b | \psi \rangle = -\vec{n}_a \cdot \vec{n}_b. \tag{6.62}$$

Therefore, if unit vector $\vec{n}_a$ is parallel to unit vector $\vec{n}_b$ the spin measurements are exactly anti-correlated, i.e

$$\langle \psi | \vec{\sigma}_a \cdot \vec{n}_a\, \vec{\sigma}_b \cdot \vec{n}_b | \psi \rangle = -1.$$

Table 6.3 Tabulation of Alice's and Bob's measuring devices and corresponding qubit values for Bell state (6.61)

Pair	1	2	3	4	5	6	7	8	9
Alice	σ_X	σ_Z	σ_Z	σ_X	σ_X	σ_Z	σ_Z	σ_X	σ_X
Result	1	1	1	1	1	1	-1	-1	1
Bob	σ_X	σ_X	σ_Z	σ_Z	σ_X	σ_X	σ_Z	σ_Z	σ_Z
Result	-1	1	-1	-1	-1	1	1	-1	-1

If Alice measures $+1$, Bob measures -1 and vice versa. If $\vec{n}_a$ is perpendicular to $\vec{n}_a$, then there is no correlation, or

$$\langle \psi | \vec{\sigma}_a \cdot \vec{n}_a \, \vec{\sigma}_b \cdot \vec{n}_b | \psi \rangle = 0.$$

So if Alice measures $+1$, Bob is just as likely to measure $+1$ or -1. After measuring their qubits, Alice's and Bob's data might look something like that shown in Table 6.3.

After measurements, Alice and Bob communicate on the telephone, and Bob announces he used the measuring devices

$$\{\sigma_X, \sigma_X, \sigma_Z, \sigma_Z, \sigma_X, \sigma_X, \sigma_Z, \sigma_Z, \sigma_Z\}.$$

Alice compares that string with her choices, and they both agree to throw out the qubit pair measurements in which their basis do not agree. In this case, it is pairs (counting from left to right) 2,4,6,8,9. So Alice has the sequence of bit values $\{1, 1, 1, -1\}$ whereas Bob's values are $\{-1, -1, -1, 1\}$. Bob knows he is anti-correlated with Alice, so he simply changes the sign of his measured values. $\{1, 1, 1, -1\}$. As in the BB84 protocol, Alice and Bob take a subset of these and compare them to determine if Eve is snooping. If they are satisfied, they use the remaining subset as a private key.

6.6.3 Quantum Dense Coding

Quantum dense coding is another interesting application of Bell pairs. It allows Alice to send two classical bits of information using only one quantum qubit, provided the latter is in a Bell state. One can think of it as a kind of *data compression* scheme. Suppose Alice wants to send the two classical bits $\{0, 0\}$ to Bob. In this protocol they share their qubits in the following Bell state

$$|\psi\rangle = \frac{1}{\sqrt{2}} \left(|00\rangle + |11\rangle \right). \qquad (6.63)$$

The first qubit belongs to Alice and the second to Bob. Alice sends her qubit to Bob who is now in possession of both qubits. Bob applies a **CNOT** gate on the qubit

Alice sent him followed by the application of a Hadamard gate, after which

$$(\mathbf{H} \otimes \mathbb{1}) \, \mathbf{CNOT} \otimes \mathbb{1} \frac{1}{\sqrt{2}} (|00\rangle + |11\rangle) = \mathbf{H} \otimes \mathbb{1} \frac{1}{\sqrt{2}} (|00\rangle + |10\rangle) =$$
$$\frac{1}{2}(|0\rangle + |1\rangle) \otimes |0\rangle + \frac{1}{2}(|0\rangle - |1\rangle) \otimes |0\rangle = |00\rangle \tag{6.64}$$

so Alice has succeeded, by giving Bob only a single qubit, in conveying the two bits $\{0, 0\}$ of information. If Alice sends $\{0, 1\}$ to Bob, she first subjects her qubit to a σ_X gate so that

$$\mathbb{1} \otimes \sigma_X \frac{1}{\sqrt{2}} (|00\rangle + |11\rangle) = \frac{1}{\sqrt{2}} (|10\rangle + |01\rangle).$$

Again, Alice gives her qubit to Bob. Bob repeats the operation with Hadamard-**CNOT** gate combination so that

$$(\mathbf{H} \otimes \mathbb{1}) \, \mathbf{CNOT} \otimes \mathbb{1} \frac{1}{\sqrt{2}} (|10\rangle + |01\rangle) = \mathbf{H} \otimes \mathbb{1} \frac{1}{\sqrt{2}} (|11\rangle + |01\rangle) =$$
$$\frac{1}{2}(|0\rangle - |1\rangle) \otimes |1\rangle + \frac{1}{2}(|0\rangle + |1\rangle) \otimes |1\rangle = |01\rangle. \tag{6.65}$$

If Alice wishes to send bits $\{1, 0\}$, she subjects her qubit to a σ_Z gate before giving it to Bob, so that

$$\mathbb{1} \otimes \sigma_Z \frac{1}{\sqrt{2}} (|00\rangle + |11\rangle) = \frac{1}{\sqrt{2}} (|00\rangle - |11\rangle).$$

Bob performs the gate operations

$$(\mathbf{H} \otimes \mathbb{1}) \, \mathbf{CNOT} \otimes \mathbb{1} \frac{1}{\sqrt{2}} (|00\rangle - |11\rangle) = \mathbf{H} \otimes \mathbb{1} \frac{1}{\sqrt{2}} (|00\rangle - |10\rangle) =$$
$$\frac{1}{2}(|0\rangle + |1\rangle) \otimes |0\rangle - \frac{1}{2}(|0\rangle - |1\rangle) \otimes |0\rangle = |10\rangle. \tag{6.66}$$

Bob's measurement apparatus yields the classic bits $\{1, 0\}$, as desired. In each of the three case, Alice was able to convey two bits of classical information using only a single qubit. The last case, in which Alice sends the bits $\{1, 1\}$, is left as an exercise.

6.7 GHZ Entaglements

The setups illustrating the Bell and CHSH inequalities require two observers making measurements that are separated by a space-like interval[4] and that perform at least three different types of measurement. As such, the arguments based on a statisti-

---

[4] Two points in space-time that cannot be bridged by a light-beam, or anything moving less than the speed of light.

cal analysis of measured data, appear somewhat convoluted and labyrinthine. Could we not just use a single qubit system to illustrate "quantum weirdness"? It turns out that a single qubit system can be described by a hidden variable model which does not display the unsettling properties associated with EPR-like systems. But is there not an alternative, perhaps a more transparent, example that does not require opaque statistical jiujitsu to illustrate a point? The answer is in the affirmative, but that development had to wait until 1989 with the introduction of the GHZ theorem, named after *Daniel Greenberger, Michael Holt*, and *Anton Zeilinger*. A GHZ state involves a quantum system in which three or more qubits are entangled. An analysis of it shows how Bell-like inequalities manifest in a way that does not require statistical ensembles. Examples of GHZ states and their properties is summarized in Mathematica Notebook 6.2.

> Mathematica Notebook 6.2: GHZ states. http://www.physics.unlv.edu/%7Ebernard/MATH_book/Chap6/chap6_link.html; See also https://bernardzygelman.github.io

Problems

6.1 All problems in this chapter are posted in the Mathematica Notebook 6.3

> Mathematica Notebook 6.3: Chapter 6 problems and exercises.

References

1. A. Aspect, P. Grangier, G. Roger, Phys. Rev. Lett. **49**, 91 (1982)
2. J.S. Bell, *Speakable and Unspeakable in Quantum Mechanics* (Cambridge University Press, 1987)
3. J.F. Clauser, M.A. Horne, A. Shimony, R.A. Holt, Phys. Rev. Lett. **23**, 880 (1969)
4. M. Giustina et al., https://doi.org/10.1103/PhysRevLett.115.250401
5. D.M. Greenberger, M.A. Horne, A. Zeilinger, arXiv:0712.0921G (2007)
6. B. Hensen et al., Loophole-free Bell inequality violation using electron spins separated by 1.3 kilometers. Nature **526**, 682 (2015)
7. M.E. Nielsen, I.L. Chuang, *Quantum Computation and Quantum Information* (Cambridge University Press, 2011)
8. C. Pacher, A. Abidin, T. Lörunser, M. Peev, R. Ursin, A. Zeilinger, J. Larsson, Quantum Inf. Process. **15**, 327 (2016)
9. S. Hawking, R. Penrose, *The Nature of Space and Time* (Princeton University Press, 1996), p. 121
10. L.K. Shalm et al., https://doi.org/10.1103/PhysRevLett.115.250402

11. A. Shimony, in *The Stanford Encyclopedia of Philosophy*, ed. by E.N. Zalta (Fall 2017 Edition), https://plato.stanford.edu/archives/fall2017/entries/bell-theorem/
12. C.S. Wu, I. Shaknov, Phys. Rev. **77**, 136 (1950)

Quantum Hardware I: Ion Trap Qubits

Abstract

The DiVincenzo criteria, a list of necessities for the construction of a quantum computer is summarized herein. In reviewing the physics underpinning the trapped ion qubit paradigm, I introduce a rotor model for atoms/ions and demonstrate how single qubit gates, such as the phase and Hadamard gates, are realized in this framework. We use the latter to illustrate how ions respond to laser radiation, the mechanism by which ion qubits are addressed. We show how the Cirac-Zoller mechanism, and its generalization, enables the realization of two-qubit control gates. I provide a survey of trapped ion qubit systems, including optical and hyperfine qubit systems.

7.1 Introduction

In the mid-1930s *Claude Shannon* introduced and utilized Boolean logic in switching circuit communication applications. Around the same time, a Boolean logic adder circuit was developed at Bell Laboratories. *Konrad Zuse* built the first binary system calculator and *John Atanasoff* conceived of and designed, a prototype for a computer based on binary electronic logic circuits. It was not until the mid-to-late 1940s that electronic digital computing machines became operational.

Fittingly, the new century saw the first laboratory demonstrations of quantum logic gates. Nevertheless, despite tremendous progress in the last two decades, quantum hardware is still in its infancy. Today, researchers routinely achieve coherent quantum states with about fifty to hundreds of qubits, but that is still a far cry from the billions of transistors crammed into a microprocessor chip. In the next couple of chapters, we delve into the heart of a quantum computer, the qubit. The blueprint for a quantum machine should address the following questions; what physical systems are viable qubit candidates? How do we write and read information stored in a register of qubits ? What are the prospects for scalability? Etc.

7.1.1 The DiVincenzo Criteria

In meeting the challenges of building an operational quantum computer, *DiVincenzo* [4] highlighted a list of essential requirements. They include,

(i) A scalable physical system that possesses well defined qubits.
(ii) The ability to initialize qubits to an initial fiducial state, such as $|000\rangle$, etc.
(iii) De-coherence times much longer than the gate operation time.
(iv) A set of universal quantum gates (i.e., phase, Hadamard and two-qubit control gates).
(v) The ability to perform qubit specific measurements and the means to read out the contents of a register.

DiVincenzo also added the following desiderata

(i) The means to transmit flying qubits (e.g., photons) between locations
(ii) Ability to inter-convert stationery and flying qubits

for enabling quantum communication capabilities.

We already introduced a couple of qubit candidates in previous chapters. The spin degrees of freedom of an electron or proton behave as qubits, and we illustrated how the polarization properties of a photon exhibit qubit properties. At the time of this writing, several qubit candidates for real-world applications are under active consideration. They include NMR (nuclear magnetic resonance) qubits, photonic qubits, Nitrogen-vacancy centers qubits and spin qubits in Silicon to name a few. Two prospective candidates have come to the fore-front; trapped ion, and superconducting qubits respectfully [1]. Trapped ion circuits are one of the first qubit technologies developed for quantum processing applications. In this chapter we focus and elaborate on the underlying physics of that technology. Unfortunately, a comprehensive discussion requires a significant background in atomic physics, quantum optics, and many-body physics. As the intended audience for this monograph is not expected to have such expertise, I introduce an accessible model and employ it to demonstrate features shared with laboratory realizations. To accomplish this goal, we first review some basic physics concepts that govern the behavior of quantum matter. We also need to understand the language used to describe classical dynamical systems and for which, I provide a lightning review below.

7.2 Lagrangian and Hamiltonian Dynamics in a Nutshell

The degrees of freedom of a mechanical system is the minimum set of numbers $q_1, q_2, \ldots q_n$, called generalized coordinates, required to characterize a system. Atomic hydrogen, the prototypical atom, consists of a single electron bound by the electrostatic Coulomb force to a proton of equal and opposite electric charge. Six generalized coordinates characterize the hydrogen atom. Three coordinates specify the location of its center of mass in space, one gives the distance of the electron from the proton, and two angle coordinates determine the orientation of the electron relative to the proton. Full analytic descriptions of an electrostatically bound system, both in the classical [5] and quantum domains [6] are available, but we focus our discussion on a simpler dynamical system, the translating rotor, characterized by only two degrees of freedom.

A free planar rotor consists of a point particle of mass m constrained to the move on a circular path of radius L (see Fig. 7.1) about a point that is allowed to translate along the horizontal axis. At first sight, the rotor seems to have no relation to the hydrogen atom, let alone complex atoms. But for our purposes, this model is adequate, as it shares many essential features of a realistic atom/ion qubit.

For a system with n degrees of freedom, a specification of the n coordinates $q_1(t), q_2(t) \ldots q_n(t)$, and their time derivatives $\dot{q}_1(t), \dot{q}_2(t) \ldots \dot{q}_n(t)$, defines its state at time t. In the Lagrangian formulation, the state is a solution to a set of n second order differential equations

$$\frac{d}{dt}\left(\frac{\partial \mathscr{L}}{\partial \dot{q}_i}\right) - \frac{\partial \mathscr{L}}{\partial q_i} = 0 \tag{7.1}$$

for $i = 1, 2 \ldots n$. $\mathscr{L}(q_1, q_2, \ldots q_n; \dot{q}_1, \dot{q}_2, \ldots \dot{q}_n)$ is the Lagrangian functional of functions $q_1(t), q_2(t), \ldots q_n(t); \dot{q}_1(t), \dot{q}_2(t), \ldots \dot{q}_n(t)$. In the mid-nineteenth century, mathematician *R.W. Hamilton* introduced an alternative description. In his formulation, a Hamiltonian functional

$$H \equiv \sum_i \dot{q}_i \frac{\partial \mathscr{L}}{\partial \dot{q}_i} - \mathscr{L} = \sum_i \dot{q}_i \, p_i - \mathscr{L}$$

$$p_i \equiv \frac{\partial \mathscr{L}}{\partial \dot{q}_i} \tag{7.2}$$

defines equations of motion given by a set of first order differential equations

$$\dot{q}_i = \frac{\partial H}{\partial p_i} \quad \dot{p}_i = -\frac{\partial H}{\partial q_i}. \tag{7.3}$$

Here, p_i are *conjugate momenta* to coordinates q_i. The Hamiltonian equations provide a state description identical to that obtained in the Lagrangian formulation.

Fig. 7.1 A translating rotor. The angle θ is the "electron's" generalized coordinate, whereas $x = |\vec{R}|$, the horizontal distance along the abscissa, is the generalized coordinate for the rotor center

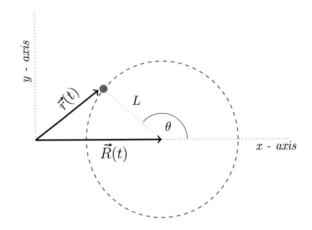

7.2.1 Dynamics of a Translating Rotor

In this model, the rotor's motion corresponds to that of electrons in a real atom/ion. We assume that a mass $M \gg m$, is located at the geometric center of the rotor, and is allowed to move along a single, horizontal, axis. The internal "electron" coordinate $\vec{r}$ along with $\vec{R}$, are as shown in Fig. 7.1

$$\vec{r} = (x + L\cos\theta)\,\hat{\mathbf{i}} + L\sin\theta\,\hat{\mathbf{j}}$$
$$\vec{R} = x\,\hat{\mathbf{i}} \qquad (7.4)$$

where $\hat{\mathbf{i}}, \hat{\mathbf{j}}$ are unit vectors along the x, y directions in that figure. The *kinetic energy* of the "electron" is

$$\frac{1}{2} m\vec{r} \cdot \vec{r} = \frac{1}{2} m(\dot{x} - L\dot{\theta}\sin\theta)^2 + \frac{1}{2} m L^2 \dot{\theta}^2 \cos^2\theta =$$
$$\frac{mL^2}{2}\dot{\theta}^2 - m\dot{\theta}\dot{x} L \sin\theta + \frac{m\dot{x}^2}{2} \qquad (7.5)$$

and

$$\frac{1}{2} M \vec{R} \cdot \vec{R} = \frac{1}{2} M \dot{x}^2. \qquad (7.6)$$

The Lagrangian functional corresponds to the total kinetic energy expressed in terms of the generalized coordinates, and so

$$\mathscr{L} = \frac{1}{2}(M+m)\dot{x}^2 + \frac{mL^2}{2}\dot{\theta}^2 - m\dot{\theta}\dot{x} L \sin\theta. \qquad (7.7)$$

With it we invoke Lagrange's equations (7.1) to obtain the equations of motion for θ, and x. We assume that $\dot{x} \ll L\dot{\theta}$ and so drop the term in (7.7) that involves cross terms between $\dot{\theta}, \dot{x}$. This approximation considerably simplifies the quantum mechanical description discussed in the next section. According to (7.2) the conjugate momenta are

$$p_\theta = \frac{\partial \mathcal{L}}{\partial \dot\theta} = m L^2 \dot\theta$$
$$p_x = \frac{\partial \mathcal{L}}{\partial \dot x} = (M + m)\dot x. \tag{7.8}$$

Using (7.8) to express generalized velocities in terms of the conjugate momenta, and inserting the latter into (7.2) we get

$$H = \frac{p_x^2}{2(m+M)} + \frac{p_\theta^2}{2 m L^2}. \tag{7.9}$$

This Hamiltonian describes two types of motion. The internal motion of the electron about the rotor center and the translational motion of the rotor. Because the two motions are de-coupled in this approximation, we study each separately. For now, let's ignore the degree of freedom associated with the translation motion and focus on the Hamiltonian for the internal coordinate θ of a stationary rotor,

$$H = \frac{p_\theta^2}{2 m L^2}. \tag{7.10}$$

> Mathematica Notebook 7.1: Lagrangian and Hamiltonian dynamics of a translating rotor. http://www.physics.unlv.edu/%7Ebernard/MATH_book/Chap7/chap7_link.html; See also https://bernardzygelman.github.io

7.3 Quantum Mechanics of a Free Rotor: A Poor Person's Atomic Model

In standard treatments [6], the route from a classical Hamiltonian description to a quantum theory proceeds by the heuristic replacement of classical variables p, q with corresponding quantum operators

$$p \Rightarrow \mathbf{P}$$
$$q \Rightarrow \mathbf{Q}.$$

The operators obey the commutation relation

$$\mathbf{PQ} - \mathbf{QP} = [\mathbf{P}, \mathbf{Q}] = -i\hbar \mathbb{1},$$

where $\hbar$ is proportional to Planck's constant and $\mathbb{1}$ is the unit operator in a Hilbert space spanned by the eigenvectors of $\mathbf{Q}$, or $\mathbf{P}$. Unlike the two-dimensional Hilbert space for a qubit, the Hilbert space describing this system is of infinite dimension [7]. A comprehensive treatment of it is beyond the domain of our discussion (e.g., see [6,7]), but we are going to proceed as before and allow the eigenstates of **Q**, or **P**, to constitute a basis for this Hilbert space. The eigenvalue equation for **Q** is

$$\mathbf{Q}|q\rangle = q|q\rangle$$

where q is the eigenvalue and $|q\rangle$ is the corresponding eigenstate. The eigenvalue $q = \theta$ spans a continuous distribution of values ranging from 0 to 2π. The quantum mechanical Hamiltonian operator

$$\mathbf{H}_0 = \frac{\mathbf{P}^2}{2mL^2}, \tag{7.11}$$

is modeled from the classical Hamiltonian by replacing p, q with the corresponding quantum operators. State $|\psi\rangle$ for the rotor evolves according to the Schrödinger equation

$$i\hbar \frac{\partial |\psi\rangle}{\partial t} = \mathbf{H}_0 |\psi\rangle, \tag{7.12}$$

and is equivalent to

$$i\hbar \frac{\partial \langle \theta | \psi \rangle}{\partial t} = \langle \theta | \mathbf{H}_0 | \psi \rangle, \tag{7.13}$$

where we have taken the inner product of the kets on both sides of (7.12) with $\langle q|$, or $\langle \theta|$. Because θ is parameterized by $0 < \theta \leq 2\pi$ and the inner product $\langle \theta | \psi \rangle$ is a complex number, $\langle \theta | \psi \rangle \equiv \psi(\theta)$ is a function of θ. Using expression (7.12) and the fact [6]

$$\langle \theta | \mathbf{P}^2 | \psi \rangle = -\hbar^2 \frac{\partial^2 \psi(\theta)}{\partial \theta^2}, \tag{7.14}$$

Schrödinger equation (7.12) takes the form of a *partial differential equation*

$$i\hbar \frac{\partial \psi(\theta, t)}{\partial t} = -\frac{\hbar^2}{2mL^2} \frac{\partial^2 \psi(\theta, t)}{\partial \theta^2}. \tag{7.15}$$

In finding solutions to it we use *the method of separation of variables*, which posits that $\psi(\theta, t) = \exp(-i Et/\hbar)\psi(\theta)$, and where E is a separation constant. Inserting this ansatz into (7.15) we obtain

$$-\frac{\hbar^2}{2mL^2} \frac{\partial^2 \psi(\theta)}{\partial \theta^2} = E\psi(\theta). \tag{7.16}$$

Solutions to (7.16) are contingent on boundary conditions (b.c.) for $\psi(\theta)$. Since $\theta = 0$ identifies with $\theta = 2\pi$ (i.e., the two values correspond to the same physical point), we

7.3 Quantum Mechanics of a Free Rotor: A Poor Person's Atomic Model

Fig. 7.2 The first four energy levels of the rotor system

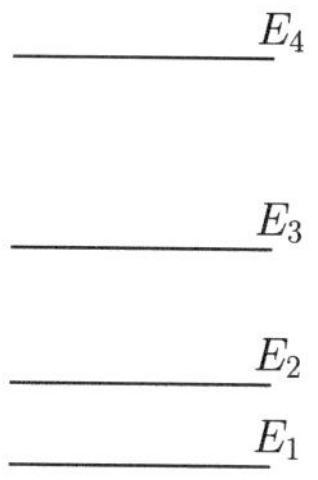

$$\underline{\qquad\qquad} E_4$$

$$\underline{\qquad\qquad} E_3$$

$$\underline{\qquad\qquad} E_2$$
$$\underline{\qquad\qquad} E_1$$

require $\psi(0) = \psi(2\pi)$.[1] In addition, we impose the condition $\psi(0) = \psi(\pi) = 0$. In that case, we arrive at a class of solutions

$$\psi_n(\theta) = \frac{1}{\sqrt{\pi}} \sin(n\theta) \quad n = 1, 2, 3 \ldots \tag{7.17}$$

provided that

$$E_n = \frac{\hbar^2 n^2}{2mL^2}. \tag{7.18}$$

The states $\psi_n(\theta)$ and the corresponding E_n are eigenstates and eigenvalues, respectively, of Hamiltonian operator $\mathbf{H}_0$. As $\mathbf{H}_0$ is hermitian, it represents an energy measurement device and its eigenvalues E_n constitute the allowed energy values obtained in a measurement. The latter form a discrete tower of allowed energy values starting from the lowest $E_1 = \hbar^2/2mL^2$ or ground energy value, followed by $E_2, E_3 \ldots$. A pictorial representation of these eigenvalues, or the energy spectrum, is illustrated in Fig. 7.2. Because (7.15) is linear, the sum

$$\psi(\theta, t) = \sum_n^\infty c_n \exp(-iE_n t/\hbar) \psi_n(\theta) \tag{7.19}$$

is a general solution to it. The coefficients c_n are determined by initial conditions $c_n(t = 0)$. Let's demand that all c_n, save for $n = 1, 2$, vanish. In that case

$$\psi(\theta, t) = c_1 \exp(-iE_1 t/\hbar)\psi_1(\theta) + c_2 \exp(-iE_2 t/\hbar)\psi_2(\theta)$$
$$|c_1|^2 + |c_2|^2 = 1. \tag{7.20}$$

It is useful to define

$$\langle \psi_n | \psi(t) \rangle \equiv \int_0^{2\pi} d\theta \, \psi_n(\theta) \, \psi(\theta, t)$$

---

[1] More precisely, we equire $\psi(0) = \psi(2\pi p)$ where p is an integer.

and

$$\underline{\psi}(t) \equiv \begin{pmatrix} \langle \psi_1 | \psi(t) \rangle \\ \langle \psi_2 | \psi(t) \rangle \end{pmatrix} = \begin{pmatrix} c_1 \exp(-iE_1 t/\hbar) \\ c_2 \exp(-iE_2 t/\hbar) \end{pmatrix}. \tag{7.21}$$

It is the matrix representation of state $\psi(\theta, t)$ with respect to the truncated basis $\psi_n(\theta)$ for $n = 1, 2$. $\underline{\psi}(t)$ is a solution to the two-dimensional matrix equation

$$i\hbar \frac{d\underline{\psi}(t)}{dt} = \underline{\mathbf{h}}_0 \, \underline{\psi}(t) \tag{7.22}$$

where $\underline{\mathbf{h}}_0$ is a 2×2 matrix whose elements are

$$\underline{\mathbf{h}}_0 \equiv \begin{pmatrix} \langle \psi_1 | \mathbf{h}_0 | \psi_1 \rangle & \langle \psi_1 | \mathbf{h}_0 | \psi_2 \rangle \\ \langle \psi_2 | \mathbf{h}_0 | \psi_1 \rangle & \langle \psi_2 | \mathbf{h}_0 | \psi_2 \rangle \end{pmatrix} = \begin{pmatrix} E_1 & 0 \\ 0 & E_2 \end{pmatrix}, \tag{7.23}$$

and where

$$\langle \psi_i | \mathbf{h}_0 | \psi_j \rangle \equiv -\frac{\hbar^2}{2L^2 m} \int_0^{2\pi} d\theta \, \psi_i(\theta) \frac{\partial^2 \psi_j(\theta)}{\partial \theta^2}.$$

It is useful to re-express $\underline{\mathbf{h}}_0$ in the form

$$\underline{\mathbf{h}}_0 = \frac{1}{2} \begin{pmatrix} E_1 + E_2 & 0 \\ 0 & E_1 + E_2 \end{pmatrix} + \frac{1}{2} \begin{pmatrix} E_1 - E_2 & 0 \\ 0 & E_2 - E_1 \end{pmatrix} =$$

$$\frac{(E_1 + E_2)}{2} \mathbb{1} - \frac{\hbar \omega_0}{2} \sigma_Z \tag{7.24}$$

where $\mathbb{1}, \sigma_Z$, are the unit and Pauli-Z matrices respectively, and

$$\hbar \omega_0 \equiv E_2 - E_1 = \frac{3\hbar^2}{2mL^2}.$$

The first term in (7.24) contributes an overall constant phase in expression (7.19) and so we neglect it.

To summarize, we found that the rotor system in the Hilbert subspace spanned by the two lowest energy eigenstates is described by Eq. (7.22) where

$$\underline{\mathbf{h}}_0 = -\frac{\hbar \omega_0}{2} \sigma_Z. \tag{7.25}$$

Note that the form of $\underline{\mathbf{h}}_0$ is identical to the *Zeeman* Hamiltonian for an electron in a homogeneous magnetic field pointed along the $\hat{z}$ direction. By truncating the Hilbert space to the first two lowest states in the rotor's energy spectrum, we mapped the rotor Hamiltonian into an equivalent Hamiltonian that describes a spin qubit in a magnetic field. Below we show how this effective qubit can be manipulated to perform gate operations.

7.3 Quantum Mechanics of a Free Rotor: A Poor Person's Atomic Model

7.3.1 Rotor Dynamics and the Hadamard Gate

Our goal is to construct a Hadamard gate from the rotor "atom" described above. We accomplish this by addressing the rotor "electron" by electromagnetic radiation, e.g., a laser field. In Chap. 2 we gave an expression for the time dependent electric field of a monochromatic field, which here we take to be

$$\vec{E}(t) = E_0(\hat{\mathbf{i}} \cos\beta \, \cos(\omega t + \delta) + \hat{\mathbf{j}} \sin\beta \, \sin(\omega t)) \quad (7.26)$$

where E_0, β, δ are constants, and ω is the frequency. The rotor "electron" has electric charge e and interacts with this electric field to induce an additional term [5]

$$\mathbf{H}_I(t) = -e\vec{r} \cdot \vec{E}(t) \quad (7.27)$$

in Hamiltonian (7.11). Here

$$\vec{r} = \hat{\mathbf{i}} L \cos\theta + \hat{\mathbf{j}} L \sin\theta \quad (7.28)$$

is the position vector for the electron, and so

$$\mathbf{H}_I(t) = -e E_0 L \cos\theta \, \cos\beta \, \cos(\omega t + \delta) - e E_0 L \sin\theta \, \sin\beta \, \sin(\omega t). \quad (7.29)$$

We project $\mathbf{H}_I(t)$ onto the qubit Hilbert space by constructing the following matrix representation

$$\underline{\mathbf{H}}_I(t) \equiv \begin{pmatrix} \langle\psi_1|\mathbf{H}_I(t)|\psi_1\rangle & \langle\psi_1|\mathbf{H}_I(t)|\psi_2\rangle \\ \langle\psi_2|\mathbf{H}_I(t)|\psi_1\rangle & \langle\psi_2|\mathbf{H}_I(t)|\psi_2\rangle \end{pmatrix} =$$

$$-\hbar\Delta \cos(\omega t + \delta) \begin{pmatrix} 0 & 1 \\ 1 & 0 \end{pmatrix} = -\hbar\Delta \, \sigma_X \cos(\omega t + \delta) \quad (7.30)$$

where

$$\langle\psi_i|\mathbf{H}_I(t)|\psi_j\rangle \equiv \int_0^{2\pi} \psi_i(\theta) \mathbf{H}_I(t) \psi_j(\theta) d\theta$$

and $\hbar\Delta = e E_0 L \cos\beta/2$. Including $\mathbf{H}_I(t)$, and in the two-state approximation, the rotor state $|\psi(t)\rangle$ obeys the following *Schrödinger equation*

$$i\hbar \frac{\partial |\psi(t)\rangle}{\partial t} = \mathbf{H}(t) |\psi(t)\rangle$$

$$\mathbf{H}(t) = -\frac{\hbar\omega_0}{2} \sigma_Z - \hbar\Delta \cos(\omega t + \delta) \sigma_X \quad (7.31)$$

Simple analytic solutions to Eq. (7.31) are not available, but for the resonance condition

$$\omega \approx \omega_0,$$

and for weak coupling $\Delta \ll \omega_0$, approximate, but accurate, solutions can be found. To that end, it is useful to express

$$\mathbf{H}_I(t) = -\hbar \Delta \cos(\omega t + \delta) \sigma_X = -\hbar \Delta \cos(\omega t + \delta) (\sigma_+ + \sigma_-) \quad (7.32)$$

where

$$\sigma_+ \equiv \begin{pmatrix} 0 & 0 \\ 1 & 0 \end{pmatrix} \quad \sigma_- \equiv \begin{pmatrix} 0 & 1 \\ 0 & 0 \end{pmatrix}, \quad (7.33)$$

and which have the properties

$$\sigma_+|0\rangle = |1\rangle \quad \sigma_+|1\rangle = 0 \quad \sigma_-|1\rangle = |0\rangle \quad \sigma_-|0\rangle = 0.$$

We attempt solution of

$$i\hbar \frac{\partial |\psi(t)\rangle}{\partial t} = \left(-\frac{\hbar \omega_0}{2} \sigma_Z + \mathbf{H}_I(t) \right) |\psi(t)\rangle \quad (7.34)$$

by replacing $|\psi(t)\rangle$ with $\exp(i\omega_0 \sigma_Z t/2)|\psi_I(t)\rangle$ and inserting that into Eq. (7.34). We get

$$i\hbar \frac{\partial |\psi_I(t)\rangle}{\partial t} = \exp(-i\omega_0 \sigma_Z t/2) \mathbf{H}_I(t) \exp(i\omega_0 \sigma_Z t/2) |\psi_I(t)\rangle, \quad (7.35)$$

the *interaction picture* Schrödinger equation. Now (see Mathematica Notebook 7.2 below)

> Mathematica Notebook 7.2: Rabi-Flopping and the Rotating Wave Approximation. http://www.physics.unlv.edu/%7Ebernard/MATH_book/Chap7/chap7_link.html; See also https://bernardzygelman.github.io

$$\exp(-i\omega_0\sigma_Z t/2)\sigma_+ \exp(i\omega_0\sigma_Z t/2) = \exp(i\omega_0 t)\sigma_+$$
$$\exp(-i\omega_0\sigma_Z t/2)\sigma_- \exp(i\omega_0\sigma_Z t/2) = \exp(-i\omega_0 t)\sigma_- \quad (7.36)$$

and so

$$i\hbar \frac{\partial |\psi_I(t)\rangle}{\partial t} =$$
$$-\hbar \Delta \cos(\omega t + \delta) \left(\exp(i\omega_0 t)\sigma_+ + \exp(-i\omega_0 t)\sigma_- \right) |\psi_I(t)\rangle. \quad (7.37)$$

Applying the resonance condition $\omega = \omega_0$, the r.h.s of (7.37) becomes

$$-\frac{\hbar \Delta}{2} (\exp(i\delta)\sigma_- + \exp(-i\delta)\sigma_+) +$$
$$-\frac{\hbar \Delta}{2} (\exp(i\delta) \exp(2i\omega_0 t)\sigma_+ + \exp(-i\delta) \exp(-2i\omega_0 t)\sigma_-) \quad (7.38)$$

7.3 Quantum Mechanics of a Free Rotor: A Poor Person's Atomic Model

The second line in (7.38) is a rapidly varying function of time, whereas the first is constant. The *rotating wave approximation (RWA)* posits that in solving Eq. (7.37) we are allowed to neglect the time varying terms so that we assume

$$i\hbar \frac{\partial |\psi_I(t)\rangle}{\partial t} = -\frac{\hbar \Delta}{2} (\exp(i\delta)\sigma_- + \exp(-i\delta)\sigma_+) |\psi_I(t)\rangle. \quad (7.39)$$

Because the interaction term on the r.h.s of (7.39) is constant we arrive at the solution

$$|\psi_I(t)\rangle = \mathbf{U}_I(t, t_0) |\psi_I(t_0)\rangle$$

$$\mathbf{U}_I(t, t_0) = \exp(i(t - t_0)\frac{\Delta}{2} (\exp(i\delta)\sigma_- + \exp(-i\delta)\sigma_+)) =$$

$$\begin{pmatrix} \cos(\frac{1}{2}\Delta(t - t_0)) & i\exp(i\delta)\sin(\frac{1}{2}\Delta(t - t_0)) \\ i\exp(-i\delta)\sin(\frac{1}{2}\Delta(t - t_0)) & \cos(\frac{1}{2}\Delta(t - t_0)) \end{pmatrix}, \quad (7.40)$$

and since $|\psi(t)\rangle = \exp(i\omega_0 \sigma_z t/2) |\psi_I(t)\rangle$

$$|\psi(t)\rangle = \mathbf{U}(t, t_0) |\psi(t_0)\rangle$$
$$\mathbf{U}(t, t_0) = \exp(i\omega_0 \sigma_z t/2) \mathbf{U}_I(t, t_0), \quad (7.41)$$

or

$$\mathbf{U}(t, t_0) =$$
$$\begin{pmatrix} \exp(i\omega_0 t/2) \cos(\frac{1}{2}\Delta(t - t_0)) & i\exp(i(\delta + \omega_0 t/2)) \sin(\frac{1}{2}\Delta(t - t_0)) \\ i\exp(-i(\delta + \omega_0 t/2)) \sin(\frac{1}{2}\Delta(t - t_0)) & \exp(-i\omega_0 t/2) \cos(\frac{1}{2}\Delta(t - t_0)) \end{pmatrix}. \quad (7.42)$$

A laser pulse with frequency $\omega = \omega_0$ is turned on at $t = t_0 = 0$, and off at $t = t_1 = 3\pi/\omega_0$ resulting in

$$\mathbf{U}(t_1, 0) = \begin{pmatrix} -i\cos(\frac{3\pi\Delta}{2\omega_0}) & -i\sin(\frac{3\pi\Delta}{2\omega_0}) \\ -i\sin(\frac{3\pi\Delta}{2\omega_0}) & i\cos(\frac{3\pi\Delta}{2\omega_0}) \end{pmatrix}, \quad (7.43)$$

where we have set the laser parameter $\delta = 3\pi/2$. Let's choose

$$\Delta = \omega_0/6$$

so that

$$\mathbf{U}(t_1) = \frac{-i}{\sqrt{2}} \begin{pmatrix} 1 & 1 \\ 1 & -1 \end{pmatrix} = -i\,\mathbf{H} \quad (7.44)$$

is the matrix representation, up to a global phase factor $-i$, of the Hadamard gate. If the rotor system was initially in some coherent state

$$|\psi\rangle = c_1 |E_1\rangle + c_2 |E_2\rangle = \begin{pmatrix} c_1 \\ c_2 \end{pmatrix}$$

at $t = t_0$, at the end of the pulse sequence it is, up to global phase factor, in state

$$\mathbf{H}|\psi\rangle,$$

where $\mathbf{H}$ is the Hadamard gate (here $\mathbf{H}$ should not be confused with $\mathbf{H}(t)$, the Hamiltonian operator).

7.3.2 Two-Qubit Gates

Consider a pair of non-interacting rotor qubits. In modeling this system, we simply incorporate Hamiltonian (7.25), for each rotor, into a two-qubit direct product operator. It is given by the expression

$$\mathbf{h}_0 = -\frac{\hbar\omega_0}{2}\sigma_Z \otimes \mathbb{1} - \frac{\hbar\omega_0}{2}\mathbb{1} \otimes \sigma_Z, \quad (7.45)$$

whose matrix representation with respect to the two-qubit basis $|00\rangle, |01\rangle, |10\rangle, |11\rangle$ is

$$\mathbf{h}_0 = \begin{pmatrix} -\hbar\omega_0 & 0 & 0 & 0 \\ 0 & 0 & 0 & 0 \\ 0 & 0 & 0 & 0 \\ 0 & 0 & 0 & \hbar\omega_0 \end{pmatrix}. \quad (7.46)$$

The *unitary time evolution operator* is given by

$$\mathbf{U}(\tau) = \exp(-i\mathbf{h}_0\tau/\hbar) = \begin{pmatrix} \exp(i\omega_0\tau) & 0 & 0 & 0 \\ 0 & 1 & 0 & 0 \\ 0 & 0 & 1 & 0 \\ 0 & 0 & 0 & \exp(-i\omega_0\tau) \end{pmatrix}. \quad (7.47)$$

Expression (7.47) is a type of *phase gate*, but it is evident that a two-qubit **CNOT** gate, whose matrix representation is

$$\begin{pmatrix} 1 & 0 & 0 & 0 \\ 0 & 1 & 0 & 0 \\ 0 & 0 & 0 & 1 \\ 0 & 0 & 1 & 0 \end{pmatrix}, \quad (7.48)$$

cannot be constructed from a pair of non-interacting qubits.

7.3 Quantum Mechanics of a Free Rotor: A Poor Person's Atomic Model

Before introducing an appropriate Hamiltonian for the latter, we note the following identity

$$\mathbf{CNOT} = (\mathbb{1} \otimes \mathbf{H}) \, \mathbf{W} \, (\mathbb{1} \otimes \mathbf{H})$$

$$\mathbf{W} \equiv \begin{pmatrix} 1 & 0 & 0 & 0 \\ 0 & 1 & 0 & 0 \\ 0 & 0 & 1 & 0 \\ 0 & 0 & 0 & -1 \end{pmatrix} \quad (7.49)$$

where **H** is a single qubit Hadamard gate.

In the previous section, we showed how to create Hadamard gates via the application of laser pulses. To realize the **CNOT** gate we also need the two-qubit phase gate **W**. One way of accomplishing the latter is by adding to $\mathbf{h}_0$ a two-qubit interaction term

$$\mathbf{H}_I = \hbar g/2 \Big((\mathbb{1} + \sigma_Z) \otimes (\mathbb{1} + \sigma_Z) \Big) - \hbar g \, \sigma_Z \otimes \sigma_Z = \hbar g \, \mathbf{W} \quad (7.50)$$

where g is a coupling constant. The time development operator for this pair of interacting qubits,

$$\mathbf{U}(\tau) = \exp(-i \, (\mathbf{h}_0 + \mathbf{H}_I) \tau / \hbar) = \begin{pmatrix} \exp(-i(g-\omega_0)\tau) & 0 & 0 & 0 \\ 0 & \exp(-ig\tau) & 0 & 0 \\ 0 & 0 & \exp(-ig\tau) & 0 \\ 0 & 0 & 0 & \exp(i(g-\omega_0)\tau) \end{pmatrix}. \quad (7.51)$$

With the coupling constant set to the value $g = \omega_0/4$

$$\mathbf{U}(\tau = 2\pi/\omega_0) = -i \, \mathbf{W}. \quad (7.52)$$

Identity (7.52) allows us to construct a **CNOT** gate as follows. First, set $g = 0$ and subject the second qubit to a laser pulse that generates a Hadamard gate. After being processed by the Hadamard gate, shut the laser off and set $g = \omega_0/4$ for a duration of $\tau = 2\pi/\omega_0$. Now adjust $g = 0$ again and repeat the Hadamard operation on the second qubit. The cumulative result from this series of pulses is the **CNOT** gate (up to an overall phase).

That scenario requires an adjustable two-qubit interaction of the form given by (7.50) but, at this point in our narrative, we have no justification for this two-qubit interaction. *Ignacio Cirac* and *Peter Zoller* [3] first showed how a series of single-qubit laser pulses applied on interacting ions generates the **W** gate and as a consequence, allows realization of the **CNOT** gate.

7.4 The Cirac-Zoller Mechanism

> Mathematica Notebook 7.3: Small vibrations and simple-harmonic motion for two interacting ions. http://www.physics.unlv.edu/%7Ebernard/MATH_book/Chap3/chap3_link.html; See also https://bernardzygelman.github.io

Consider two-rotors, situated side-by-side, in the plane of this page, that are allowed to move along the horizontal axis. Including the kinetic energy of the internal motion, their dynamics are governed by the Hamiltonian

$$h_0 = \frac{p_a^2}{2M} + \frac{p_b^2}{2M} + v(|R_a - R_b|) + \frac{p_{\theta_a}^2}{2mL^2} + \frac{p_{\theta_b}^2}{2mL^2}. \tag{7.53}$$

where p_a, p_b are the momenta of the rotor cores a, b respectively. We allow a rotor-rotor interaction energy $v(|R_a - R_b|)$ that depends only on the relative separation between them. In a trapped ion set-up $v(|R_a - R_b|)$ is a consequence of the repulsive Coulomb force between the positively charged ions. The repulsive force is typically balanced by a trap potential (not shown here) so that the ions assume an equilibrium position at values of $R_a(0), R_b(0)$. Instead of coordinates R_a, R_b it is convenient to define a new set of coordinates

$$R = \frac{R_a + R_b}{2}$$
$$u = R_a - R_b, \tag{7.54}$$

and the corresponding conjugate momenta p_R, p_u. Noting that

$$p_b = p_R/2 - p_u$$
$$p_a = p_R/2 + p_u \tag{7.55}$$

we obtain

$$h_0 = \frac{p_R^2}{4M} + \frac{p_u^2}{2\mu} + V(|u|) + \frac{p_{\theta_a}^2}{2mL^2} + \frac{p_{\theta_b}^2}{2mL^2} \tag{7.56}$$

by replacing R_a, R_b, p_a, p_b in (7.53) with (7.54) and (7.55). $\mu = M/2$ is the reduced mass for the relative motion, R represents the center of mass (CM), or bulk, motion of the rotors and coordinate u their relative motion. We ignore the CM motion[2] and focus our attention on the Hamiltonian that governs the relative motion of the two rotors. We assume that this motion undergoes small deviations from the equilibrium

---

[2] In the original Cirac-Zoller protocol [3], the CM motion is the medium that negotiates ion-ion interactions.

7.4 The Cirac-Zoller Mechanism

value $u_0 \equiv R_a(0) - R_b(0)$. In this approximation

$$V = V(u_0) + \frac{\partial V}{\partial u}\bigg|_{u_0}(u - u_0) + \frac{\partial^2 V}{\partial u^2}\bigg|_{u_0}(u - u_0)^2/2 + \cdots \quad (7.57)$$

By definition, at the equilibrium position u_0, $\partial V/\partial u$ vanishes and so

$$V(z) - V(u_0) \approx \frac{k}{2}z^2 \quad k \equiv \frac{\partial^2 V}{\partial u^2}\bigg|_{u_0} \quad z \equiv u - u_0 \quad (7.58)$$

Using (7.58), a coordinate transformation from u to $z = u - u_0$, and neglecting the constant $V(u_0)$, we obtain Hamiltonian

$$H_{SHO} \equiv \frac{p_z^2}{2\mu} + \frac{k}{2}z^2, \quad (7.59)$$

which describes the relative motion of the two rotors. The presence of the potential energy $kz^2/2$ term results in a restoring force

$$F = -kz$$

that localizes the relative motion of the ion pair about their mutual equilibrium positions. This type of motion is called *simple harmonic motion* and plays an important role in many areas of physics.

7.4.1 Quantum Theory of Simple Harmonic Motion

A quantum theory of simple harmonic motion (SHO) follows by employing the prescription used to quantize the free rotor. Classical variable z and its conjugate momentum p_z are replaced by non-commuting operators,

$$z \Rightarrow \mathbf{Q} \quad p_z \Rightarrow \mathbf{P}$$

so that

$$\mathbf{H}_{SHO} = \frac{\mathbf{P}^2}{2\mu} + \frac{k}{2}\mathbf{Q}^2 \quad (7.60)$$

where $[\mathbf{Q}, \mathbf{P}] = i\hbar \mathbf{1}$. Instead of $\mathbf{P}, \mathbf{Q}$, it is desirable to define new quantum operators

$$\mathbf{a} \equiv \sqrt{\frac{\mu \nu}{2\hbar}}\left(\mathbf{Q} + \frac{i}{\mu \nu}\mathbf{P}\right)$$

$$\mathbf{a}^\dagger \equiv \sqrt{\frac{\mu \nu}{2\hbar}}\left(\mathbf{Q} - \frac{i}{\mu \nu}\mathbf{P}\right)$$

$$\nu \equiv \sqrt{\frac{k}{\mu}}. \quad (7.61)$$

Using the commutation relations for **P**, **Q**, we find that

$$[\mathbf{a}, \mathbf{a}^\dagger] = 1 \quad [\mathbf{a}, \mathbf{a}] = [\mathbf{a}^\dagger, \mathbf{a}^\dagger] = 0$$
$$\mathbf{H}_{SHO} = \hbar\nu\left(\mathbf{a}^\dagger\mathbf{a} + \frac{1}{2}\right). \tag{7.62}$$

The last identity is obtained by expressing **P**, **Q** in terms of **a** and $\mathbf{a}^\dagger$ in expression (7.60). We now seek the eigenstates and eigenvalues of $\mathbf{H}_{SHO}$. Suppose there is a state $|\emptyset\rangle$, commonly called the *vacuum state*, that has the property

$$\mathbf{a}|\emptyset\rangle = 0. \tag{7.63}$$

I claim that $|\emptyset\rangle$ is an eigenstate of $\mathbf{H}_{SHO}$ with eigenvalue $\frac{1}{2}\hbar\nu$, and that $|\nu\rangle \equiv \mathbf{a}^\dagger|\emptyset\rangle$ is an eigenstate with eigenvalue $\frac{3}{2}\hbar\nu$.

Proof Operate $\mathbf{H}_{SHO}$ on the vacuum state to get

$$\mathbf{H}_{SHO}|\emptyset\rangle = \hbar\nu\left(\mathbf{a}^\dagger\mathbf{a} + \frac{1}{2}\right)|\emptyset\rangle = \frac{1}{2}\hbar\nu|\emptyset\rangle \tag{7.64}$$

where I used the fact that $\mathbf{a}|\emptyset\rangle = 0$. Likewise

$$\mathbf{H}_{SHO}|\nu\rangle = \hbar\nu\left(\mathbf{a}^\dagger\mathbf{a} + \frac{1}{2}\right)\mathbf{a}^\dagger|\emptyset\rangle$$
$$= \hbar\nu\left(\mathbf{a}^\dagger\mathbf{a}\mathbf{a}^\dagger|\emptyset\rangle\right) + \frac{1}{2}\hbar\nu|\nu\rangle. \tag{7.65}$$

The term $\mathbf{a}\mathbf{a}^\dagger$ in the second line can be re-written $\mathbf{a}\mathbf{a}^\dagger = [\mathbf{a}, \mathbf{a}^\dagger] + \mathbf{a}^\dagger\mathbf{a} = \mathbf{a}^\dagger\mathbf{a} + 1$, therefore

$$\mathbf{a}\mathbf{a}^\dagger|\emptyset\rangle = (\mathbf{a}^\dagger\mathbf{a} + 1)|\emptyset\rangle = |\emptyset\rangle,$$

and

$$\hbar\nu\left(\mathbf{a}^\dagger\mathbf{a} + \frac{1}{2}\right)\mathbf{a}^\dagger|\emptyset\rangle = \hbar\nu\mathbf{a}^\dagger|\emptyset\rangle + \frac{1}{2}\hbar\nu|\nu\rangle = \frac{3}{2}\hbar\nu|\nu\rangle, \tag{7.66}$$

or

$$\mathbf{H}_{SHO}|\nu\rangle = \frac{3}{2}\hbar\nu|\nu\rangle. \tag{7.67}$$

This result generalizes [6] so that that $|2\nu\rangle \equiv \mathbf{a}^\dagger|\nu\rangle = \mathbf{a}^\dagger\mathbf{a}^\dagger|\emptyset\rangle$, or $(\mathbf{a}^\dagger)^2|\emptyset\rangle$ is an eigenstate of $\mathbf{H}_{SHO}$ with eigenvalue $\frac{5}{2}\hbar\nu$, and in general

$$|n\nu\rangle \equiv (\mathbf{a}^\dagger)^n|\emptyset\rangle \tag{7.68}$$

is an eigenstate with eigenvalue $(2n+1)/2\,\hbar\nu$ where $n = 0, 1, 2\ldots$. The energy spectrum of $\mathbf{H}_{SHO}$ constitutes an infinite tower of levels, of constant $\hbar\nu$ increment, starting from the lowest vacuum energy $\hbar\nu/2$. Because states of higher energy are

7.4 The Cirac-Zoller Mechanism

generated by the repeated application $\mathbf{a}^\dagger$ on the vacuum state, the latter are called *creation*, or raising operators. Its adjoint $\mathbf{a}$ is called a *destruction* or lowering operator. The above analysis reveals that the eigenvalues of the *number operator*

$$\mathbf{N} \equiv \mathbf{a}^\dagger \mathbf{a} \qquad (7.69)$$

are the integers $0, 1, 2 \ldots$. The normalized eigenstates of $\mathbf{H}_{SHO}$ and $\mathbf{N}$ are [6]

$$|nv\rangle = \frac{(\mathbf{a}^\dagger)^n}{\sqrt{n!}} |\emptyset\rangle \qquad (7.70)$$

and have the properties

$$\mathbf{a}^\dagger |nv\rangle = \sqrt{n+1}|(n+1)v\rangle$$
$$\mathbf{a}|nv\rangle = \sqrt{n}|(n-1)v\rangle$$
$$\langle nv|mv\rangle = \delta_{mn} \qquad (7.71)$$

where $n, m = 0, 1, 2 \ldots$

7.4.2 A Phonon—Qubit Pair Hamiltonian

According to the above analysis the quantum Hamiltonian for two interacting rotors is

$$\mathbf{H}_0 = \mathbf{h}_0 + \mathbf{H}_{SHO} = -\frac{\hbar\omega_0}{2}\sigma_Z \otimes \mathbb{1} - \frac{\hbar\omega_0}{2}\mathbb{1} \otimes \sigma_Z + \hbar v\, \mathbf{a}^\dagger \mathbf{a}, \qquad (7.72)$$

where we have ignored an overall constant $\hbar v/2$. Here $\mathbf{h}_0$ represents the internal motion of the rotors and $\mathbf{H}_{SHO}$ describes the relative motion of the rotor pair. The Hilbert space in which $\mathbf{H}_0$ resides, is a direct product of the two-qubit Hilbert space for the internal rotor motion, with that of an infinite dimensional Hilbert space spanned by the eigenstates of $\mathbf{H}_{SHO}$. The states $|nv\rangle$ represent quantized excitations of simple-harmonic motion and are called *single-mode phonon states*. So the vacuum $|\emptyset\rangle$ is a state containing no phonons, $|v\rangle$ is a state containing one phonon, $|2v\rangle$ contains two phonons, and so on. A typical eigenstate for $\mathbf{H}_0$ might look like

$$|nv\rangle \otimes |0\rangle \otimes |0\rangle$$

with eigenvalue $-\hbar\omega_0/2 - \hbar\omega_0/2 + \hbar n v$.

Previously, we truncated the rotor Hilbert space to a two-dimensional qubit subspace, and here we do the same in the phonon Hilbert space. That is, we truncate the latter to a sub-space spanned by the vectors $|\emptyset\rangle$, $|v\rangle$. The direct product of these states with the four-dimensional two-qubit rotor Hilbert space is spanned by the following basis vectors,

$$|k\rangle_3 \equiv |n\nu\rangle \otimes |\psi\rangle \tag{7.73}$$

where $|\psi\rangle$ are the two-qubit basis vectors for the internal motion of the two rotors, and the phonon basis states $|n\nu\rangle \equiv |0\rangle, |1\rangle$, for $n = 0, 1$ respectively. Explicitly

$$|0\rangle_3 = |000\rangle \quad |1\rangle_3 = |001\rangle \quad |2\rangle_3 = |010\rangle \quad |3\rangle_3 = |011\rangle$$
$$|4\rangle_3 = |100\rangle \quad |5\rangle_3 = |101\rangle \quad |6\rangle_3 = |110\rangle \quad |7\rangle_3 = |111\rangle. \tag{7.74}$$

It turns out that the *Cirac-Zoller gate* requires an additional quantum state. We label it

$$|\emptyset\rangle \otimes |\phi\rangle \equiv |8\rangle \tag{7.75}$$

where $|\phi\rangle = |0\rangle \otimes |2\rangle$ is a two-qubit state in which rotor (a) is in state $|0\rangle$ and rotor (b) in state $|2\rangle$ (beyond, the two-state approximation). Assuming that the energy eigenvalue for state $|2\rangle$ is $\hbar(\omega_1 + \omega_0/2)$, the total energy for state $|8\rangle$ is $\hbar\omega_1$. Joining state $|8\rangle$ with set (7.74) leads to a nine-dimensional Hilbert space. In it, the matrix representation of $\mathbf{H}_0$ is

$$\begin{pmatrix}
-\hbar\omega_0 & 0 & 0 & 0 & 0 & 0 & 0 & 0 & 0 \\
0 & 0 & 0 & 0 & 0 & 0 & 0 & 0 & 0 \\
0 & 0 & 0 & 0 & 0 & 0 & 0 & 0 & 0 \\
0 & 0 & 0 & \hbar\omega_0 & 0 & 0 & 0 & 0 & 0 \\
0 & 0 & 0 & 0 & \hbar\nu - \hbar\omega_0 & 0 & 0 & 0 & 0 \\
0 & 0 & 0 & 0 & 0 & \hbar\nu & 0 & 0 & 0 \\
0 & 0 & 0 & 0 & 0 & 0 & \hbar\nu & 0 & 0 \\
0 & 0 & 0 & 0 & 0 & 0 & 0 & \hbar\nu + \hbar\omega_0 & 0 \\
0 & 0 & 0 & 0 & 0 & 0 & 0 & 0 & \hbar\omega_1
\end{pmatrix}. \tag{7.76}$$

The upper left hand block in (7.76), deliminated by the dashed lines, represents a sub-space in which the two qubits form direct products with phonon state $|0\nu\rangle$. The middle block represents the sub-space in which the qubits form products with state $|1\nu\rangle$.

7.4.3 Light-Induced Rotor-Phonon Interactions

We now investigate the effect of a collimated laser beam on a rotor electron. Suppose a plane polarized beam is incident on rotor *a* only. In the vicinity of the rotor's

7.4 The Cirac-Zoller Mechanism

electron, the electric field is given by

$$\vec{E}(t) = E_a \exp(i\omega t) \exp(i\,\delta_a) \exp(i\vec{k}\cdot\vec{r}_a)\,\hat{\boldsymbol{\varepsilon}} + h.c.$$
$$\vec{k} = \hat{\mathbf{i}}\,k_x + \hat{\mathbf{j}}\,k_y \quad \vec{k}\cdot\hat{\boldsymbol{\varepsilon}} = 0 \tag{7.77}$$

Here h.c. stands for complex conjugate, E_a is a real constant, $\vec{k}$ determines the direction of propagation of the laser beam, and $\hat{\boldsymbol{\varepsilon}}$ is a unit vector in the plane of the page but perpendicular to $\vec{k}$. Because $\vec{k}$ is also in the plane of this page, we include the modulating factor $\exp(i\vec{k}\cdot\vec{r}_a)$ as required by Maxwell's equations [5]. The $\vec{E}$-field-electron interaction introduces the term (Here we neglect any direct interactions with the charge of the rotor's core.)

$$H_I(a) = -e\,\vec{r}_a \cdot \vec{E}(t) \tag{7.78}$$

and for the geometry shown in Fig. 7.1 in which the translational coordinate for ion a is given by R_a.

$$\vec{r}_a = \hat{\mathbf{i}}\,R_a + \hat{\mathbf{i}}\,L\cos\theta_a + \hat{\mathbf{j}}\,L\sin\theta_a. \tag{7.79}$$

Assuming that $|\vec{k}\,L| \ll 1$ and keeping only the lowest order terms of the latter we find

$$H_I(a) \approx$$
$$2eE_a(\sin\phi(R_a + L\cos\theta_a) - \cos\phi L\sin\theta_a)\cos(\omega t + \delta_a + k_x R_a) \tag{7.80}$$

where ϕ is the angle between $\vec{k}$ and the x-axis. Using (7.54) to express R_a in terms of R, u, and constructing the matrix representation of (7.80) with respect to the internal rotor functions, we posit that

$$\underline{\mathbf{H}}_I(a) = \begin{pmatrix} 0 & 2\hbar\,\Delta_a \cos(\omega t + \delta_a + k_x\,\mathbf{u}/2) \\ 2\hbar\,\Delta_a \cos(\omega t + \delta_a + k_x\,\mathbf{u}/2) & 0 \end{pmatrix}.$$

We have neglected coupling to the CM motion, elevated u to a quantum variable, defined $\hbar\,\Delta_a = e\,E_a \sin\phi\,L/2$, and ignored the diagonal elements to $\mathbf{H}_I(a)$ as they are a correction to $\mathbf{h}_0$. Therefore,

$$\mathbf{H}_I(a) = 2\hbar\,\Delta_a \cos(\omega t + \delta_a + k_x\mathbf{u}/2)\,\sigma_X(a) \tag{7.81}$$

where $\sigma_X(a)$ is the Pauli-X qubit operator for rotor (a). We use definitions (7.61) to express $\mathbf{u}$ in terms of the raising and lowering operators

$$\mathbf{u} = \sqrt{\frac{\hbar}{2\mu\nu}}\,(\mathbf{a} + \mathbf{a}^\dagger), \tag{7.82}$$

assume the factor $\eta \equiv \sqrt{k_x^2\hbar/8\mu\nu} < 1$ to get

$$\cos(\omega t + \delta_a + k_x\mathbf{u}/2) \approx \cos(\omega t + \delta_a) - \eta\sin(\omega t + \delta_a)(\mathbf{a} + \mathbf{a}^\dagger) + \mathcal{O}(\eta^2) + \cdots.$$

Up to terms linear in η

$$\mathbf{H}_I(a) = \mathbf{H}_I^{(1)}(a) + \mathbf{H}_I^{(2)}(a)$$
$$\mathbf{H}_I^{(1)}(a) = 2\hbar\Delta_a \cos(\omega t + \delta_a)\sigma_X(a)$$
$$\mathbf{H}_I^{(2)}(a) = -2\hbar\Delta_a \eta \sin(\omega t + \delta_a)\sigma_X(a)(\mathbf{a} + \mathbf{a}^\dagger). \qquad (7.83)$$

We construct the matrix representations of the phonon creation and destruction operators with respect to states $|\emptyset\rangle$, $|\nu\rangle$, so that

$$\underline{\mathbf{a}} = \begin{pmatrix} \langle\emptyset|\mathbf{a}|\emptyset\rangle & \langle\emptyset|\mathbf{a}|\nu\rangle \\ \langle\nu|\mathbf{a}|\emptyset\rangle & \langle\nu|\mathbf{a}|\nu\rangle \end{pmatrix} = \begin{pmatrix} 0 & 1 \\ 0 & 0 \end{pmatrix}$$
$$\underline{\mathbf{a}^\dagger} = \begin{pmatrix} \langle\emptyset|\mathbf{a}^\dagger|\emptyset\rangle & \langle\emptyset|\mathbf{a}^\dagger|\nu\rangle \\ \langle\nu|\mathbf{a}^\dagger|\emptyset\rangle & \langle\nu|\mathbf{a}^\dagger|\nu\rangle \end{pmatrix} = \begin{pmatrix} 0 & 0 \\ 1 & 0 \end{pmatrix}. \qquad (7.84)$$

In this representation $(\mathbf{a} + \mathbf{a}^\dagger) = \sigma_X(\nu)$, the Pauli-X operator in the Hilbert space of the *vibrational qubit*.

If the laser on rotor (a) is turned on, the total three-qubit interaction Hamiltonian has the form

$$\mathbf{H}_I(t) = \mathbb{1} \otimes \mathbf{H}_I^{(1)}(a) \otimes \mathbb{1} - 2\eta\hbar\Delta_a \sin(\omega t + \delta_a)\sigma_X(\nu) \otimes \sigma_X(a) \otimes \mathbb{1}. \quad (7.85)$$

The first term in expression (7.85) was discussed in the previous section where it was noted that only for the resonance condition $\hbar\omega = E_1 - E_0$, does it play a role. We adjust the laser frequency ω so that the latter is not satisfied and are justified in ignoring that term. Therefore

$$\mathbf{H}_I(t) \approx -2\eta\hbar\Delta_a \sin(\omega t + \delta_a)\sigma_X(\nu) \otimes \sigma_X(a) \otimes \mathbb{1}. \qquad (7.86)$$

or

$$i\eta\hbar\Delta_a \exp(i\delta_a)\exp(i\omega t) \times$$
$$\Big(\sigma_+(\nu)\sigma_+(a) + \sigma_+(\nu)\sigma_-(a) + \sigma_-(\nu)\sigma_+(a) + \sigma_-(\nu)\sigma_-(a)\Big) \otimes \mathbb{1}$$
$$-i\eta\hbar\Delta_a \exp(-i\delta_a)\exp(-i\omega t) \times$$
$$\Big(\sigma_+(\nu)\sigma_+(a) + \sigma_+(\nu)\sigma_-(a) + \sigma_-(\nu)\sigma_+(a) + \sigma_-(\nu)\sigma_-(a)\Big) \otimes \mathbb{1}.$$
$$(7.87)$$

For the sake of economy in notation, we ignored in (7.87) the implicit direct product operator between the vibrational and rotor qubits. In the interaction picture we are allowed the substitutions

$$\sigma_\pm(a) \to \exp(\pm i\omega_0 t)\sigma_\pm(a)$$
$$\sigma_\pm(b) \to \exp(\pm i\omega_0 t)\sigma_\pm(b)$$
$$\sigma_\pm(\nu) \to \exp(\pm i\nu t)\sigma_\pm(\nu). \qquad (7.88)$$

7.4 The Cirac-Zoller Mechanism

We impose the resonance condition

$$\omega = \omega_0 - \nu, \tag{7.89}$$

and using (7.88) construct the interaction picture phonon-qubit coupling

$$\mathbf{H}_I = i\eta\hbar\Delta_a\Big(\exp(i\delta_a)\sigma_+(\nu) \otimes \sigma_-(a) - \exp(-i\delta_a)\sigma_-(\nu) \otimes \sigma_+(a)\Big) \otimes \mathbb{1}. \tag{7.90}$$

In deriving (7.90) we took advantage of the RWA approximation by ignoring all terms that contain time dependent phases. For the interval $\tau_1 = t_1 - t_0$ the RWA solution to the *interaction picture* Schroedinger equation is

$$\psi_I(\tau_1) = \mathbf{U}^{(I)}(t_1, t_0)\psi_I(0)$$
$$\mathbf{U}^{(I)}(t_1, t_0) = \exp(-i\,\mathbf{H}_I\tau_1/\hbar) \tag{7.91}$$

or

$$\mathbf{U}^{(I)}(t_1, t_0) = \begin{pmatrix} 1 & 0 & 0 & 0 \\ 0 & \cos(\eta\Delta_a\tau_1) & -\exp(-i\delta_a)\sin(\eta\Delta_a\tau_1) & 0 \\ 0 & \exp(i\delta_a)\sin(\eta\Delta_a\tau_1) & \cos(\eta\Delta_a\tau_1) & 0 \\ 0 & 0 & 0 & 1 \end{pmatrix} \otimes \mathbb{1} \tag{7.92}$$

$\mathbf{U}^{(I)}$ describes unitary evolution in the three-qubit sub-space. In the nine-dimensional Hilbert space that includes state $|8\rangle$, the full unitary evolution matrix has the block form

$$\mathbf{U}_a^{(I)}(\tau_1) = \left(\begin{array}{c|c} \mathbf{U}^{(I)}(\tau_1) & 0 \\ \hline 0 & \mathbb{1} \end{array}\right). \tag{7.93}$$

In interval $\tau_2 = t_2 - t_1$, the pulse on rotor (a) is turned off (i.e. $\Delta_a = 0$) and a second pulse is applied on ion (b). The frequency, ω, of that pulse is chosen so that levels $|100\rangle = |4\rangle_3$, and $|002\rangle = |8\rangle$ are in resonance. In that case we assume that the interaction Hamiltonian, in the RWA approximation, has the form

$$(\mathbf{H}_I)_{ij} = i\,\exp(i\delta_b)\Delta_b\eta_b$$

for matrix indices $i = 5, j = 9$, and for $i = 9, j = 5$ its complex conjugate. The parameters $\Delta_b, \delta_b, \eta_b$ determine the strength and phase of this interaction. The remaining entries for $i, j \notin 5, 9$ are set to the null value. The unitary evolution matrix, $\exp(-i\,\mathbf{H}_I\,\tau_2)$ is given by,

$$\mathbf{U}_b^{(I)}(\tau_2) = \begin{pmatrix} 1 & 0 & 0 & 0 & 0 & 0 & 0 & 0 & 0 \\ 0 & 1 & 0 & 0 & 0 & 0 & 0 & 0 & 0 \\ 0 & 0 & 1 & 0 & 0 & 0 & 0 & 0 & 0 \\ 0 & 0 & 0 & 1 & 0 & 0 & 0 & 0 & 0 \\ \hline 0 & 0 & 0 & 0 & \cos(\Delta_b\eta_b\tau) & 0 & 0 & 0 & \sin(\Delta_b\eta_b\tau) \\ 0 & 0 & 0 & 0 & 0 & 1 & 0 & 0 & 0 \\ 0 & 0 & 0 & 0 & 0 & 0 & 1 & 0 & 0 \\ 0 & 0 & 0 & 0 & 0 & 0 & 0 & 1 & 0 \\ \hline 0 & 0 & 0 & 0 & -\sin(\Delta_b\eta_b\tau) & 0 & 0 & 0 & \cos(\Delta_b\eta_b\tau) \end{pmatrix} \quad (7.94)$$

where we set the parameter $\delta_b = 0$.

Consider a series of 3 pulses of the type described above. The first, with time evolution operator (7.93), is of duration $\tau_1 = \pi/2\eta\Delta_a$. Evaluating (7.93) we find

$$\mathbf{U}_a^{(I)}(\tau_1) = \begin{pmatrix} 1 & 0 & 0 & 0 & 0 & 0 & 0 & 0 & 0 \\ 0 & 1 & 0 & 0 & 0 & 0 & 0 & 0 & 0 \\ 0 & 0 & 0 & 0 & -\exp(-i\delta_a) & 0 & 0 & 0 & 0 \\ 0 & 0 & 0 & 0 & 0 & -\exp(-i\delta_a) & 0 & 0 & 0 \\ \hline 0 & 0 & \exp(i\delta_a) & 0 & 0 & 0 & 0 & 0 & 0 \\ 0 & 0 & 0 & \exp(i\delta_a) & 0 & 0 & 0 & 0 & 0 \\ 0 & 0 & 0 & 0 & 0 & 0 & 1 & 0 & 0 \\ 0 & 0 & 0 & 0 & 0 & 0 & 0 & 1 & 0 \\ \hline 0 & 0 & 0 & 0 & 0 & 0 & 0 & 0 & 1 \end{pmatrix} \quad (7.95)$$

Subsequently to it, a pulse of duration $\tau_2 = \pi/\eta_b\Delta_b$ on ion b described by gate (7.94) or

$$\begin{pmatrix} 1 & 0 & 0 & 0 & 0 & 0 & 0 & 0 & 0 \\ 0 & 1 & 0 & 0 & 0 & 0 & 0 & 0 & 0 \\ 0 & 0 & 1 & 0 & 0 & 0 & 0 & 0 & 0 \\ 0 & 0 & 0 & 1 & 0 & 0 & 0 & 0 & 0 \\ \hline 0 & 0 & 0 & 0 & -1 & 0 & 0 & 0 & 0 \\ 0 & 0 & 0 & 0 & 0 & 1 & 0 & 0 & 0 \\ 0 & 0 & 0 & 0 & 0 & 0 & 1 & 0 & 0 \\ 0 & 0 & 0 & 0 & 0 & 0 & 0 & 1 & 0 \\ \hline 0 & 0 & 0 & 0 & 0 & 0 & 0 & 0 & -1 \end{pmatrix}. \quad (7.96)$$

7.4 The Cirac-Zoller Mechanism

> Mathematica Notebook 7.4 The Cirac-Zoller Mechanism. http://www.physics.unlv.edu/%7Ebernard/MATH_book/Chap7/chap7_link.html; See also https://bernardzygelman.github.io

Finally, a pulse with evolution operator (7.95) is again impressed on ion a. The trio of pulses leads to

$$U^{(I)}(t_f) = U_a^{(I)}(\tau_1) U_b^{(I)}(\tau_2) U_a^{(I)}(\tau_1) =$$

$$\left(\begin{array}{cccc|cccc} 1 & 0 & 0 & 0 & 0 & 0 & 0 & 0 \\ 0 & 1 & 0 & 0 & 0 & 0 & 0 & 0 \\ 0 & 0 & 1 & 0 & 0 & 0 & 0 & 0 \\ 0 & 0 & 0 & -1 & 0 & 0 & 0 & 0 \\ \hline 0 & 0 & 0 & 0 & -1 & 0 & 0 & 0 \\ 0 & 0 & 0 & 0 & 0 & -1 & 0 & 0 \\ 0 & 0 & 0 & 0 & 0 & 0 & 1 & 0 \\ 0 & 0 & 0 & 0 & 0 & 0 & 0 & 1 \\ 0 & 0 & 0 & 0 & 0 & 0 & 0 & -1 \end{array} \right). \quad (7.97)$$

where $t_f = \tau_1 + \tau_2 + \tau_1$. Inspection of the upper left hand block in gate (7.97) shows that the subspace spanned by the direct product of vibrational state $|\emptyset\rangle$ with rotor states $|00\rangle, |01\rangle, |10\rangle, |11\rangle$, is acted on by a control-phase gate $\mathbf{W}$. The gates were constructed in the interaction picture and so from the definition of the interaction picture

$$\mathbf{U}(t_f, t_0) = \exp(-i\,\mathbf{H}_0\,t_f) \mathbf{U}^{(I)}(t_f, t_0) \exp(i\,\mathbf{H}_0\,t_0) \quad (7.98)$$

where $\mathbf{U}(t_f, t_0)$ is the time development operator in the space spanned by the computational basis. Therefore, assuming $t_0 = 0$, the series of pulses result in the computational basis gate

$$\exp(-i\,\mathbf{H}_0\,t_f) \mathbf{U}^{(I)}(t_f), \quad (7.99)$$

and it is necessary to compensate for this phase shift by allowing the system to evolve for an additional time period, from t_f to t_1' so that

$$\mathbf{U}(t_1', 0) = \exp(-i\,\mathbf{H}_0(t_1' - t_f)) \exp(-i\,\mathbf{H}_0 t_f) \mathbf{U}^{(I)}(t_f) =$$
$$\exp(-i\mathbf{H}_0\,t_1') \mathbf{U}^{(I)}(t_f) \quad (7.100)$$

provided that $t_1' > t_f$. If $t_1' = 2n\pi/\omega_0$, where n is an integer, operator $\exp(-i\mathbf{H}_0 t_1')$ is the identity operator in the phonon vacuum subspace. Thus, in this subspace, we realize a control-phase gate $\mathbf{W}$ for the computational basis states. With it and the action of a pair of Hadamard gates, a **CNOT** gate for the two rotors is realized.

7.5 Trapped Ion Qubits

Because of our limited command of the quantum mechanics of real atoms, we introduced and relied on a rotor model to describe qubits interacting with each other and with external radiation fields. With that model, we proceeded to construct the Hadamard and **CNOT** gates. The latter, with phase gates, serve as *universal gates* from which we are able to build general logic gates [8]. Despite its simplicity, the rotor model incorporates many essential properties of laboratory ion-qubits. For example, it features discrete levels whose energy defects (i.e. energy difference between two adjacent levels), are not uniform. This property allows the rotor to behave like a two-level system, or qubit. If the energy spacings were constant, as in a simple harmonic oscillator, a laser pulse of resonant frequency would be able to access all levels simultaneously thereby destroying the binary character of the proposed qubit system.

However, any resemblance between the rotor model and real atoms/ions end there. Atoms and ions contain states that are not stable, as they decay by a process called *spontaneous emission*. In that event, an atom in a state with energy E_b cascades to a state of lower energy $E_a < E_b$. Energy is conserved by the emission of a photon with energy $\Delta E = E_b - E_a$. Spontaneous emission results in unwanted heating and is a leading cause of decoherence. On the other hand, photon emission via cascade does have its advantages, as it provides a mechanism that allows qubit interrogation. For example, Fig. 7.3 illustrates the energy spectrum for the first few levels of the Ca^+ optical qubit. The solid lines identify the qubit states. They are metastable states, which means that spontaneous emission is so slow that it does not hinder the coherence of the qubit during a gate operation. However nearby levels spontaneously emit photons readily. So if those states are populated, they emit and scatter photons which can be monitored by a detection device. If the qubit is populated in its $|1\rangle$ (the upper level in the diagram) state, a laser beam with a frequency in resonance with the energy of the unstable level excites the latter and triggers photon detectors. We then know that the qubit must have been in state $|1\rangle$. If no photons are detected, the qubit must have been in state $|0\rangle$ as the frequency of the applied field is not in resonance with the energy defect between $|0\rangle$ and the unstable state.

Unlike a rotor, atoms/ions possess degrees of freedom that lead to a complicated quantized energy structure. With decades of laboratory efforts and advances, physicists have learned to control an ion's environment in such a way that it behaves like a two-level system, or qubit. Ion-traps take advantage of the Coulomb repulsion of individual ions in competition with a trapping potential so that an equilibrium configuration is reached. For sufficient jostling motion or kinetic energy, ions overcome the trapping potential and exit the trap. Since temperature is a measure of average kinetic energy, this loss mechanism is avoided by employing laser cooling techniques to "freeze" the ions in place. The laser-cooled ions not only prevent trap loss but inhibit excitation of unwanted internal energy levels of the ion, thereby constraining the ion to navigate qubit Hilbert space. Both laser cooling and ion trapping technologies allow laboratory realizations of ion crystals. In a one-dimensional crystal, ions are separated in space on the order of several microns. Though the ions are

7.5 Trapped Ion Qubits

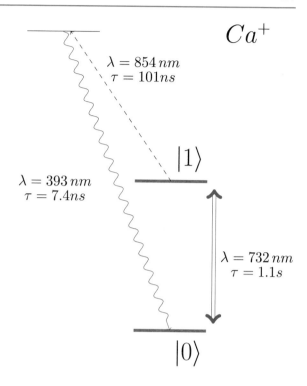

Fig. 7.3 The Ca$^+$ optical qubit. The $|1\rangle$ qubit state is metastable having a lifetime of 1.1 s. Nearby states undergo rapid spontaneous photon emission which can be monitored for qubit readout purposes

"frozen" they can exhibit bulk motion about their equilibrium position. Phonons are quantized excitations of that motion and, in an ion trap, serve as a "phonon bus", the medium by which two ions communicate. The Cirac-Zoller mechanism exploits ion-phonon coupling thus allowing multi-qubit gates. Once the ions have been trapped and cooled one should, according to the DiVincenzo criteria, be able to individually address individual or pairs of ions for the purpose of gate operations, and qubit read-out capabilities. Present day laser technology allows beams that are only on the order of a few microns (10^{-6} m) in diameter. The trapping potential is adjusted so that inter-ion spacings accommodate laser addressing of single ions.

The magnitude of the level energy separation or defect, defines two classes of qubits. In the Ca$^+$ ion, the energy separation ΔE of the qubit states is on the order of a 1.6 electron-volts (about 10^{-19} J). The Einstein energy relation $\Delta E = \hbar\omega$ relates, through Planck's constant $\hbar$, an angular frequency ω associated with this energy defect. So ΔE for the Ca$^+$ corresponds to frequencies on order of 10^{15} Hz, a frequency in the optical region of the electromagnetic spectrum; hence the term *optical qubit*.

In addition to the strong Coulomb interaction between electrons and the nucleus of an atom, the spin of the electron interacts with internal magnetic fields produced by the spin of the nucleus. Magnetic interactions, also called hyperfine interactions, are much weaker than Coulomb interactions between electrons and nucleus and so the corresponding energy splittings are much smaller. A pair of states split by hyperfine interactions are called *hyperfine qubits*. Energy defects in hyperfine qubits are on the order of 10^{10} Hz, about 100,000 times smaller than the frequencies associated with optical qubits. Hyperfine qubits are typically driven and accessed by *microwave radiation*. They do not succumb as readily to spontaneous emission but are sensitive to stray magnetic fields.

> Mathematica Notebook 7.5 The Paul ion trap. http://www.physics.unlv.edu/%7Ebernard/MATH_book/Chap7/chap7_link.html; See also https://bernardzygelman.github.io; See also https://bernardzygelman.github.io

> Mathematica Notebook 7.6 Doppler Cooling. http://www.physics.unlv.edu/%7Ebernard/MATH_book/Chap7/chap7_link.html; See also https://bernardzygelman.github.io

Because ions are separated on the order of a few microns, each can be individually addressed by a laser beam. The expression

$$H_{ion} = -\sum_i^N \frac{\hbar \omega_0}{2} (\sigma_Z)_i, \qquad (7.101)$$

where N is the number of ions in the crystal, is the effective ion internal Hamiltonian for the configuration. In this notation, operator $(\sigma_Z)_i$ represents the action of σ_Z on qubit i, with all other qubits acted on by the identity so that

$$\begin{aligned}(\sigma_Z)_1 &\equiv \mathbb{1}_N \otimes \cdots \otimes \sigma_Z \\ (\sigma_Z)_2 &\equiv \mathbb{1}_N \otimes \cdots \otimes \sigma_Z \otimes \mathbb{1}_1 \\ (\sigma_Z)_3 &\equiv \mathbb{1}_N \otimes \cdots \otimes \sigma_Z \otimes \mathbb{1}_2 \otimes \mathbb{1}_1\end{aligned} \qquad (7.102)$$

and so on.

In the Yb$^+$ system the qubit levels are separated by a frequency of ≈ 12.64 GHz ($1G = 10^9$). *Magnetic dipole radiation* can be used to drive transitions between these states, instead *Stimulated Raman Excitation (SRE)* is the preferred method for inducing Rabi-like flopping in this hyperfine qubit. Though a rigorous discussion of the physics behind SRE is beyond the scope of this text, for our purposes it is sufficient to think of *SRE* as the application of two coherent laser beams of frequencies ω_1, ω_2 (see Fig. 7.4) on the ion. The difference in frequency of the lasers, or the beat frequency,

$$\omega = \omega_1 - \omega_2 \quad \omega_i \equiv k_i c$$

7.5 Trapped Ion Qubits

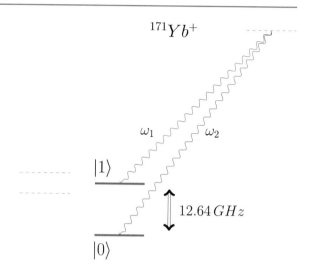

Fig. 7.4 The Yb$^+$ hyperfine qubit. The curly lines represent the Raman laser beams and the two dashed lines near state $|1\rangle$ are additional hyperfine levels shifted by an external magnetic field. The super-script 171 represents the atomic number (the total number of neutrons and protons in the nucleus) of that isotope

is chosen so that ω is nearly resonance with the energy difference $\Delta E = \hbar \omega_0$ between the levels of the hyperfine qubit. It is useful to define a de-tuning parameter

$$\delta \equiv \omega - \omega_0,$$

and the wave number parameter $k = |\vec{k}_1 - \vec{k}_2|$ where $\vec{k}_i, i = 1, 2$ are the propagation vectors for beams 1 and 2 respectively. With an appropriate choice of $\vec{k}_1, \vec{k}_2$, the ion experiences (in the interaction picture) an interaction Hamiltonian of the form [9]

$$\mathbf{H}_I(t) = \hbar \Omega \, \sigma_X(t) \cos(k \, \mathbf{x}(t) - \omega t + \phi)$$
$$\sigma_X(t) = \sigma_+ \exp(i\omega_0 t) + \sigma_- \exp(-i\omega_0 t) \quad (7.103)$$

where

$$\Omega \equiv \frac{g_1 g_2}{\Delta}$$

is an effective Rabi-frequency proportional to the Rabi couplings g_1, g_2 and Δ is a de-tuning parameter, of the two Raman components.

$$\mathbf{x}(t) = \sqrt{\frac{\hbar}{2m\nu}} (\mathbf{a} \exp(-i\nu t) + \mathbf{a}^\dagger \exp(i\nu t))$$

is a position operator, and ϕ a phase constant. The position operator is parameterized by the mass m of the ion and ν the vibrational frequency[3] (for the center of mass motion) of the ions in the trap. Expression (7.103) can be expanded in powers of the

---

[3] For the sake of simplicity, we assume coupling to only one vibrational mode.

Lamb-Dicke parameter

$$\eta \equiv \sqrt{\frac{\hbar k^2}{2m\nu}} < 1,$$

provided that the inequality holds. For detuning $\delta = 0$ the leading order term, in the RWA approximation, of that expansion is

$$\mathbf{H}_C = \frac{\hbar \Omega}{2}(\sigma_+ \exp(i\phi) + \sigma_- \exp(-i\phi)). \tag{7.104}$$

Hamiltonian (7.104) induces Rabi transitions between the two-levels of the hyperfine qubit and, by convention, is called the *carrier transition*. In the case where the detuning is toward the red side of the *carrier frequency*, that is $\omega - \omega_0 = -\delta$ ($\delta > 0$), and fixing the phase ϕ to the value $-\pi/2$, the term in the expansion proportional to η, and that survives in a RWA approximation, is

$$\mathbf{H}_R = \frac{\hbar \Omega \eta}{2} \left(\mathbf{a} \exp(-i\nu t) + \mathbf{a}^\dagger \exp(i\nu t) \right) (\sigma_+ \exp(i\delta t) + \sigma_- \exp(-i\delta t)). \tag{7.105}$$

On resonance with $\delta = \nu$, we find

$$\mathbf{H}_R = \frac{\hbar \Omega \eta}{2} \left(\sigma_+ \mathbf{a} + \sigma_- \mathbf{a}^\dagger \right). \tag{7.106}$$

> Mathematica Notebook 7.7 Mølmer-Sørenson Coupling. http://www.physics.unlv.edu/%7Ebernard/MATH_book/Chap7/chap7_link.html; See also https://bernardzygelman.github.io

Because Hamiltonian (7.106) contains harmonic oscillator creation and destruction operators, it couples qubit rotations with excitations of the vibrational degrees of freedom of the ion. Hamiltonian (7.106) is the interaction term in the *Jaynes-Cummings model*, which we discuss in more detail, and in a different context, in the next chapter. Analogously, when the laser beat frequency is shifted to the blue side of the carrier frequency so that $\omega - \omega_0 = \delta$, the leading order term in η of (7.103) is

$$\mathbf{H}_B = \frac{\hbar \Omega \eta}{2} \left(\mathbf{a} \exp(-i\nu t) + \mathbf{a}^\dagger \exp(i\nu t) \right) (\sigma_+ \exp(-i\delta t) + \sigma_- \exp(i\delta t)). \tag{7.107}$$

At resonance $\delta = \nu$ it reduces to

$$\mathbf{H}_B = \frac{\hbar \Omega \eta}{2} \left(\sigma_+ \mathbf{a}^\dagger + \sigma_- \mathbf{a} \right), \tag{7.108}$$

an example of an *anti-Jaynes-Cummings Hamiltonian*. $\mathbf{H}_B$ induces so-called *blue-sideband transitions*. With pulses of the carrier, blue and *red-sideband*, gate

7.5 Trapped Ion Qubits

operations similar to those illustrated in our rotor model are possible. For example, $\mathbf{H}_R$ induces transitions from the ion excited state $|1\rangle$, in the $n = 0$ phonon bus state, into the ground $|0\rangle$ ion state and $n = 1$ phonon bus state, i.e. $|1\rangle \otimes |\emptyset\rangle \to |0\rangle \otimes |\nu\rangle$. Because $\mathbf{H}_R |0\rangle \otimes |\emptyset\rangle = 0$, a red-detuned pulse drives transitions from "spin-up" (the excited ion state) to "spin down" (the ground ion state), but not the other way around. If the ion is in a superposition state

$$(\alpha|0\rangle + \beta|1\rangle) \otimes |\emptyset\rangle,$$

a red detuned pulse can induce the mapping

$$(\alpha|0\rangle + \beta|1\rangle) \otimes |\emptyset\rangle \to |0\rangle \otimes (\alpha|\emptyset\rangle + \beta|1\rangle), \tag{7.109}$$

or the ion superposition state is mapped into a superposition of phonon bus states. This scenario features conditional dynamics keeping in the spirit with the Cirac-Zoller mechanism. Additional pulses can alter, retrieve and map the quantum state of the phonon bus into different ions [2].

7.5.1 Mølmer-Sørenson Coupling

In Sect. 7.4, we employed the Cirac-Zoller mechanism to construct a multi-qubit gate. More recent laboratory efforts have explored alternative methods built upon this foundation. One of the more popular is the Mølmer-Sørenson (MS) procedure. In it, two pairs of lasers are employed that are detuned to, but not in resonance with, the red and blue side band regions of the spectrum. One member of the Raman-laser pair is tuned to a beat frequency in the blue region of the carrier and the other in the red. In this set-up, the Mølmer-Sørenson Hamiltonian

$$\mathbf{H}_{MS} = \hbar \Omega \, \eta \, \sigma_X \, \cos(\delta \, t) \left(\mathbf{a} \exp(-i\nu t) + \mathbf{a}^\dagger \exp(i\nu t) \right), \tag{7.110}$$

the sum of (7.105) and (7.107), governs ion-qubit dynamics.

Using (7.110), we seek to address two adjacent ions, labeled i, j, in a two-qubit ion string. To that end each ion is subjected to bi-chromatic Raman beams, and the Hamiltonian, in the Lamb-Dicke limit, for the ion pair is the sum of $\mathbf{H}_{MS}$ for each ion, i.e.,

$$\mathbf{H}_{MS}(t) = \cos(\delta t) \left(\mathbf{a} \exp(-i\nu t) + \mathbf{a}^\dagger \exp(i\nu t) \right) \otimes$$
$$(\hbar \eta_i \Omega_i (\sigma_X)_i \otimes \mathbb{1} + \hbar \eta_j \Omega_j \mathbb{1} \otimes (\sigma_X)_j), \tag{7.111}$$

where η_i, Ω_i are, respectively, the Lamb-Dicke, and Rabi-coupling laser parameters for ion i.

According to (3.37), the time evolution operator is

$$\mathbf{U}_{MS}(t, t_0) = T \, \exp(-i/\hbar \int_{t_0}^{t} dt' \, \mathbf{H}_{MS}(t')) \tag{7.112}$$

which can be expanded so that it has the form

$$\mathbf{U}_{MS}(t,t_0) = \exp(-i/\hbar \int_{t_0}^{t} dt' \mathbf{H}_{MS}(t')) \exp(-\frac{1}{2\hbar^2} \int_{t_0}^{t} dt_1 \int_{t_0}^{t_1} dt_2 [\mathbf{H}_{MS}(t_1), \mathbf{H}_{MS}(t_2)]) \quad (7.113)$$

provided that all higher commutators (e.g $[\mathbf{H}_{MS}(t), [\mathbf{H}_{MS}(t_1), \mathbf{H}_{MS}(t_2)]]$) vanish. The argument of the first exponential integrates to an ion-phonon operator that has the generic form

$$\frac{(\alpha(t)\mathbf{a} + \beta(t)\mathbf{a}^\dagger)}{\nu^2 - \delta^2} \otimes (\eta_i \Omega_i (\sigma_X)_i \otimes \mathbb{1} + \eta_j \Omega_j \mathbb{1} \otimes (\sigma_X)_j), \quad (7.114)$$

where the coefficients $\alpha(t)$, $\beta(t)$ are complicated functions of t, ν and δ. In applications it is desirable to adjust the laser parameters in such way, e.g. making the de-tuning δ large, so that for all practical purposes operator (7.114) can be ignored. The exponential proportional to the commutator in expression (7.113) is more interesting. Evaluating the commutator

$$[\mathbf{H}_{MS}(t_1), \mathbf{H}_{MS}(t_2)] =$$
$$\hbar^2 \cos(\delta t_1)\cos(\delta t_2)[\mathbf{a}\exp(-i\nu t_1) + \mathbf{a}^\dagger \exp(i\nu t_2), \mathbf{a}\exp(-i\nu t_1) + \mathbf{a}^\dagger \exp(i\nu t_2)] \otimes$$
$$(\eta_i \Omega_i (\sigma_X)_i \otimes \mathbb{1} + \eta_j \Omega_j \mathbb{1} \otimes (\sigma_X)_j)^2 =$$
$$\hbar^2 \cos(\delta t_1)\cos(\delta t_2) 2i \sin((t_2 - t_1)\nu) \mathbb{1} \otimes (\eta_i \Omega_i (\sigma_X)_i \otimes \mathbb{1} + \eta_j \Omega_j \mathbb{1} \otimes (\sigma_X)_j)^2. \quad (7.115)$$

Integrating the argument of the exponential in (7.113) over t_1, t_2 we get

$$-\frac{1}{2\hbar^2} \int_{t_0}^{t} dt_1 \int_{t_0}^{t_1} dt_2 [\mathbf{H}_{MS}(t_1), \mathbf{H}_{MS}(t_2)] =$$
$$\left(\frac{-i\nu t}{2(\delta^2 - \nu^2)} + f(t)\right) \mathbb{1} \otimes (\eta_i \Omega_i (\sigma_X)_i \otimes \mathbb{1} + \eta_j \Omega_j \mathbb{1} \otimes (\sigma_X)_j)^2 \quad (7.116)$$

where the function $f(t)$ contain rapidly oscillating terms. For larger values of t and de-tuning, those terms are insignificant compared to the contribution from the term linear in t. Therefore, in the qubit sector,

$$\mathbf{U}_{MS} \approx \exp(-i\tilde{\mathbf{H}}t/\hbar) \exp(-i\mathbf{H}_0' t/\hbar)$$
$$\tilde{\mathbf{H}} \equiv J_{ij}(\sigma_X)_i \otimes (\sigma_X)_j$$
$$J_{ij} \equiv \hbar\nu \frac{\eta_i \eta_j \Omega_i \Omega_j}{(\delta^2 - \nu^2)} \quad (7.117)$$

Here $\mathbf{H}_0'$ is an effective two-qubit operator proportional to $\mathbb{1} \otimes \mathbb{1}$. It contributes an overall phase to the time evolution of the qubit pair. $\tilde{\mathbf{H}}$ is a two-qubit operator that "spin"-flips both qubits labeled i and j. The line of reasoning leading to (7.117) can be generalized, by incorporating different phases in (7.103), to generate terms such as $\sigma_Y \otimes \sigma_Y$, $\sigma_Z \otimes \sigma_Z$, etc. Therefore, with the MS procedure, it's possible to induce effective time-independent Hamiltonians that have the generic form

$$\mathbf{H}_{eff} = \sum_{j<i} J_{ij}^{XX} (\sigma_X)_i \otimes (\sigma_X)_j +$$
$$\sum_{j<i} J_{ij}^{YY} (\sigma_Y)_i \otimes (\sigma_Y)_j + \sum_{j<i} J_{ij}^{ZZ} (\sigma_Z)_i \otimes (\sigma_Z)_j, \quad (7.118)$$

also known as the *Heisenberg spin model* and which has important applications in quantum statistical physics. By adjusting the various laser parameters one can tailor Hamiltonian (7.118), and by extension, quantum gate

$$\mathbf{U} = \exp(-i/\hbar\, \mathbf{H}_{eff}(t - t_0)).$$

For example, with a judicious choice of laser parameters, it is possible to engineer a form given by expression (7.50), which as shown in the previous section, leads to gate (7.52). Hamiltonian (7.118) is also a starting point for *Adiabatic Quantum Computing*, an alternative to the circuit model of quantum computing (see Chap. 10).

Problems

7.1 Derive expression (7.21) by evaluating the integrals $\langle \psi_n | \psi(t) \rangle$.

7.2 Derive expression (7.23) by evaluating the integrals $\langle \psi_i | \mathbf{h}_0 | \psi_j \rangle$.

7.3 Derive the matrix representation (7.30) by evaluating the integrals $\langle \psi_i | \mathbf{H}_I(t) | \psi_j \rangle$ for all $i, j = 1, 2$.

7.4 Prove identities (7.36).

7.5 Use Mathematica's matrix exponentiation function to exponentiate the Hamiltonian in expression (7.39) in order to obtain $\mathbf{U}_I(t)$ in form given by (7.40).

7.6 Verify that the **CNOT** gate can be expressed in the form given by (7.49).

7.7 Verify identity (7.50).

7.8 Verify identity (7.51).

7.9 Using Hamiltonian (7.60), the commutation relations for **P**, **Q** and definitions (7.61), derive expression (7.62).

7.10 Derive matrix representation (7.76) of Hamiltonian (7.72) with respect to basis states (7.74) and (7.75).

7.11 Derive expression (7.87) from definition (7.86).

7.12 Show that the interaction picture operators,

$$\exp(i\mathbf{H}_0 t/\hbar)\mathbb{1} \otimes \sigma_\pm(a) \otimes \mathbb{1} \exp(-i\mathbf{H}_0 t/\hbar),$$
$$\exp(i\mathbf{H}_0 t/\hbar)\mathbb{1} \otimes \mathbb{1} \otimes \sigma_\pm(b) \exp(-i\mathbf{H}_0 t/\hbar),$$
$$\exp(i\mathbf{H}_0 t/\hbar)\sigma_\pm(\nu) \otimes \mathbb{1} \otimes \mathbb{1} \exp(-i\mathbf{H}_0 t/\hbar),$$

where $\mathbf{H}_0$ is given by (7.72), satisfy mappings (7.88).

7.13 Construct, using Mathematica, the matrix representation of Hamiltonian (7.90). Exponentiate it to derive the matrix representing gate (7.92).

7.14 Using Mathematica, evaluate the matrix representation of gate

$$\mathbf{U}_a^{(I)}(\tau_1)\mathbf{U}_b^{(I)}(\tau_2)\mathbf{U}_a^{(I)}(\tau_1)$$

and verify identity (7.97).

7.15 Derive the Mølmer-Sørenson Hamiltonian (7.110).

7.16 Evaluate $\exp(-i/\hbar \int_{t_0}^{t} dt' \mathbf{H}_{MS}(t'))$ and find $\alpha(t), \beta(t)$ given in (7.114).

7.17 Derive identity (7.116).

7.18 Find the expression for $f(t)$ given in (7.116) and determine the value of t in which $f(t)$ is a factor 10^2 smaller than the term linear in t.

7.19 Determine the laser parameters required to generate the gate

$$J_{ij}^{XY}(\sigma_X)_i \otimes (\sigma_Y)_j.$$

References

1. A. Fruchtman, I. Chois, *Technical Roadmap for Tolerant Quantum Computing*, NQIT Report, https://nqit.ox.ac.uk/content/technical-roadmap-fault-tolerant-quantum-computing
2. F. Schmidt-Kaler, H. Haffner, M. Riebe, S. Gulde, G.P.T. Lancaster, T. Deuschle, C. Becher, C. Roos, J. Eschner, R. Blatt, Nature, **422**, 408EP (2003)
3. J.I. Cirac, P. Zoller, Phys. Rev. Lett. **74**, 4091 (1995)

References

4. D.P. DiVincenzo, arXiv:quanr-ph/0002077
5. R.P. Feynman, R.B Leighton, M. Sands, *Feynman Lectures in Physics, Vol. II* (Addison-Wesley, 1965)
6. K. Gottfried, T.-M. Yan, *Quantum Mechanics: Fundamentals* (Springer, 2003)
7. F.T. Jordan, *Linear Operators for Quantum Mechanics* (Dover Publications Inc., Mineola, New York, 1997)
8. M.E. Nielsen, I.L. Chuang, *Quantum Computation and Quantum Information* (Cambridge University Press, 2011)
9. D. Leibfried, R. Blatt, C. Monroe, D. Wineland, Rev. Mod. Phys. **75**, 281 (2003)

Quantum Hardware II: cQED and cirQED

Abstract

We introduce the vacuum Maxwell equations and use them to describe electromagnetic fields trapped in a cavity. Boundary conditions for the cavity are shown to lead to standing wave solutions which we quantize to construct cavity QED, a quantum theory for those fields. We insert a rotor into the cavity and are led to a quantum description of a rotor(atom) qubit coupled to a quantized electromagnetic field. We derive the Jaynes-Cummings Hamiltonian and find its approximate eigenvalues and eigenvectors in the strong atom-radiation coupling regime. We show how artificial atoms, composed of superconducting Josephson junctions, interact with microwave line-resonator photons, thus allowing a circuit analog of cavity QED. We discuss how the latter is described with electrical circuit terminology.

8.1 Introduction

In a vacuum, which contains no free charges or currents, electric $\vec{E}(x, y, z, t)$ and magnetic $\vec{B}(x, y, z, t)$ fields obey the following Maxwell equations (in Gaussian units):

$$
\begin{aligned}
&(I) && \frac{\partial E_x}{\partial x} + \frac{\partial E_y}{\partial y} + \frac{\partial E_z}{\partial z} = 0 \\
&(II) && \frac{\partial B_x}{\partial x} + \frac{\partial B_y}{\partial y} + \frac{\partial B_z}{\partial z} = 0 \\
&(III) && \frac{\partial E_z}{\partial y} - \frac{\partial E_y}{\partial z} + \frac{\dot{B}_x}{c} = \frac{\partial E_x}{\partial z} - \frac{\partial E_z}{\partial x} + \frac{\dot{B}_y}{c} = \frac{\partial E_y}{\partial x} - \frac{\partial E_x}{\partial y} + \frac{\dot{B}_z}{c} = 0 \\
&(IV) && \frac{\partial B_z}{\partial y} - \frac{\partial B_y}{\partial z} - \frac{\dot{E}_x}{c} = \frac{\partial B_x}{\partial z} - \frac{\partial B_z}{\partial x} - \frac{\dot{E}_y}{c} = \frac{\partial B_y}{\partial x} - \frac{\partial B_x}{\partial y} - \frac{\dot{E}_z}{c} = 0.
\end{aligned}
\qquad (8.1)
$$

Here c is the speed of light in the vacuum, and we used Newton's dot notation to denote time derivatives. Let's consider the following ansatz for the electric $\vec{E}(z,t) = \hat{i} E(z,t)$ and magnetic $\vec{B}(z,t) = \hat{j} B(z,t)$ fields

$$B(z,t) = a(t) k \exp(i(kz - \pi/2)) + h.c.$$
$$E(z,t) = \dot{a}(t)/c \exp(ikz) + h.c \qquad (8.2)$$

where $a(t)$ is a complex function of time t, k is a real number and $h.c.$ is the complex conjugate of the latter term. Plugging (8.2) into (8.1) we find that equations (I) and (II) are immediately satisfied as both $\vec{E}$ and $\vec{B}$ are functions of z and have no vector components in the z-direction. Condition (III) is also satisfied but (IV) requires that

$$\ddot{a}(t) + k^2 c^2 a(t) = 0. \qquad (8.3)$$

We recognize (8.3) as the equation of motion for a simple harmonic oscillator. Its solutions are $a(t) = a_\omega \exp(\pm i \omega t)$ where a_ω is a complex constant and $\omega = kc$. The time average, $<\vec{S}>$, over a single period $2\pi/\omega$, of the Poynting vector [1]

$$\vec{S} \equiv \frac{c}{4\pi} \vec{E} \times \vec{B} \qquad (8.4)$$

denotes an energy current (i.e., it has units of energy/area/time), and if $a(t) = a_\omega \exp(-i\omega t)$,

$$<\vec{S}> = |a_\omega|^2 \frac{\omega}{c} \hat{k}. \qquad (8.5)$$

Thus (8.2) represents an electromagnetic wave that transmits energy along the z-axis. For that wave, k is the wavenumber, and it is related to its wavelength $\lambda = 2\pi/k$; the distance by which the phase changes, at a single instance of time, from 0 to 2π. The rate of change of the phase at a given point in space is called the phase velocity and is here given by the speed of light c.

Now let's explore how these fields are modified in a setup in which two large parallel (perfectly) conducting plates in the xy plane are situated at $z = 0$ and $z = d$ on the propagation axis. Though fields (8.2) satisfy Maxwell's equations, they do not satisfy boundary conditions (b.c.) at the plates. We require b.c. so that $\vec{E}(0,t) = \vec{E}(d,t) = 0$ [1]. Consider the condition at $z = 0$, (8.2) stipulates that

$$E(0,t) \equiv E^+(0,t) = -i\frac{\omega}{c}\left(a_\omega \exp(-i\omega t) - a_\omega^* \exp(i\omega t)\right)$$

and so the boundary condition is not met for arbitrary values of t. Now,

$$E^-(z,t) = i\frac{\omega}{c}\left(b_\omega \exp(i\omega t)\exp(ikz) - b_\omega^* \exp(-i\omega t)\exp(-ikz)\right) \qquad (8.6)$$

is also a possible solution to Maxwell's equations. Its Poynting vector is directed along the negative z-axis. Choosing $b_\omega = a_\omega$, and using the fact that Maxwell's equations are linear,

8.1 Introduction

$$E^+(z,t) + E^-(z,t) = -2\frac{\omega}{c}\sin(\omega t)\Big(a_\omega \exp(ik\,z) + a_\omega^* \exp(-ik\,z)\Big), \quad (8.7)$$

is also a possible solution. Choosing a_ω to be pure imaginary (i.e., $a_\omega = i|a_\omega|$), $a_\omega + a_\omega^* = 0$, the b.c. at $z = 0$ is satisfied since,

$$E^+(0,t) + E^-(0,t) = -2\frac{\omega}{c}\sin(\omega t)\Big(a_\omega + a_\omega^*\Big) = 0. \quad (8.8)$$

At $z = d$

$$E^+(d,t) + E^-(d,t) = -2\omega/c \sin(\omega t)\Big(a_\omega \exp(ik\,d) + a_\omega^* \exp(-ik\,d)\Big) \quad (8.9)$$

which, in general, does not vanish unless the exponential factors reduce to unity. The latter condition is satisfied if $k\,d = \pi n$ where n is an integer, and so the wavenumber

$$k_n = \frac{\pi n}{d} \quad (8.10)$$

and the angular frequency $\omega_n = k_n c$ assume discrete values determined by the value of index n. Equation (8.7) is a *standing wave* solution illustrated in Fig. 8.1. The values of index n determine the *modes* of the cavity. The lowest frequency $\omega_0 = \pi c/d$ corresponds to mode index $n = 1$. It is called the fundamental frequency, whereas integer products of ω_0 are *harmonics* of the fundamental frequency. By adjusting the dimensions of the cavity, it is possible to tune the mode structure of the standing waves.

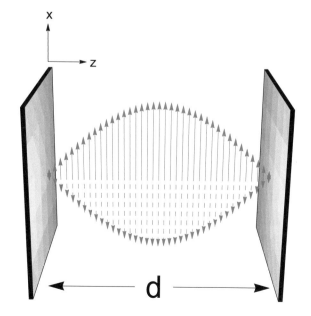

Fig. 8.1 Standing wave for the $n = 1$ mode of electric field $\vec{E}$, plane polarized along x direction. The solid line vectors represent $\vec{E}$ at $t = \pi/\omega_1$. The dashed line vectors describe the field at $t = 3\pi/\omega_1$. The standing wave oscillates in time with period $2\pi/\omega_1$ between these two extremes

8.2 Cavity Quantum Electrodynamics (cQED)

Standing wave (8.9), with wavenumber (8.10), satisfies both Maxwell's equations and boundary conditions. Our goal is a quantum version of this classical description, a theory commonly called *quantum electrodynamics* or QED. Specifically, we are interested in a quantum theory for radiation trapped in the cavity, hence the moniker cavity QED, or cQED for short.

We follow the method proposed in [4] and express the standing wave solutions $E(z,t)$, $B(z,t)$, for a given mode, by the following relations

$$E(z,t) = -\sqrt{\frac{8\pi}{L^2 d}}\, P(t)\, \sin(k_n z)$$

$$B(z,t) = \sqrt{\frac{8\pi}{L^2 d}}\, \omega_n\, Q(t)\, \cos(k_n z)$$

$$k_n = \frac{n\pi}{d} \qquad \omega_n = k_n c \tag{8.11}$$

where $P(t)$, $Q(t)$ are real parameters. With this ansatz, Maxwell's equations require that

$$\dot{Q}(t) = P(t)$$
$$\dot{P}(t) = -\omega_n^2 Q(t), \tag{8.12}$$

where the first line in (8.12) follows from condition (III) in (8.1), and the second from condition (IV).

They are identical to the equations obtained from Hamilton's principle if

$$H = \frac{1}{2}\left(P^2 + \omega^2 Q^2\right), \tag{8.13}$$

since

$$\frac{\partial H}{\partial P} = \dot{Q} \qquad \frac{\partial H}{\partial Q} = -\dot{P}.$$

The energy content of an electromagnetic field in a box of volume $V = L^2 d$ is given by the expression [1]

$$E = \frac{1}{8\pi}\int dV \left(\vec{E}\cdot\vec{E} + \vec{B}\cdot\vec{B}\right), \tag{8.14}$$

where the integral is over the volume of the box. In the region bounded by the capacitor plates, whose dimension L is much greater than the spacing distance d between the plates, we find that, using (8.11),

$$E = \frac{1}{2}\left(P^2(t) + \omega_n^2 Q^2(t)\right) = \frac{\omega_n^2}{2} A^2. \tag{8.15}$$

8.2 Cavity Quantum Electrodynamics (cQED)

The second identity follows from the SHO solutions to (8.12)

$$Q(t) = A \cos(\omega_n t + \phi) \quad P(t) = -\omega_n A \sin(\omega_n t + \phi),$$

where A, ϕ are constants. The total energy in this box is constant and whose value is given by Hamiltonian (8.13), In other words, parameters P, Q are conjugate variables whose time development is governed by Hamiltonian (8.13).

> Mathematica Notebook 8.1: Standing electromagnetic waves in a cavity and the Fabry-Perot interferometer. http://www.physics.unlv.edu/%7Ebernard/MATH_book/Chap8/chap8_link.html; See also https://bernardzygelman.github.io

We arrive at a quantum theory by elevating the canonical variables P, Q to quantum variables so that

$$Q \to \mathbf{Q} \quad P \to \mathbf{P}$$
$$[\mathbf{Q}, \mathbf{P}] = i\hbar. \tag{8.16}$$

With commutation relation (8.16), operators

$$\mathbf{a}_n \equiv \frac{1}{\sqrt{2\hbar\omega_n}} (\mathbf{P} - i\omega_n \mathbf{Q})$$
$$\mathbf{a}_n^\dagger \equiv \frac{1}{\sqrt{2\hbar\omega_n}} (\mathbf{P} + i\omega_n \mathbf{Q}) \tag{8.17}$$

obey

$$\left[\mathbf{a}_n, \mathbf{a}_n^\dagger\right] = 1 \tag{8.18}$$

and Hamiltonian,

$$\mathbf{H}_0 = \frac{1}{2} \left(\mathbf{P}^2 + \omega_n^2 \mathbf{Q}^2\right) = \hbar\omega_n \left(\mathbf{a}_n^\dagger \mathbf{a}_n + 1/2\right). \tag{8.19}$$

Once again, we are led to the quantum theory of a SHO. Operators $\mathbf{a}_n, \mathbf{a}_n^\dagger$ are destruction and creation operators for a quantum excitation of the electromagnetic field. We call this excitation a *cavity photon*, as it corresponds to a well defined energy $\hbar\omega_n$. Hamiltonian (8.19) and the commutation relations for $\mathbf{a}_n, \mathbf{a}_n^\dagger$ are identical to that of phonon excitations described in the previous chapter. A single cavity photon, in mode n, is described by state

$$\mathbf{a}_n^\dagger |\emptyset\rangle,$$

and N photons by the state

$$|\Psi_n\rangle = \frac{1}{\sqrt{N!}}(\mathbf{a_n}^\dagger)^N |\emptyset\rangle = \frac{1}{\sqrt{N!}} \underbrace{\mathbf{a}_n^\dagger \ldots \mathbf{a}_n^\dagger}_{N \text{ times}} |\emptyset\rangle, \qquad (8.20)$$

where $\mathbf{a}_n|\emptyset\rangle = 0$. $|\Psi_n\rangle$ is an eigenstate of $\mathbf{H}_0$ so that

$$\mathbf{H}_0 |\Psi_n\rangle = \hbar\omega_n(N + 1/2)|\Psi_n\rangle.$$

Because Maxwell's equations are linear, the most general Hamiltonian is a sum of Hamiltonians for each mode, i.e.

$$\mathbf{H}_{em} = \sum_m \hbar\omega_m(\mathbf{a}_m^\dagger \mathbf{a}_m + 1/2) \qquad (8.21)$$

where $\mathbf{a}_m^\dagger$, $\mathbf{a}_m$ are the corresponding creation and destruction operators for mode m, and obey commutation relations

$$\left[\mathbf{a}_n, \mathbf{a}_m^\dagger\right] = \delta_{nm} \quad \left[\mathbf{a}_n^\dagger, \mathbf{a}_m^\dagger\right] = [\mathbf{a}_n, \mathbf{a}_m] = 0. \qquad (8.22)$$

The vacuum state $|\emptyset\rangle$ is defined so that

$$\mathbf{a}_m |\emptyset\rangle = 0 \qquad (8.23)$$

for all values of m. Ket

$$|\Psi\rangle = \frac{1}{\sqrt{N_n! N_m! \ldots N_k!}} (\mathbf{a_n}^\dagger)^{N_n} (\mathbf{a_m}^\dagger)^{N_m} \ldots (\mathbf{a_k}^\dagger)^{N_k} |\emptyset\rangle \qquad (8.24)$$

describes a state where N_m cavity photons are in mode m, N_n in mode n and N_k in mode k.

In applications, it is desirable to have a single photon occupy the cavity. In that case the system is in state $|\nu_n\rangle \equiv \mathbf{a}_n^\dagger|\emptyset\rangle$, for mode n, and the mean square value of the electric field, is given by the expectation value

$$<\vec{\mathbf{E}} \cdot \vec{\mathbf{E}}> \equiv \langle\nu_n|\vec{\mathbf{E}} \cdot \vec{\mathbf{E}}|\nu_n\rangle. \qquad (8.25)$$

Using the expression (8.11) for $\vec{E}$ and the relation

$$\vec{\mathbf{E}} = -\sqrt{\frac{8\pi}{L^2 d}} \vec{\mathbf{P}} \sin k_n z, \quad \vec{\mathbf{P}} = \sqrt{\frac{\hbar\omega_n}{2}} \left(\mathbf{a}_n^\dagger + \mathbf{a}_n\right) \hat{\mathbf{i}}$$

(8.25) reduces to

$$\hbar\omega_n \frac{4\pi}{L^2 d} \sin^2(k_n z)\langle\nu_n|(\mathbf{a}_n^\dagger + \mathbf{a}_n)^2|\nu_n\rangle = \hbar\omega_n \frac{12\pi}{L^2 d} \sin^2(k_n z), \qquad (8.26)$$

and is proportional to the average electric field energy density (energy/volume). For the lowest frequency mode $n = 1$, where $\omega_n \equiv \omega$, the electric field energy density

8.2 Cavity Quantum Electrodynamics (cQED)

is proportional to $\sin^2(\pi z/d)$ and has its maximum value at the center $z = d/2$ of the plates. For the sake of economy in notation, and since we restrict our discussion to a single photon mode, we ignore the mode subscripts in the expressions for the photon destruction and creation operators $\mathbf{a}, \mathbf{a}^\dagger$.

A qubit, typically a two-level system such as an atom, is placed at the center $z = d/2$ of the plates where it interacts with the quantized electric field. With $\vec{r}$ the position coordinate of the electric charge, the interaction energy is given by the expression

$$\Delta W \equiv -e\vec{r} \cdot \vec{E} = -e\mathbf{x}\,\mathbf{E}(d/2) = e\mathbf{x}\sqrt{\frac{4\pi}{L^2 d}}\sqrt{\hbar\omega}\,(\mathbf{a}+\mathbf{a}^\dagger) \qquad (8.27)$$

where $\mathbf{x}$ is the quantum operator associated with the x coordinate of the electron. Let's assume that the atom is represented by a rotor situated along the xz plane of Fig. 8.1. Using the matrix representation for the rotor-electron operator $\mathbf{x} = R/2\sigma_X$, we obtain

$$\Delta W = \hbar\Omega\,\sigma_X\left(\mathbf{a}+\mathbf{a}^\dagger\right) =$$

$$\Omega = eR\sqrt{\frac{\pi\omega}{V\hbar}} \qquad (8.28)$$

where R is the rotor radius and $V = dL^2$ is the volume of the cavity confined by the capacitor plates. Including Hamiltonian (7.25) for the rotor, and (8.19) for the single mode cavity photons, the interacting atom (rotor)—cavity photon Hamiltonian is

$$\mathbf{H} = \mathbf{h}_0 + \hbar\omega(\mathbf{a}^\dagger\mathbf{a}+1/2) + \hbar\Omega\,\sigma_X\left(\mathbf{a}+\mathbf{a}^\dagger\right)$$

$$\mathbf{h}_0 = -\frac{\hbar\omega_0}{2}\sigma_Z. \qquad (8.29)$$

In the interaction picture, these time-independent operators become time-dependent operators via the prescription

$$\mathbf{a} \rightarrow \exp(i\,\mathbf{H}_0\,t/\hbar)\,\mathbf{a}\,\exp(-i\,\mathbf{H}_0\,t/\hbar) = \mathbf{a}\,\exp(-i\omega t)$$
$$\mathbf{a}^\dagger \rightarrow \exp(i\,\mathbf{H}_0\,t/\hbar)\,\mathbf{a}^\dagger\,\exp(-i\,\mathbf{H}_0\,t/\hbar) = \mathbf{a}^\dagger\,\exp(i\omega t) \qquad (8.30)$$

also

$$\sigma_X \rightarrow \exp(i\,\mathbf{h}_0\,t/\hbar)\,\sigma_X\,\exp(-i\,\mathbf{h}_0\,t/\hbar) = \exp(i\,\mathbf{h}_0\,t/\hbar)(\sigma_+ + \sigma_-)\exp(-i\,\mathbf{h}_0\,t/\hbar)$$
$$\rightarrow \sigma_+\exp(i\omega_0 t) + \sigma_-\exp(-i\omega_0 t), \qquad (8.31)$$

and the product

$$\sigma_X\left(\mathbf{a}+\mathbf{a}^\dagger\right) \rightarrow$$

$$(\sigma_+\exp(i\omega_0 t) + \sigma_-\exp(-i\omega_0 t))\left(\mathbf{a}\exp(-i\omega t) + \mathbf{a}^\dagger\exp(i\omega t)\right). \qquad (8.32)$$

Expanding (8.32) we find terms proportional to $\exp(\pm i(\omega + \omega_0)t)$ and $\exp(\pm i(\omega - \omega_0)t)$. Close to resonance where $\omega \approx \omega_0$, it is only the latter terms that contribute in the RWA approximation. Therefore, we are allowed to replace (8.28) with operator

$$\hbar \Omega \left(\mathbf{a}\sigma_+ + \mathbf{a}^\dagger \sigma_- \right).$$

In this approximation we obtain the *Jaynes-Cummings Hamiltonian* [4], a work-horse of cQED,

$$\mathbf{H}_{JC} = -\frac{\hbar\omega_0}{2}\sigma_Z + \hbar\omega(\mathbf{a}^\dagger \mathbf{a} + 1/2) + \hbar \Omega \left(\mathbf{a}\sigma_+ + \mathbf{a}^\dagger \sigma_- \right). \quad (8.33)$$

Indeed, we already met $\mathbf{H}_{JC}$ in the trapped-ion scenario discussed in Chap. 7. There, a SHO Hamiltonian describes the spectrum of phonon excitations, i.e., the "phonon bus", and here, cavity photons assume the role of the "bus". With photon-atom coupling, we can shuttle quantum information from qubits to cavity photons, and vice-versa [2].

8.2.1 Eigenstates of the Jaynes-Cummings Hamiltonian

We seek energy eigenstates of Hamiltonian (8.33), i.e., solutions to

$$\mathbf{H}_{JC}|\phi\rangle = E_n |\phi\rangle. \quad (8.34)$$

To that end it is useful to define the states

$$|n, 0\rangle \equiv |n\rangle \otimes |0\rangle \quad |n, 1\rangle \equiv |n\rangle \otimes |1\rangle \quad (8.35)$$

where $|0\rangle, |1\rangle$ are atomic(rotor) qubit states, $|n\rangle$ the eigenstates of Hamiltonian $\hbar\omega(\mathbf{a}^\dagger \mathbf{a} + 1/2)$, and n is the photon occupation number. The cavity is tuned to near resonance so that $\hbar\omega \approx \hbar\omega_0$ and states $|\phi_0\rangle \equiv |n, 0\rangle, |\phi_1\rangle \equiv |n-1, 1\rangle$ are nearly degenerate. For approximate solutions to (8.34), we posit the ansatz

$$|\phi\rangle = c_1|\phi_0\rangle + c_2|\phi_1\rangle, \quad (8.36)$$

and construct the matrix representation of the Jaynes-Cummings Hamiltonian with basis $|\phi_0\rangle, |\phi_1\rangle$. Thus

$$\underline{\mathbf{H}}_{JC} = \begin{pmatrix} \langle\phi_0|\mathbf{H}_{JC}|\phi_0\rangle & \langle\phi_0|\mathbf{H}_{JC}|\phi_1\rangle \\ \langle\phi_1|\mathbf{H}_{JC}|\phi_0\rangle & \langle\phi_1|\mathbf{H}_{JC}|\phi_1\rangle \end{pmatrix} = \begin{pmatrix} \hbar\omega_0 n & \hbar\Omega\sqrt{n} \\ \hbar\Omega\sqrt{n} & \hbar\omega_0 n \end{pmatrix}, \quad (8.37)$$

where we used the fact that $\mathbf{a}^\dagger|n\rangle = \sqrt{n+1}|n\rangle$. With ansatz (8.36) we obtain the matrix representation of eigenvalue Eq. (8.34)

$$\begin{pmatrix} \hbar\omega_0 n & \hbar\Omega\sqrt{n} \\ \hbar\Omega\sqrt{n} & \hbar\omega_0 n \end{pmatrix} \begin{pmatrix} c_1 \\ c_2 \end{pmatrix} = E \begin{pmatrix} c_1 \\ c_2 \end{pmatrix} \quad (8.38)$$

8.2 Cavity Quantum Electrodynamics (cQED)

whose eigenvalues are

$$E_\pm = \hbar\omega_0 n \pm \hbar\Omega\sqrt{n} \tag{8.39}$$

corresponding to eigenstates

$$|\phi_+\rangle = \frac{1}{\sqrt{2}}(|n, 0\rangle + |n-1, 1\rangle)$$

$$|\phi_-\rangle = \frac{1}{\sqrt{2}}(|n, 0\rangle - |n-1, 1\rangle) \tag{8.40}$$

respectively. Suppose a qubit in the ground state $|0\rangle$ is introduced into a cavity containing n-single mode photons at $t = 0$. The Jaynes-Cummings Hamiltonian predicts that the composite system evolves, for $t > 0$, according to

$$|\psi(t)\rangle = \exp(-i\mathbf{H}_{JC}t/\hbar)|n, 0\rangle =$$
$$\exp(-i\omega_0 n t)\left(\cos(\Omega\sqrt{n}\,t)|n, 0\rangle - i\sin(\Omega\sqrt{n}\,t)|n-1, 1\rangle\right). \tag{8.41}$$

Equation (8.41) predicts a probability to find n photons in the cavity, and which oscillates between the photon number n and $n - 1$ with a period determined by Ω and $\sqrt{n}$. It is yet another manifestation of coherent Rabi-flopping. Energy quanta are exchanged between the qubit and the electromagnetic field. Unlike Rabi-flopping of a single qubit, cQED features flopping of entangled states of the qubit and photons.

Laboratory demonstrations of *Rydberg atom* qubits interacting with cavity photons have verified the existence of the predicted oscillations. In *S. Haroche's* laboratory [2], a cavity comprised of a high-Q reflecting material confined single mode photons for as long as 130 ms. It translates, given the dimensions of the cavity, to a total transit distance of 40,000 km as the photon bounces back and forth billions of times. At the same time Rydberg atoms whose qubits states are separated by an energy defect corresponding to $2\pi\omega_0 = 51$ GHz, are introduced into the nearly resonant cavity. By using a method called non-demolition measurements, the group was able to measure the cavity photon number after transit of the Rydberg atom qubit. Measurements revealed the predicted oscillations in photon number as a function of atom-cavity transit time t and confirmed the predicted strong-coupling of an atom/ion with a quantized field. S. Haroche and D. Wineland shared the 2012 Nobel prize in physics for their pioneering work in the domain of cavity QED and ion-trap demonstrations respectively. In the Nobel committee statement, they were cited *"for ground-breaking experimental methods that enable measuring and manipulation of individual quantum systems"*, and *"Their ground-breaking methods have enabled this field of research to take the very first steps towards building a new type of super fast computer based on quantum physics."*

8.3 Circuit QED (cirQED)

A descendant of cavity QED, circuit QED or cirQED, shows great promise as a quantum computing and information platform. In a span of a dozen years or so, cirQED has positioned itself from a dark-horse to a leading contender. Instead of atoms, cirQED employs "artificial atoms" for its qubits and which are best described with electronic circuit terminology. In cirQED, microwave photons supported by planar superconducting transmission lines, or excitations of a superconducting circuit operating at the quantum limit, serve the role of the cavity photon "bus" in cQED.

8.3.1 Quantum LC Circuits

Let's consider an electrical circuit that consists of conducting elements, such as a *capacitor* and *inductor* connected in series (see Fig. 8.2 panel (a)). A capacitor consists of two conductors that are separated, on which equal but opposite signed charges $\pm Q$ reside. Those charges support an electric field in the space between the conductors and, in turn, leads to a voltage difference ΔV_C between the positive and negative charged plates. It turns out that the ratio of charge Q and ΔV_C is always constant so that

$$Q/\Delta V_C = C \tag{8.42}$$

where C is called the capacitance. The *SI* unit of capacitance is called the *Farad*, or F, after the nineteenth century electro-magnetism pioneer *Michael Faraday*. Capacitors store electric field energy and are found in a wide array of electric circuit applications.

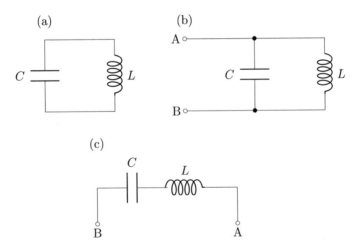

Fig. 8.2 Circuit diagrams for an LC. Circuit. Panel **a** is a stand-alone LC circuit. Panels **b** and **c** illustrate LC circuits containing ports that allow coupling to an external voltage source

8.3 Circuit QED (cirQED)

Another common component in electric circuits, an inductor, stores magnetic energy. The generic inductor is a conducting wire configured into a solenoid loop. As current winds around the loop, a magnetic field is set-up inside and along the axis of the solenoid. If the magnetic field changes in time due a time dependent current, the *Faraday induction law* stipulates that the circuit responds with an induced electric field. The latter sets up a voltage difference between the endpoints, or ports, of the inductor. Faraday's law provides a precise relationship between the voltage ΔV_L difference and the time derivative of the current passing through the solenoid. If $I_{ab}(t)$ is the current traversing from terminal a to terminal b of a solenoid

$$\Delta V_L = -L \frac{dI_{ab}}{dt}$$
$$\Delta V_L \equiv V_b - V_a. \tag{8.43}$$

The constant L is called the *inductance* of the circuit and is expressed in units of *Henry*, or H, after *Joseph Henry*. When the conducting leads of each capacitor plate are connected, so that the circuit is closed, current flows to neutralize the charge separation between the capacitor plates. Without an inductor, neutralization occurs in a minuscule fraction of a second. With an inductor along the current path, a counteracting potential prevents neutralization of the circuit. In the initial stages of the discharge, the current increases in the inductor thereby increasing the magnetic flux. The increasing magnetic flux reverses, because of induction, the voltage polarity of the inductor. This scenario, in which electric and magnetic energy is being shuttled back and forth between capacitor and inductor, is best expressed mathematically. Since the current in the wire is the negative of the time rate of change of the charge Q on the capacitor

$$I(t) = -\frac{dQ(t)}{dt}$$

and assuming that outside each lumped circuit element the magnetic flux vanishes, we appeal to the *Kirchoff loop law* [1], $\Delta V_C + \Delta V_L = 0$, or

$$\frac{Q(t)}{C} = L\dot{I}(t)$$
$$\dot{Q}(t) = -I(t), \tag{8.44}$$

The second equation follows from the definition of current and we used the dot notation for time derivatives. It is convenient to express (8.44) in terms of $\Phi(t) \equiv -LI(t)$, the magnetic flux associated with the inductor, and so

$$\frac{Q(t)}{C} = -\dot{\Phi}(t)$$
$$\frac{\Phi(t)}{L} = \dot{Q}(t). \tag{8.45}$$

Let's define the Hamiltonian

$$H = \frac{\Phi^2}{2L} + \frac{Q^2}{2C} \tag{8.46}$$

where Φ is the conjugate momentum to variable Q. Hamilton's equations lead to the expressions

$$\begin{aligned} \frac{\partial H}{\partial \Phi} &= \dot{Q} \Rightarrow \frac{\Phi}{L} = \dot{Q} \\ \frac{\partial H}{\partial Q} &= -\dot{\Phi} \Rightarrow \frac{Q}{C} = -\dot{\Phi}, \end{aligned} \tag{8.47}$$

that are in harmony with (8.45).

Quantization of (8.46) proceeds by replacing the classical variables Φ, Q with their quantum operators that obey the quantization rule $[\mathbf{\Phi}, \mathbf{Q}] = -i\hbar$. Defining the operators

$$\begin{aligned} \mathbf{a} &= \sqrt{\frac{Z_0}{2\hbar}} \left(\mathbf{Q} + i \frac{\mathbf{\Phi}}{Z_0} \right) \\ \mathbf{a}^\dagger &= \sqrt{\frac{Z_0}{2\hbar}} \left(\mathbf{Q} - i \frac{\mathbf{\Phi}}{Z_0} \right) \\ Z_0 &\equiv \sqrt{\frac{L}{C}} \end{aligned} \tag{8.48}$$

we find that $\left[\mathbf{a}, \mathbf{a}^\dagger\right] = 1$ and

$$\begin{aligned} \mathbf{H} &= \hbar\omega \left(\mathbf{a}^\dagger \mathbf{a} + 1/2 \right) \\ \omega &= \frac{1}{\sqrt{LC}}. \end{aligned} \tag{8.49}$$

We arrived at yet another example involving a simple harmonic oscillator model description. In Chap. 7, a SHO Hamiltonian (7.62) described vibrational excitation about an equilibrium configuration of interacting ions. Equation (8.19) posits a SHO model for quantum excitations of the electromagnetic field in a cavity. Above, (8.49) describes the quantum excitation of a current in an LC circuit. Three disparate physical systems are depicted by the SHO model, underscoring its universal utility.

With advances in the microelectronic engineering of *superconducting* elements, LC circuits exhibiting quantum behavior have been realized in the laboratory. An essential feature of superconducting micro LC circuits is the lack of dissipation, i.e., the circuit does not contain resistive elements that drain and convert electric and magnetic energy into heat. Under ordinary conditions, perfect conductors (without dissipation) do not exist, but a superconductor can sustain a current without significant dissipation. Therefore, microelectronic superconducting LC circuits operating in the quantum regime can serve as a "bus" in quantum processing applications. Because quantum LC circuits, photons, and phonons are all described by the SHO model, a

8.3 Circuit QED (cirQED)

common theoretical framework is available. The converse is also true; we can model cavity photons as excitations of an LC circuit.

Superconductivity, a phenomenon characterized by the apparent loss of resistivity that allows the persistence of electric current, was discovered in 1911 by *Heike Kamerlingh Onnes*. Superconductivity also displays the *Meissner effect*, the rejection of magnetic fields in the transition from an ordinary conduction state to a superconductor. Materials that exhibit super-conduction do so at extremely low temperatures, typically at temperatures where helium gas becomes liquid, i.e., at around 4 K. One K is equivalent to about $-272.15°$ C. In the past 30 years or so a new class of materials displaying superconductivity at around 100 K had been discovered. Fittingly, the phenomenon is called high-T_c superconductivity. However, those materials have not yet found widespread application in electronic circuitry.

In a micro-circuit with capacitance $C = 10^{-11}$ F and inductance $L = 10^{-9}$ H, the quanta of energy is given by the expression

$$\hbar\omega = \hbar\frac{1}{\sqrt{LC}} = 1.05 \times 10^{-24} \, J$$

where J denotes the Joule, the SI unit of energy. Lets compare this figure of merit with $k_B T$, where k_B is the *Boltzmann constant*, the ambient energy that characterizes the environment. Or

$$T = \hbar\omega/k_B \approx 80 \text{ mK}.$$

At temperatures above this threshold exposure of the LC oscillator with the environment threatens coherence and tends to classical behavior. Clearly, a viable QCI platform based on present day superconducting technology requires a very cold environment.

To build quantum gates we need to couple the LC circuits to qubits. In the terminology of electronic circuit theory, this is accomplished by terminals or ports, connected, as illustrated in Fig. 8.2, to the LC circuit. Typically, a time-dependent voltage difference $V(t)$ across the ports drives current $I(t)$. We assume the time dependence is sinusoidal so that

$$V(t) = Re(v \exp(j\omega t)) \quad I(t) = Re(i \exp(j\omega t)), \tag{8.50}$$

(Note: in this section, we use the electrical engineering convention in which the imaginary number i is replaced by the symbol j) where v, i are complex numbers, ω is a driving frequency, and Re represent the real part of these expressions. For linear circuits, v is related to i via the relationship

$$v = i Z \tag{8.51}$$

where Z is a complex number and is called the *circuit impedance*. In a lumped circuit $\partial \vec{B}/\partial t$ is assumed to vanish in regions of the circuit that are external to the circuit elements, including capacitors, inductors, and resistors. Each is characterized by an element impedance, defined according to the rules [1]

$$Z_C = \frac{1}{j\omega C} \quad Z_L = j\omega L \quad Z_R = R. \tag{8.52}$$

Here Z_C, Z_L, Z_R are impedances for a capacitor with capacitance C, inductor with inductance L, and resistor with resistance R. For a circuit with elements connected in series, as in panel (c) of Fig. 8.2, the effective impedance Z of the circuit is the sum

$$Z = Z_1 + Z_2 + Z_3 + \cdots$$

where Z_i is the impedance of element i. If the elements are connected in parallel, as in panel (b) of Fig. 8.2,

$$\frac{1}{Z} = \frac{1}{Z_1} + \frac{1}{Z_2} + \frac{1}{Z_3} + \cdots$$

So for an LC circuit, with negligible resistance, connected in series

$$Z = \frac{1}{j\omega C} + j\omega L = -j\frac{1-\omega^2 LC}{\omega C} = (X_C - X_L)\exp(-j\pi/2)$$

where $X_C \equiv 1/\omega C$, $X_L \equiv \omega L$ are the capacitance and inductive reactance respectively. Inserting this expression into (8.51) we find that

$$I(t) = \frac{1}{X_C - X_L}\cos(\omega t + \pi/2 + \phi_0) \tag{8.53}$$

where $v = V\exp(j\phi_0)$ and V is real. If the driving frequency ω is close to $\omega_0 = 1/\sqrt{LC}$, the natural or resonance frequency of the LC circuit, the current $I(t)$ approaches its maximum value. At resonance $\omega = \omega_0$, expression (8.53) is undefined, but if we include a non-vanishing resistance R, we obtain

$$I(t) = V/R \cos(\omega_0 t + \phi_0)$$

at resonance.

> Mathematica Notebook 8.2: An Introduction to Transmission Line resonators. http://www.physics.unlv.edu/%7Ebernard/MATH_book/Chap8/chap8_link.html; See also https://bernardzygelman.github.io

8.3.2 Artificial Atoms

In one version of cirQED [3], a microwave transmission line resonator serves as the analog of a cQED cavity. But, we also need the circuit analog of a qubit to develop QCI capabilities. Unlike an atom/ion, the quantum energy defects in an LC circuit are uniform and so are not suitable qubit candidates. In cirQED, a qubit is realized with circuits containing *Josephson junctions* (*JJ*), after *Brian Josephson* who first predicted their behavior in the 1960s. Circuits containing Josephson junctions are

8.3 Circuit QED (cirQED)

built to exhibit atom/ion-like features and, therefore, have also been called artificial atoms.

Superconductivity is an inherently quantum mechanical effect, but unlike atoms, molecules and other familiar systems that exhibit quantum behavior, superconductivity is a macroscopic phenomenon. A conductor supports about 10^{22} conduction electrons per cubic cm, but in the superconducting state, they exhibit coherent behavior by forming so-called *Cooper pairs*, after F. Cooper, who along with colleagues *Bardeen*, and *Schrieffer* developed the modern low-temperature theory of superconductivity, also know as the BCS theory. The theory posits that electrons behave as a coherent collective entity much like a wave, rather than individual classical billiard-like objects. Though the BCS theory is beyond the scope of our discussion, we will rely on a more accessible phenomenological description that allows us to predict the behavior of currents and charges near a Josephson junction.

At its most fundamental level, a Josephson junction consists of two superconducting wires separated by a small gap filled by some non-conducting material that cannot be bridged by ordinary currents. That sounds like a description for a capacitor, and the Josephson junction does have an intrinsic capacitance but, unlike a standard capacitor, Cooper pairs with charge $2e$ can bridge the gap by a process called *tunneling*. The latter is a common feature in quantum systems, but for our purposes, we need not concern ourselves with details of tunneling theory, as long as we accept it as a phenomenological fact.

8.3.3 Superconducting Qubits

Let n_a, n_b represent the number of Cooper pairs on superconducting wires a, b that form the boundary of the junction. We define phase parameters δ_a, δ_b for the corresponding regions. These parameters arise from the need to describe the electron gas by a quantum mechanical wave amplitude. For example, in our discussion of the quantum rotor system, the wave amplitude was written in the form (7.17) which includes a magnitude and a phase. Here the amplitude $\psi \sim \sqrt{n} \exp(i\delta)$ describes the collective behavior of electron pairs in a superconductor. The JJ is characterized by the variables

$$\delta \equiv \delta_b - \delta_a \quad Q = 2e(n_a - n_b)$$

where Q represents the excess charge and δ the phase difference across a junction. In applying the BCS theory, Brian Josephson derived the following equations

$$\dot{\delta} = \frac{2e\,V}{\hbar}$$
$$I = I_0 \sin \delta \tag{8.54}$$

where $I(t)$ is the current (or supercurrent) flowing across the junction, $V(t)$ the voltage difference across the junction and $I_0 = E_J\, 2e/\hbar$ is a constant. The quantity

E_J is the Josephson energy, a measure of junction characteristics. Since the junction also acts as a capacitor, $V = Q/C$ and $I = -\dot{Q}$, and we re-write (8.54) in the form

$$\dot{\delta} = \frac{2e}{\hbar} \frac{Q}{C}$$
$$\dot{Q} = -I_0 \sin \delta \qquad (8.55)$$

where C is the capacitance of the junction. In typical junctions, C is on the order of 10^{-12} F and I_0 on the order of $10\,\mu\text{A}$ (A is an Ampere, the SI unit for current). We define an effective Hamiltonian $H_{JJ}(Q, \Phi)$

$$H_{JJ}(Q, \Phi) = E_C \left(\frac{Q}{e}\right)^2 + E_J(1 - \cos(\Phi/\Phi_0)).$$

$$E_C \equiv \frac{e^2}{2C} \quad \Phi_0 = \frac{\hbar}{2e} \qquad (8.56)$$

where $\Phi = \Phi_0 \delta$ is a generalized coordinate and Q its conjugate momentum. From Hamilton's equations

$$\frac{\partial H_{JJ}}{\partial Q} = \dot{\Phi} \quad \frac{\partial H_{JJ}}{\partial \Phi} = -\dot{Q}$$

it follows that

$$\frac{2 E_C Q}{e^2} = \Phi_0 \dot{\delta}$$

$$\frac{E_J}{\Phi_0} \sin(\Phi/\Phi_0) = -\dot{Q}, \qquad (8.57)$$

which when substituting the definitions for E_C, Φ_0 is identical to the Josephson equations (8.55).

> Mathematica Notebook 8.3: Eigenstates of a Josephson junction. http://www.physics.unlv.edu/%7Ebernard/MATH_book/Chap8/chap8_link.html; See also https://bernardzygelman.github.io

Suppose we are in the regime in which $\Phi/\Phi_0 < 1$ and it is legitimate to express $\cos(\Phi/\Phi_0)$ as a power series in Φ/Φ_0. We then obtain

$$H_{JJ} = H_{JJ}^0 + H_{NL}$$
$$H_{JJ}^0 = \frac{Q^2}{2C} + \frac{\Phi^2}{2L_J} \qquad (8.58)$$

where H_{NL} consists of all terms beyond second order in the power expansion of the argument, and

$$L_J \equiv \frac{\Phi_0}{I_0}$$

8.3 Circuit QED (cirQED)

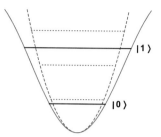

Fig. 8.3 Artificial atom qubit energy spectrum for Hamiltonian (8.58). States $|0\rangle$, $|1\rangle$ denote the qubit states, whereas the dotted lines are, uniformly spaced, energy eigenstates of H_{JJ}^0. The dashed curve represents V_{SHO}, whereas the solid curve includes the anharmonic contribution H_{NL}

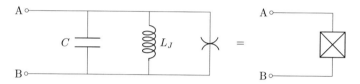

Fig. 8.4 Equivalent circuit for the JJ qubit shown by the crossed boxed symbol. We designate the non-linear component, arising from H_{NR}, with the spider symbol

is a self-inductance. H_{JJ}^0 describes a SHO whereas H_{NL} is an anharmonic correction. The latter term allows JJ circuits to function as viable qubits. Because H_{NL} introduces anharmonicity, as shown in Fig. 8.3, to the harmonic potential

$$V_{SHO} = \frac{\Phi^2}{2L_J}$$

the energy eigenvalue defects of H_{JJ} are not equally spaced.

We express a system governed by H_{JJ} with the circuit shown in Fig. 8.4. In that figure the elements corresponding to the JJ capacitance and self-inductance are connected in parallel to the element denoted by the spider symbol. The latter is a non-linear circuit component that represents the anharmonic contribution H_{NL} in Hamiltonian (8.58). Together they constitute an equivalent circuit containing only a single element, represented by the boxed cross symbol in Fig. 8.4.

> Mathematica Notebook 8.4: Charge, Phase and Flux artificial atom qubits. http://www.physics.unlv.edu/%7Ebernard/MATH_book/Chap8/chap8_link.html; See also https://bernardzygelman.github.io

In QCI applications it is desirable to have the JJ qubit coupled to a LC "bus." There are several ways of doing this, one of which is illustrated in Fig. 8.5. In that figure, a JJ qubit is connected to a LC circuit via two capacitors. It can be shown [6] that the Hamiltonian describing it is similar to the Jaynes-Cummings Hamiltonian discussed in the previous section. There we noted how the eigenstates

Fig. 8.5 Josephson junction qubit coupled to a resonator circuit

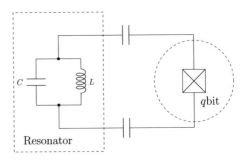

Fig. 8.6 A BBQ circuit

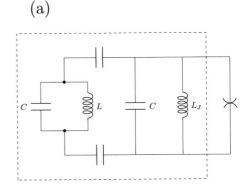

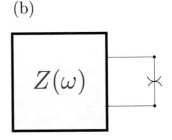

of the latter exhibit entanglement between qubit and photon states. Analogously, we expect entanglement of the JJ qubit with resonator quanta.

The system described by the circuit illustrated in (8.5) is somewhat simplistic. In laboratory realizations, circuit involving non-linear JJ elements are coupled to a complex network of linear elements, e.g., capacitors, resistors, and inductors. As shown in the previous section, such a network is conveniently described by its impedance. Knowing that a Josephson junction is equivalent to the circuit shown in Fig. 8.4, we re-express the circuit in Fig. 8.5 with that shown in panel (a) of Fig. 8.6. In it, we collected all linear elements, circumscribed by the dashed line, and replaced them with a "black box" characterized by a single parameter, the black box impedance $Z(\omega)$ expressed as a function of the driving frequency ω. The non-linear

8.3 Circuit QED (cirQED)

JJ element is coupled, as shown in panel (b) of Fig. 8.6, to the box. In a procedure called *black-box quantization* [5], or BBQ for short, knowledge of $Z(\omega)$ allows one to predict the energy spectrum, and basis vectors associated with it, of the black box. The basis vectors can then be used to form a matrix representation, which can be diagonalized, of the nonlinear term coupled to the black box. In this way, a systematic procedure is available to find the spectrum and eigenstates for a cirQED system.

Problems

8.1 Demonstrate that ansatz (8.2) satisfies conditions (I), (II), (III) in (8.1).

8.2 Show that ansatz (8.2) satisfies condition (IV) in (8.1), only if $a(t)$ is a solution to (8.3).

8.3 Verify that $a(t) = a_\omega \exp(\pm i \omega t)$ are solutions to (8.3), provided that $\omega = kc$.

8.4 Using definition (8.4) for the Poynting vector, and ansatz (8.2), derive relation (8.5) for the time average of the Poynting vector.

8.5 Find the time average of the Poynting vector, for field configurations (8.6) and (8.7). Comment.

8.6 Show that $\vec{E}(t) = \hat{\mathbf{i}} E(z,t)$ and $\vec{B}(t) = \hat{\mathbf{j}} B(z,t)$, where $E(z,t)$, $B(z,t)$ are given by (8.11), satisfy the Maxwell equations in a vacuum.

8.7 Consider the single mode (mode index not shown, and non-normalized) photon state
$$|\psi\rangle = |\emptyset\rangle + \mathbf{a}^\dagger \mathbf{a}^\dagger |\emptyset\rangle.$$
Evaluate $\langle \psi | \psi \rangle$. (Hint: to evaluate the expectation value of operator $\mathbf{a}\mathbf{a}\mathbf{a}^\dagger \mathbf{a}^\dagger$, use commutation relations (8.22) to move the destruction operators toward the right so that you can exploit (8.23) and "destroy the vacuum".)

8.8 Consider the single mode (mode index not shown) photon state
$$|\psi\rangle = \frac{1}{\sqrt{2}} \left(|\emptyset\rangle + \mathbf{a}^\dagger |\emptyset\rangle \right).$$
Find the expectation value $\langle \psi | N | \psi \rangle$, where $N = \mathbf{a}^\dagger \mathbf{a}$ is a photon number operator.

8.9 For the state given in problem (8.8), calculate the variance

$$\Delta \mathbf{N}^2 \equiv \langle \psi | \mathbf{N}^2 | \psi \rangle - \langle \psi | \mathbf{N} | \psi \rangle^2.$$

8.10 For the fields defined in problem (8.6), evaluate expression (8.14) and verify relation (8.15).

8.11 Using definition (8.25) verify relation (8.26).

8.12 For $\vec{\mathbf{E}}$ given in problem 8.6, find the expectation value

$$\langle \psi | \vec{\mathbf{E}} \cdot \vec{\mathbf{E}} | \psi \rangle$$

for the, single mode, state

$$|\psi\rangle = \frac{1}{\sqrt{2}} \mathbf{a}^\dagger |\emptyset\rangle + \frac{1}{2} \mathbf{a}^\dagger \mathbf{a}^\dagger |\emptyset\rangle.$$

The mode index of the creation operator is not shown.

8.13 Verify identities (8.30).

8.14 Using the basis states (8.35), evaluate the matrix representation of the Jaynes-Cummings Hamiltonian and verify (8.37).

8.15 Using definitions (8.40) evaluate the expectation values

$$\langle \phi_\pm | \mathbf{H}_{JC} | \phi_\pm \rangle.$$

8.16 Evaluate the expectation value

$$\langle \psi(t) | \mathbf{H}_{JC} | \psi(t) \rangle$$

where $|\psi(t)\rangle$ is given by (8.41).

8.17 Find the mean number of photons $\mathbf{N}$, as a function of time, for state (8.41). Evaluate the variance $\Delta \mathbf{N}^2$ for this state.

8.18 Find the matrix representation $\underline{\psi}(t)$, with respect to basis vectors (8.35), of ket (8.41). Show that it satisfies the Schroedinger equation

$$i\hbar \frac{\partial \underline{\psi}(t)}{\partial t} = \underline{\mathbf{H}}_{JC} \underline{\psi}(t),$$

where $\underline{\mathbf{H}}_{JC}$ is given by (8.37).

8.19 Consider the circuit comprised of linear elements and shown within the dashed line in panel (a) of Fig. 8.6. Assume the unlabeled coupling capacitors in that circuit have the value $C/2$. Find the impedance of this circuit as a function of a driving term with angular frequency ω.

8.20 Evaluate the exercises in Mathematica Notebook 8.3.

References

1. R.P. Feynman, R.B Leighton, M. Sands, *Feynman Lectures in Physics, Vol. II* (Addison-Wesley, 1965)
2. S. Haroche, Rev. Mod. Phys. **85**, 1083 (2013)
3. S. Girvin, in *Quantum Machines: Measurement and Control of Engineered Quantum Systems: Lecture Notes of the Les Houches Summer School*, ed. by M. Devoret, B. Huard, R. Schoelkopf, L.F. Cugliandolo, vol. 96 (Oxford Scholarship Online, 2014)
4. E.T. Jaynes, F.W. Cummings, Proc. IEEE **51**, 89 (1963)
5. S.E. Nigg, H. Paik, B. Vlastakis, G. Kirchmair, S. Shankar, L. Frunzio, M.H. Devoret, R.J. Schoelkopf, S.M. Girvin, Phys. Rev. Lett. **108**, 240502 (2012)
6. U. Vool, M. Devoret, arXiv:1610.0348v2 (2017)

9. Computare Errare Est: Quantum Error Correction

Abstract

We show how to diagnose and rehabilitate bit-flip errors in a classical error correcting model. We compare logical versus physical bits, define codewords, and introduce a classical error-correcting model. We illustrate elements of the latter, including logical versus physical bits, codewords and the notion of an error correcting code. Quantum codes must consider the no-cloning theorem, the collapse hypothesis, and the possibility of continuous errors. We present encoding, syndrome measurement, and recovery circuits for single qubit bit-flip and phase shift errors. We review the Shor code, introduce the stabilizer formalism, and illustrate stabilizers role in its implementation. We demonstrate the use of the stabilizer formalism in the analysis of quantum error-detection in the Laflamme and Steane codes and the development of surface codes. We discuss the threshold theorem and its role in allowing for fault-tolerant quantum computing.

9.1 Introduction

So far, we have primarily considered qubits that evolve uncorrupted in time and space. In other words, these ideal qubits navigate noiseless communication *channels*. In the real world, qubits and their classical cousins, bits, are regularly subjected to corrupting environmental perturbations. If those disturbances alter the value of a bit, let's say changing a bit assignment from 0 to 1, a *bit-flip error* has been incurred. Qubits are subjected to those and other types of errors. The goal of error correction is to mitigate them in such a way that we can be confident, within acceptable bounds, that a computation proceeds in an intended manner.

In this chapter, before discussing quantum error correction, we first investigate how bit-flip errors are diagnosed and corrected in the classical circuit model. Let's define p as the probability that a bit occupies the wrong state, i.e., it has flipped, and $1-p$ the probability that it has not. We also assume that p is independent of the bit state. Consider a classical channel in which we wish to transfer the following information encoded in a series of 8 bits,

$$11001010. \qquad (9.1)$$

If $p = 0.15$ we are almost certain to incur one error during its transmission. One way to mitigate this error is to include extra bits to construct *logical bits*. Up to now we assigned the Boolean logical values 0, 1 to a single bit, instead let's use three bits in state 000, to denote logical value 0, and 111 logical value 1. We replace the eight-bit string (9.1) with the 24-bit string

$$111111000000111000111000. \qquad (9.2)$$

Suppose, at the receiver, we find

$$\begin{array}{c} 1\ 1\ 1\ 1\ 1\ 0\ 0\ 0\ 0\ 0\ 0\ 1\ 1\ 1\ 0\ 0\ 0\ 1\ 0\ 1\ 0\ 0\ 0 \\ 24\ 23\ 22\ 21\ 20\ 19\ 18\ 17\ 16\ 15\ 14\ 13\ 12\ 11\ 10\ 9\ 8\ 7\ 6\ 5\ 4\ 3\ 2\ 1 \end{array} \qquad (9.3)$$

where the first line is the received bit string and the second line the bit label. Reading from right to left, we notice that all bit values, except for bits 5 and 19, appear in threes. The latter are the exceptions, and so we suspect their presence is due to bit-flip errors. We recover the original message (9.2) if the value of bit 5 is changed to 1, and the value of bit 19 to 1. Strings of three identical digits, such as 000 representing the logical 0, is called a *codeword*. The set of all codewords constitute a *code*, and in the example given here, the 24-letter code (9.2) represents the logical value 11001010 (or 202 in base-10).

The process in which a logical binary string is converted into a codeword is called encoding. The inverse of encoding is *error recovery* or decoding. To implement error recovery, we investigate all possible errors and their probabilities. Let's take the three-bit code word 000 as an example. The probability for the occurrence of the single bit (e.g., 000 → 001) error is $p(1-p)(1-p)$ since p is the probability for a flip, as the other two bits in the code word are correct. Table 9.1 itemizes all possible errors and probabilities. We restrict our discussion to cases in which there is, at most,

Table 9.1 Bit flip errors in a three-bit codeword

Codeword error	Total probability
000 → 000	$(1-p)^3$
000 → 001	$p(1-p)^2$
000 → 010	$p(1-p)^2$
000 → 100	$p(1-p)^2$
000 → 011	$p^2(1-p)$
000 → 101	$p^2(1-p)$
000 → 110	$p^2(1-p)$
000 → 111	p^3

Table 9.2 Error syndrome of three-bit codewords 000 and 111. P_{ij} denotes the parity of bit i with respect to bit j

Codewords	P_{12}	P_{13}	Corrupt bit	Codewords	P_{12}	P_{13}	Corrupt bit
000	1	1	None	111	1	1	None
001	−1	−1	1	110	−1	−1	1
010	−1	1	2	101	−1	1	2
100	1	−1	3	011	1	−1	3
011	1	−1	na	100	1	−1	na
101	−1	1	na	010	−1	1	na
110	−1	−1	na	001	−1	−1	na
111	1	1	na	000	1	1	na

one-bit flip. Those errors are itemized in the first four rows of Table 9.1. We use a majority vote to recover the errors by identifying the "odd man out" in a trio of bits. So in the codeword 010, the odd man out is the second bit, and we replace its value with a zero, the silent majority. Obviously, this procedure only works for single-bit codeword errors. Alternatively, we can calculate the relative *parity* of a bit pair in the codeword. Let's assign a positive sign (+1) to bit value 0 and a negative sign (−1) to bit value 1. If the product of signs of two bits is positive (even parity), both bits must have the same value. Likewise, if the product is negative (odd parity), we know that the corresponding bit values differ. In Table 9.2 we itemize parity assignments for every bit pair in a codeword. Note that single-bit errors are uniquely identified in this table, which is also called an *error syndrome*. So code word 101 has the parity signature $\{-1\ 1\}$ which implies that bit 2 is corrupted, regardless of the logical bit codeword 000, or 111. Of course, that parity signature is also evident in row 6 of both tables and signifies a 2-bit flip error, i.e., 000 → 101 or 111 → 010. So in syndrome decoding, parity assignments tell us that either bit 2 is corrupted or a double bit flip has occurred. According to Table 9.1 the probability for the latter is $3p^2(1-p)$, and so we reduce the probability of error from p to order p^2.

This discussion scratches only the surface of classical error correction theory, but it hints at a strategy for achieving the ultimate goal of *fault-tolerant computing*.

Mathematica Notebook 9.1: Simulating a classical bit flip code http://www.physics.unlv.edu/%7Ebernard/MATH_book/Chap9/chap9_link.html

9.2 Quantum Error Correction

Can we borrow strategies from the classical error correcting model for application in quantum error correcting codes? To answer this question we must be made aware of the novel challenges presented by quantum physics. They include,

(I) The no-cloning theorem, discussed in Chap. 6, posits that quantum states cannot be copied, or cloned. That is, for any qubit state $|\psi\rangle = \alpha|0\rangle + \beta|1\rangle$ the following operation,

$$\mathbb{1} \otimes \mathbb{1} \otimes |\psi\rangle \to |\psi\rangle \otimes |\psi\rangle \otimes |\psi\rangle \tag{9.4}$$

is disallowed. At first sight, this restriction appears to throw a major wrench in a quantum encoding program.

(II) According to the collapse postulate, any measurement of a ket, let's say with an occupation operator **n**, collapses the state to an eigenstate of **n**. In that collapse information contained in the initial state is lost. Since syndrome measurements require bit interrogation, the collapse hypothesis seems to preclude a similar protocol in the quantum domain.

(III) In a classical bit, a bit flip from $0 \to 1$ is discrete and requires, depending on the bit hardware, significant energy in the form of noise. In practice, present-day classical micro-circuits are very robust, and the chance for a bit flip in a commercial chip is one part in 10^{17}. In quantum mechanics, errors are inflicted in continuous increments.

For example, the state $|\psi\rangle = \cos\theta|0\rangle + \sin\theta|1\rangle$ is parameterized by the angle θ. Because the parameter θ is continuous, even small changes in θ can alter $|\psi\rangle$ in an adverse manner. The cited obstacles offer slim hope for the possibility of a robust quantum error correction scheme. Indeed, this was the prevailing view until 1995 when P. Shor introduced a nine-qubit code that opened a path to fault-tolerant quantum computing.

Consider circuit diagram Fig. 9.1. The first qubit is in state $|\psi\rangle = \alpha|0\rangle + \beta|1\rangle$ and the direct product state $|0\rangle \otimes |0\rangle \otimes |\psi\rangle$ enters the gates on the left. After passage through the first **CNOT** gate,

$$|0\rangle \otimes |0\rangle \otimes |\psi\rangle = \alpha|000\rangle + \beta|001\rangle \Rightarrow \alpha|000\rangle + \beta|011\rangle. \tag{9.5}$$

Fig. 9.1 Circuit for encoding state $|\psi\rangle = \alpha|0\rangle + \beta|1\rangle$ into $\alpha|000\rangle + \beta|111\rangle$

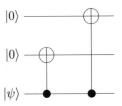

9.2 Quantum Error Correction

A pass through the second **CNOT** gate in Fig. 9.1 delivers the desired state $\alpha|000\rangle + \beta|111\rangle$. The physical qubits $|0\rangle, |1\rangle$ have been effectively replaced by the logical states $|0\rangle_L \equiv |000\rangle, |1\rangle_L \equiv |111\rangle$. We have not violated the no-cloning theorem as the gate in Fig. 9.1 does not carry out the cloning transformation (9.4). Instead, we have embedded our state in a two-dimensional sub-space, spanned by the kets $|000\rangle, |111\rangle$ in a three-qubit, or eight-dimensional, Hilbert space. From this point of view, we can think of a bit flip error as the rotation of a vector in the two-dimensional sub-space of the three-qubit Hilbert space, into an orthogonal sub-space. Having performed the encoding operation, we need to carry out a syndrome type measurement without information loss. Suppose a qubit suffers a bit-flip error so that

$$|\Phi\rangle = \alpha|000\rangle + \beta|111\rangle \rightarrow \mathbf{X}_1(\alpha|000\rangle + \beta|111\rangle) \equiv$$
$$|\Phi'\rangle = \alpha|001\rangle + \beta|110\rangle \qquad (9.6)$$

where $\mathbf{X}_1 = \mathbb{1} \otimes \mathbb{1} \otimes \sigma_X$ represents a bit-flip error in the first qubit. We are treating the error as the passage of a qubit state through an unwanted gate, in this case $\mathbf{X}_1$. Our goal is to "undo" the effect of the phantom gate $\mathbf{X}_1$. Figure 9.2 illustrates an error syndrome measurement and recovery circuit for the first qubit in state $|\psi_3\psi_2\psi_1\rangle$, $\psi_i \in \{0, 1\}$. First, suppose the incoming state is not corrupted and is in state $|000\rangle$; or state $|0\rangle_{anc} \otimes |000\rangle$ if we include the *ancillary qubit*, in state $|0\rangle_{anc}$. The time-line for this state, as it navigates the circuit from left to right is,

$$|0\rangle_{anc} \otimes |000\rangle \rightarrow |0\rangle_{anc} \otimes |000\rangle \rightarrow |0\rangle_{anc} \otimes |000\rangle \rightarrow |0\rangle_{anc} \otimes |000\rangle$$

where each arrow denotes passage through one of the three **CNOT** gates. Now suppose that the first qubit entering the circuit is corrupted. In that case,

$$|0\rangle_{anc} \otimes |001\rangle \rightarrow |0\rangle_{anc} \otimes |001\rangle \rightarrow |1\rangle_{anc} \otimes |001\rangle \rightarrow |1\rangle_{anc} \otimes |000\rangle,$$

and we recover the original state $|000\rangle$. Similarly, for the uncorrupted state

$$|0\rangle_{anc} \otimes |111\rangle \rightarrow |1\rangle_{anc} \otimes |111\rangle \rightarrow |0\rangle_{anc} \otimes |111\rangle \rightarrow |0\rangle_{anc} \otimes |111\rangle,$$

and for the corrupted state

$$|0\rangle_{anc} \otimes |110\rangle \rightarrow |1\rangle_{anc} \otimes |110\rangle \rightarrow |1\rangle_{anc} \otimes |110\rangle \rightarrow |1\rangle_{anc} \otimes |111\rangle.$$

In all four cases, the circuit diagnosed and corrected the error. Because quantum mechanics is a linear theory, a bit flip error in the first qubit in the superposition state $|\psi\rangle = \alpha|000\rangle + \beta|111\rangle$ is also corrected. In general, any of the input qubits $|\psi_i\rangle$ could suffer a bit-flip error. In that case, the circuit in Fig. 9.2 is inadequate. A generalization of it, which requires two ancillary qubits to perform syndrome measurements, is shown in Fig. 9.3. In it, we include an "error box" that represents each of the possible bit-flip errors $\mathbf{X}_1, \mathbf{X}_2, \mathbf{X}_3$. To construct that circuit we used the identity shown in Fig. 9.4.

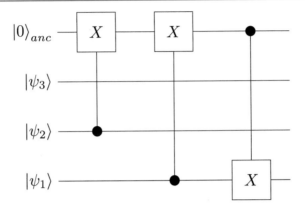

Fig. 9.2 Syndrome and error recovery for a flip error in the first qubit of the codeword

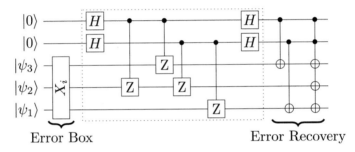

Fig. 9.3 A three-qubit error correcting code for a single flip error. The gates inside the dotted line constitute the syndrome measurements. The error is generated by phantom gates $X_1 = \mathbb{1} \otimes \mathbb{1} \otimes X$, $X_2 = \mathbb{1} \otimes X \otimes \mathbb{1}$, or $X_3 = X \otimes \mathbb{1} \otimes \mathbb{1}$. the error recovery segment includes a sequence of Toffoli gates discussed in Mathematica Notebook 9.2

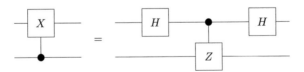

Fig. 9.4 Identity between a control X gate with a circuit containing the $Z = \sigma_Z$ phase flip operator and Hadamard gates H

Let's call the set of gates, comprising the syndrome and recovery components, in Fig. 9.3 $\mathbf{R}_{bf}$. As it is a product of unitary operators, $\mathbf{R}_{bf}$ is unitary. Consider the three-qubit, eight-dimensional, sub-space in which the codewords $|000\rangle$, $|111\rangle$ reside. In that sub-space, $\mathbf{R}_{bf}$ induces the mapping $\tilde{\mathbf{R}}$, defined by

$$|000\rangle \mapsto |000\rangle \quad |001\rangle \mapsto |000\rangle \quad |010\rangle \mapsto |000\rangle \quad |011\rangle \mapsto |111\rangle$$
$$|100\rangle \mapsto |000\rangle \quad |101\rangle \mapsto |111\rangle \quad |110\rangle \mapsto |111\rangle \quad |111\rangle \mapsto |111\rangle. \quad (9.7)$$

9.2 Quantum Error Correction

For any vector $|\psi\rangle$ in the codeword subspace,

$$\tilde{\mathbf{R}}\, \mathbf{E_X}\,|\psi\rangle \to |\psi\rangle$$

where $\mathbf{E}_X$ represents an error box that includes any of the following operators

$$\mathbf{E_0} \equiv \mathbb{1} \otimes \mathbb{1} \otimes \mathbb{1},\ \mathbf{X_3} \equiv \mathbf{X} \otimes \mathbb{1} \otimes \mathbb{1},\ \mathbf{X_2} \equiv \mathbb{1} \otimes \mathbf{X} \otimes \mathbb{1},\ \mathbf{X_1} \equiv \mathbb{1} \otimes \mathbb{1} \otimes \mathbf{X}. \quad (9.8)$$

Alternatively, in Hilbert space that includes the ancillary qubits,

$$\mathbf{R}_{bf}\,|00\rangle \otimes \mathbf{E_X}\,|\psi\rangle = |\phi_1\phi_2\rangle \otimes |\psi\rangle \quad (9.9)$$

where $|\phi_1\phi_2\rangle$ is a state in the sub-space of ancillary qubits. Prior to measurement the density operator is

$$\rho_{in} = |00\rangle\langle 00| \otimes |\psi\rangle\langle\psi|$$

whose partial trace over the Hilbert space of the ancillary qubits give

$$\rho^{\psi} \equiv Tr_A(|00\rangle\langle 00| \otimes |\psi\rangle\langle\psi|) = |\psi\rangle\langle\psi|. \quad (9.10)$$

According to (9.9) the recovery process maps

$$\rho_{in} \mapsto \mathbf{R}_{bf}|00\rangle \otimes \mathbf{E_X}|\psi\rangle\langle\psi|\mathbf{E_X^\dagger} \otimes \langle 00|\mathbf{R}_{bf}^\dagger = |\phi_1\phi_2\rangle \otimes |\psi\rangle\langle\psi| \otimes \langle\phi_2\phi_1| \quad (9.11)$$

and since $Tr(|\phi_1\phi_2\rangle\langle\phi_2\phi_1|) = 1$, we find

$$\rho^{\psi} \mapsto \rho^{\psi}.$$

9.2.1 Phase Flip Errors

As long as we are restricting the errors generated to that by $\mathbf{X_1}, \mathbf{X_2}, \mathbf{X_3}$, a three-qubit error correcting code is adequate. But this is not the only error. For example, either $|0\rangle, |1\rangle$ can undergo a sign change, or phase flip, as in

$$\alpha|0\rangle + \beta|1\rangle \Rightarrow \alpha|0\rangle - \beta|1\rangle.$$

Obviously, phase errors do not arise in classical circuits, but they are common and destructive in quantum circuits. Since $\mathbf{Z}|0\rangle = |0\rangle, \mathbf{Z}|1\rangle = -|1\rangle$, we define the sign-change operator $\mathbf{Z} = \sigma_Z$. Let's define the states

$$|+\rangle = \frac{1}{\sqrt{2}}(|0\rangle + |1\rangle)$$
$$|-\rangle = \frac{1}{\sqrt{2}}(|0\rangle - |1\rangle), \quad (9.12)$$

and note that

$$\mathbf{Z}|+\rangle = |-\rangle \quad \mathbf{Z}|-\rangle = |+\rangle.$$

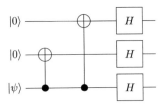

Fig. 9.5 Encoding circuit for $|0\rangle \mapsto |+++\rangle$, and $|1\rangle \mapsto |---\rangle$

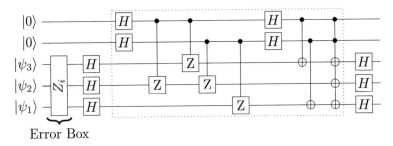

Fig. 9.6 Phase flip error correcting circuit

Vectors $|+\rangle$, $|-\rangle$ behave as qubit flips under the action of error **Z**. Consider the mapping,

$$\alpha|0\rangle + \beta|1\rangle \mapsto \alpha|+++\rangle + \beta|---\rangle, \tag{9.13}$$

whose encoding circuit is illustrated in Fig. 9.5. In the circuit illustrated in Fig. 9.6, kets $|\psi_3\rangle$, $|\psi_2\rangle$, $|\psi_1\rangle$ are qubit states expressed in the $|+\rangle$, $|-\rangle$ basis. As they enter and exit the error box, they are subjected to any one of the three possible errors $\mathbf{Z}_3 \otimes \mathbb{1} \otimes \mathbb{1}$, $\mathbb{1} \otimes \mathbf{Z}_2 \otimes \mathbb{1}$, $\mathbb{1} \otimes \mathbb{1} \otimes \mathbf{Z}_1$. Suppose an initial state $|+\rangle \otimes |+\rangle \otimes |+\rangle$ exits the error box in state $|-\rangle \otimes |+\rangle \otimes |+\rangle$, that is, a phase flip error occurred in qubit 3. Noting that

$$\mathbf{H}|+\rangle = |0\rangle, \quad \mathbf{H}|-\rangle = |1\rangle,$$

the first three qubits are in state $|1\rangle \otimes |0\rangle \otimes |0\rangle$ as they enter the gates within the dotted line of Fig. 9.6. The latter series of gates (including the ancillary states) are represented by $\mathbf{R}_{bf}$, and so the first three qubits are mapped by operator $\tilde{\mathbf{R}}$. Since $\tilde{\mathbf{R}}|1\rangle \otimes |0\rangle \otimes |0\rangle = |0\rangle \otimes |0\rangle \otimes |0\rangle$, the first three qubits enter the final three Hadamard gates in that figure, and from which they exit in the corrected state $|+\rangle \otimes |+\rangle \otimes |+\rangle$. In summary, under the action of

$$\mathbf{H} \otimes \mathbf{H} \otimes \mathbf{H}\,\tilde{\mathbf{R}}\,\mathbf{H} \otimes \mathbf{H} \otimes \mathbf{H} \tag{9.14}$$

single qubit phase errors in the code word $|\psi_3\rangle \otimes |\psi_2\rangle \otimes |\psi_1\rangle$ are corrected.

The general form of a single qubit operator, or error, is

$$\alpha \mathbb{1} + \beta\,\sigma_X + \gamma\,\sigma_Y + \varepsilon\,\sigma_Z \tag{9.15}$$

where $\alpha, \beta, \gamma, \varepsilon$ are real numbers. Having illustrated the quantum error correction model for single qubit bit-flip and phase flip errors, we proceed to explore how errors of the type generated by expression (9.15) are diagnosed and corrected. The first quantum code that corrects for all possible single qubit errors is called the Shor code, after Peter Shor who first published it in 1995 [1].

9.3 The Shor Code

The Shor code is a nine-qubit error correction code whose encoding scheme is defined by the following,

$$|0\rangle_{9L} = \frac{1}{\sqrt{8}} (|000\rangle + |111\rangle) \otimes (|000\rangle + |111\rangle) \otimes (|000\rangle + |111\rangle)$$
$$|1\rangle_{9L} = \frac{1}{\sqrt{8}} (|000\rangle - |111\rangle) \otimes (|000\rangle - |111\rangle) \otimes (|000\rangle - |111\rangle) \quad (9.16)$$

where $|0\rangle_{9L}, |1\rangle_{9L}$ correspond to the logical qubit values for 0, 1 respectively. The nine-qubit Hilbert space is of dimension $2^9 = 512$, but the logical qubits reside in a sub-space of dimension 2. An error induces a rotation of a vector in this subspace to its orthogonal compliment. Let's define,

$$|0\rangle_{3L} \equiv |000\rangle \quad |1\rangle_{3L} \equiv |111\rangle$$
$$|+\rangle_{3L} \equiv \frac{1}{\sqrt{2}} (|0\rangle_{3L} + |1\rangle_{3L}) \quad |-\rangle_{3L} \equiv \frac{1}{\sqrt{2}} (|0\rangle_{3L} - |1\rangle_{3L}) \quad (9.17)$$

and so

$$|0\rangle_{9L} = |+\rangle_{3L} \otimes |+\rangle_{3L} \otimes |+\rangle_{3L}$$
$$|1\rangle_{9L} = |-\rangle_{3L} \otimes |-\rangle_{3L} \otimes |-\rangle_{3L}. \quad (9.18)$$

The encoding scheme (9.16) is implemented by the circuit diagram illustrated in Fig. 9.7. Suppose the first qubit suffers a bit flip error. It induces the rotations

$$|0\rangle_L = \frac{1}{\sqrt{8}} (|000\rangle + |111\rangle) \otimes (|000\rangle + |111\rangle) \otimes (|000\rangle + |111\rangle) \rightarrow$$
$$\frac{1}{\sqrt{8}} (|000\rangle + |111\rangle) \otimes (|000\rangle + |111\rangle) \otimes (|001\rangle + |110\rangle) \quad (9.19)$$

and

$$|1\rangle_L = \frac{1}{\sqrt{8}} (|000\rangle - |111\rangle) \otimes (|000\rangle - |111\rangle) \otimes (|000\rangle - |111\rangle) \rightarrow$$
$$\frac{1}{\sqrt{8}} (|000\rangle - |111\rangle) \otimes (|000\rangle - |111\rangle) \otimes (|001\rangle - |110\rangle). \quad (9.20)$$

Fig. 9.7 Shor nine-qubit encoding circuit. If $|\Psi\rangle = |0\rangle$ the register is mapped into $|0\rangle_{9L}$, if it is equal to $|1\rangle$ the register is mapped into $|1\rangle_{9L}$

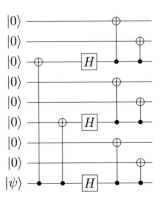

The diagnosis of that error requires 2-ancillary qubits for the cluster of the first three qubits. A circuit that is similar to the gates enclosed by the dotted line in Fig. 9.3 does the job, and also corrects bit-flip errors in qubit 2 and qubit 3. Similarly, for the cluster comprised of qubits 4,5,6, we require an additional pair of ancillary qubits to perform syndrome measurement. The same is true for the final cluster of qubits 7,8,9. Thus at least six ancillary qubits are required for analyzing bit-flip errors. How about phase flip errors? If the first qubit suffers a phase flip error,

$$|+\rangle_{3L} = \frac{1}{\sqrt{2}}(|000\rangle + |111\rangle) \mapsto \frac{1}{\sqrt{2}}(|000\rangle - |111\rangle) = |-\rangle_{3L}$$

or

$$|+\rangle_{3L} \otimes |+\rangle_{3L} \otimes |+\rangle_{3L} \mapsto |+\rangle_{3L} \otimes |+\rangle_{3L} \otimes |-\rangle_{3L}.$$

The same is true for qubits 2 and 3. Similarly, if one of the qubits 4,5,6 experience a phase flip error

$$|+\rangle_{3L} \otimes |+\rangle_{3L} \otimes |+\rangle_{3L} \mapsto |+\rangle_{3L} \otimes |-\rangle_{3L} \otimes |+\rangle_{3L}.$$

Likewise,

$$|+\rangle_{3L} \otimes |+\rangle_{3L} \otimes |+\rangle_{3L} \mapsto |-\rangle_{3L} \otimes |+\rangle_{3L} \otimes |+\rangle_{3L}$$

for qubits 9,8,7. In total, eight ancillary qubits are needed for syndrome measurements in the nine-qubit code. A circuit for the latter is given in Fig. 9.8.

Mathematica Notebook 9.2: Simulating a quantum error correcting code (QEC) http://www.physics.unlv.edu/%7Ebernard/MATH_book/Chap9/chap9_link.html; See also https://bernardzygelman.github.io

9.4 Stabilizers

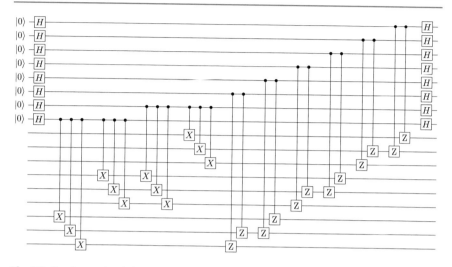

Fig. 9.8 Syndrome circuit for Shor nine-qubit code. The code words $|0\rangle_{9L}, |1\rangle_{9L}$ are subjected to an error box consisting of all possible single-qubit errors. The nine-qubits are represented by the first nine wires (starting from the bottom of the figure), and the eight ancillary qubits by the last eight wires

9.4 Stabilizers

Consider the cluster of the first three $\mathbf{X}$ gates in Fig. 9.8. For the state $|000\rangle + |111\rangle = \sqrt{2}|+\rangle_{3L}$ in the Hilbert sub-space of the first three qubits,

$$\mathbf{X}_3 \otimes \mathbf{X}_2 \otimes \mathbf{X}_1 (|000\rangle + |111\rangle) = \mathbf{X}_3 \otimes \mathbf{X}_2 (|001\rangle + |110\rangle) = $$
$$\mathbf{X}_3 (|011\rangle + |100\rangle) = (|111\rangle + |000\rangle). \tag{9.21}$$

In the same manner we find that, for $\sqrt{2}|-\rangle_{3L}$,

$$\mathbf{X}_3 \otimes \mathbf{X}_2 \otimes \mathbf{X}_1 (|000\rangle - |111\rangle) = -(|000\rangle - |111\rangle). \tag{9.22}$$

Thus the state $|+\rangle_{3L}$ is an eigenstate of $\mathbf{X}_3 \otimes \mathbf{X}_2 \otimes \mathbf{X}_1$ with eigenvalue $+1$, and $|-\rangle_{3L}$ is an eigenstate with eigenvalue -1. Similar relations hold for the operators $\mathbf{X}_6 \otimes \mathbf{X}_5 \otimes \mathbf{X}_4$ and $\mathbf{X}_9 \otimes \mathbf{X}_8 \otimes \mathbf{X}_7$ in the corresponding Hilbert spaces. Let's define the following operators

$$\mathbf{M}_1 = \mathbb{1} \otimes \mathbb{1} \otimes \mathbb{1} \otimes \mathbf{X}_6 \otimes \mathbf{X}_5 \otimes \mathbf{X}_4 \otimes \mathbf{X}_3 \otimes \mathbf{X}_2 \otimes \mathbf{X}_1$$
$$\mathbf{M}_2 = \mathbf{X}_9 \otimes \mathbf{X}_8 \otimes \mathbf{X}_7 \otimes \mathbf{X}_6 \otimes \mathbf{X}_5 \otimes \mathbf{X}_4 \otimes \mathbb{1} \otimes \mathbb{1} \otimes \mathbb{1}. \tag{9.23}$$

Using relations (9.21) and (9.22) we find that

$$\mathbf{M}_1|0\rangle_{9L} = (+1)|0\rangle_{9L} \quad \mathbf{M}_1|1\rangle_{9L} = (+1)|1\rangle_{9L}$$
$$\mathbf{M}_2|0\rangle_{9L} = (+1)|0\rangle_{9L} \quad \mathbf{M}_2|1\rangle_{9L} = (+1)|1\rangle_{9L} \tag{9.24}$$

and

$$[\mathbf{M}_1, \mathbf{M}_2] = 0. \tag{9.25}$$

Now we investigate how the various $\mathbf{Z}_i$ gates appearing in Fig. 9.8 rotate the codeword qubits. For the first pair $\mathbf{Z}_2 \otimes \mathbf{Z}_1$ acting on $|000\rangle \pm |111\rangle$,

$$\mathbb{1} \otimes \mathbf{Z}_2 \otimes \mathbf{Z}_1 |\pm\rangle_{3L} = |\pm\rangle_{3L}. \tag{9.26}$$

Also

$$\mathbf{Z}_3 \otimes \mathbf{Z}_2 \otimes \mathbb{1} |\pm\rangle_{3L} = |\pm\rangle_{3L}. \tag{9.27}$$

We define and itemize the following

$$\begin{aligned}
\mathbf{M}_3 &\equiv \mathbb{1} \otimes \mathbb{1} \otimes \mathbb{1} \otimes \mathbb{1} \otimes \mathbb{1} \otimes \mathbb{1} \otimes \mathbb{1} \otimes \mathbf{Z}_2 \otimes \mathbf{Z}_1 \\
\mathbf{M}_4 &\equiv \mathbb{1} \otimes \mathbb{1} \otimes \mathbb{1} \otimes \mathbb{1} \otimes \mathbb{1} \otimes \mathbb{1} \otimes \mathbf{Z}_3 \otimes \mathbf{Z}_2 \otimes \mathbb{1} \\
\mathbf{M}_5 &\equiv \mathbb{1} \otimes \mathbb{1} \otimes \mathbb{1} \otimes \mathbb{1} \otimes \mathbf{Z}_5 \otimes \mathbf{Z}_4 \otimes \mathbb{1} \otimes \mathbb{1} \otimes \mathbb{1} \\
\mathbf{M}_6 &\equiv \mathbb{1} \otimes \mathbb{1} \otimes \mathbb{1} \otimes \mathbf{Z}_6 \otimes \mathbf{Z}_5 \otimes \mathbb{1} \otimes \mathbb{1} \otimes \mathbb{1} \otimes \mathbb{1} \\
\mathbf{M}_7 &\equiv \mathbb{1} \otimes \mathbf{Z}_8 \otimes \mathbf{Z}_7 \otimes \mathbb{1} \otimes \mathbb{1} \otimes \mathbb{1} \otimes \mathbb{1} \otimes \mathbb{1} \otimes \mathbb{1} \\
\mathbf{M}_8 &\equiv \mathbf{Z}_9 \otimes \mathbf{Z}_8 \otimes \mathbb{1} \otimes \mathbb{1} \otimes \mathbb{1} \otimes \mathbb{1} \otimes \mathbb{1} \otimes \mathbb{1} \otimes \mathbb{1}.
\end{aligned} \tag{9.28}$$

With definitions (9.23) and (9.28) we find that

$$[\mathbf{M}_i, \mathbf{M}_j] = 0 \tag{9.29}$$

and

$$\mathbf{M}_i |0\rangle_{9L} = |0\rangle_{9L}, \quad \mathbf{M}_i |1\rangle_{9L} = |1\rangle_{9L} \tag{9.30}$$

for all $i, j \in \{1, 8\}$. In other words, the codewords $|0\rangle_{9L}, |1\rangle_{9L}$ are invariant under the action of $\mathbf{M}_i$ on them. The set of operators $\mathbf{M}_i$ are called *stabilizers*. The code words of the Shor code are simultaneous $(+1)$ eigenstates of $\mathbf{M}_i$ and are said to be stabilized by the stabilizers. The operators $\mathbf{M}_i$ are members of a structure called a *group*. Though we cannot do full justice to the theory of groups in this text, below we present a short introduction to the theory of groups [2], in particular, the Pauli group.

9.4.1 A Short Introduction to the Pauli Group

Consider the set S_0 of single qubit operators

$$\{\mathbb{1}, \mathbf{X}, \mathbf{Z}, \mathbf{Y}\},$$

which are shorthand for the unit and Pauli matrices $\sigma_X, \sigma_Z, \sigma_Y$ respectively. Multiplying $\mathbb{1}$ with itself and with $\mathbf{X}, \mathbf{Z}, \mathbf{Y}$ reproduces S_0. Also, we know that $\mathbf{XX} = \mathbf{ZZ} = \mathbf{YY} = \mathbb{1}$ and so they also belong to S_0. However the products $\mathbf{XZ} = -\mathbf{ZX} = -i\mathbf{Y}$,

9.4 Stabilizers

$XY = -YX = iZ$, $YZ = -ZY = iX$ are not included in S_0 and so we amend S_0 to construct a new set S_1 consisting of

$$\{\mathbb{1}, X, Z, Y, iX, -iX, iY, -iY, iZ, -iZ\}.$$

Products of members, or elements, of set S_1 spawn new operators, such as $Y(iZ) = -X$, and $(iX)(iX) = -\mathbb{1}$. We include them in set S_2, containing 16 elements, given by

$$S_2 \equiv \mathscr{P}_1 = \{\pm\mathbb{1}, \pm i\mathbb{1}, \pm X, \pm iX, \pm Z, \pm iZ, \pm Y, \pm iY\}. \tag{9.31}$$

Multiplying any two members of S_2 gives a product that is also a member of S_2. In math jargon, we say that the set is closed under matrix multiplication. An arbitrary sequence of multiplications among the various group members always results in an object contained in the original set. This property is one of the defining requirements for the set of objects to constitute a group. There are other requirements including (i) given group members a, b, c group multiplication must be associative i.e.

$$(ab)c = a(bc).$$

Associativity is certainly satisfied by matrix multiplication. (ii) There exists a unique element e in the group, called the identity element, that has the property $ea = ae$ for every element a in the group. It is clear that for $\mathscr{P}_1$, $\mathbb{1}$ is the identity element. (iii) Each element a is mirrored by an element a', contained in the group and called the inverse of a, that has the property

$$aa' = a'a = e.$$

It can be verified that each element of $\mathscr{P}_1$ does indeed possess an inverse.

Within a group, one can always find at least one sub-set of group elements that satisfy all conditions itemized above. For example, let's take the elements $\mathbb{1}, X, -X, -\mathbb{1}$ from $\mathscr{P}_1$ and form the multiplication table

	$\mathbb{1}$	X	$-X$	$-\mathbb{1}$
$\mathbb{1}$	$\mathbb{1}$	X	$-X$	$-\mathbb{1}$
X	X	XX	$-XX$	$-X$
$-X$	$-X$	$-XX$	XX	X
$-\mathbb{1}$	$-\mathbb{1}$	$-X$	X	$\mathbb{1}$

where the entries contain matrix products of the elements in the set. Because $XX = \mathbb{1}$ we recognize that set $p = \{\mathbb{1}, X, -X, -\mathbb{1}\}$ is closed under matrix multiplication. We find that $\mathbb{1}$ is the identity, and each element in p is its inverse. Therefore p is also a group, but since it is a sub-set of $\mathscr{P}_1$, it is a sub-group of $\mathscr{P}_1$.

We define the multi-qubit Pauli group [3]

$$\mathscr{P}_n \equiv \{\pm 1, \pm i\} \times \{\mathbb{1}, X, Z, Y\}^{\otimes n}. \tag{9.32}$$

The superscript $\otimes n$ implies all possible permutations of n products of the arguments. For example, $\{A, B\}^{\otimes 3}$ denotes the set of 2^3 elements $\{AAA, AAB, ABA, \ldots\}$. The number of members, or elements, of a group, is called the *order* of the group. So $\mathscr{P}_1$ is of order 16, and from the definition (9.32) we surmise that the order of $\mathscr{P}_n$ is 4^{n+1}. The Pauli group has some interesting properties. Notably,

(i) Two members g_i, g_j of $\mathscr{P}_n$ either commute $g_i g_j - g_j g_i \equiv [g_i, g_j] = 0$, or anti-commute $g_i g_j + g_j g_i \equiv \{g_i, g_j\}_+ = 0$.
(ii) An element g_i of $\mathscr{P}_n$ is unitary. Since g_i is a matrix, unitarity requires that $g_i g_i^\dagger = g_i^\dagger g_i = \mathbb{1}$.
(iii) Each element g_i in $\mathscr{P}_n$ has the property that $g_i g_i = \pm \mathbb{1}$.

A group in which all its elements commute is called an *Abelian group*. A *stabilizer group* is an Abelian subgroup of the Pauli group, that does not include the element $-\mathbf{I}$ [3]. Stabilizers, the elements of a stabilizer group $\mathscr{S}$, share (+1) eigenstates that are said to be stabilized, or fixed, by the stabilizers. The eight operators $\mathbf{M}_i$, in $\mathscr{P}_9$, stabilize the codewords $|0\rangle_{9L}, |1\rangle_{9L}$, and because every element in $\mathscr{S}$ can be expressed as products of them, they constitute the *generators* of $\mathscr{S}$. $\mathscr{P}_1$ contains 16 elements, but we can fill in its multiplication table by using only the elements $\mathbf{X}, \mathbf{Y}, \mathbf{Z}$. In other words, all members of $\mathscr{P}_1$ can be expressed as finite strings of products of the latter. So instead of explicitly listing all elements of a group, it suffices to list its generators. Typically, a bracket convention is used to characterize a group by its generators. In that convention the equality

$$\mathscr{P}_1 = \langle \mathbf{X}, \mathbf{Y}, \mathbf{Z} \rangle,$$

implies that group $\mathscr{P}_1$ is generated from the elements itemized within the bracket. It is important that we do not confuse this bracket notation with that of the Dirac notation. Thus a stabilizer sub-group $\mathscr{S}$ of $\mathscr{P}_9$ is defined by [3]

$$\mathscr{S} = \langle \mathbf{M}_1, \mathbf{M}_2, \ldots \mathbf{M}_8 \rangle. \tag{9.33}$$

To get a firmer grip on the stabilizer formalism, let's focus on the Pauli group $\mathscr{P}_3$. Consider the subset of operators, $\mathscr{S}$, in that group,

$$\mathbf{III} \equiv \mathbb{1} \otimes \mathbb{1} \otimes \mathbb{1} \quad \mathbf{IZZ} \equiv \mathbb{1} \otimes \mathbf{Z} \otimes \mathbf{Z}$$
$$\mathbf{ZZI} \equiv \mathbf{Z} \otimes \mathbf{Z} \otimes \mathbb{1} \quad \mathbf{ZIZ} \equiv \mathbf{Z} \otimes \mathbb{1} \otimes \mathbf{Z}. \tag{9.34}$$

Multiplication table

	III	IZZ	ZZI	ZIZ
III	III	IZZ	ZZI	ZIZ
IZZ	IZZ	III	ZIZ	ZZI
ZZI	ZZI	ZIZ	III	IZZ
ZIZ	ZIZ	ZZI	IZZ	III

9.4 Stabilizers

demonstrates that $\mathscr{S}$ is a sub-group of $\mathscr{P}_3$. The computation basis kets $|000\rangle, |001\rangle \ldots |111\rangle$, are eigenstates of each element in $\mathscr{S}$, but only two basis vectors $|000\rangle, |111\rangle$, are $(+1)$ eigenstates common to all members of $\mathscr{S}$. Therefore, $\mathscr{S}$ is a stabilizer group that stabilize the vectors

$$|\psi\rangle = \alpha |000\rangle + \beta |111\rangle$$

in three-qubit sub-space spanned by $|000\rangle, |111\rangle$. We can represent $\mathscr{S}$ by generators gleaned from the multiplication table, so that

$$\mathscr{S} = \langle \mathbf{ZZI}, \mathbf{IZZ} \rangle.$$

An operator $\mathbf{E}$ in $\mathscr{P}_3$ either commutes or anti-commutes with elements in $\mathscr{S}$. If it commutes with the generators of $\mathscr{S}$ it commutes with all its members. If $\mathbf{E}$ anti-commutes with at least one element in $\mathscr{S}$ it anti-commutes with at least one generator of $\mathscr{S}$. Suppose operator $\mathbf{E}$ anti-commutes with $\mathbf{ZZI}$. Then,

$$(\mathbf{ZZI})(\mathbf{E}|\psi\rangle) = -\mathbf{E}(\mathbf{ZZI})|\psi\rangle = -\mathbf{E}|\psi\rangle,$$

and the state $\mathbf{E}|\psi\rangle$ is an eigenstate of $\mathbf{ZZI}$ with eigenvalue (-1). Likewise, if $\mathbf{E}$ commutes with $\mathbf{ZZI}$, $\mathbf{E}|\psi\rangle$ is an eigenstate $\mathbf{ZZI}$ with eigenvalue $(+1)$. $\mathbf{E}$ either commutes or anti-commutes with the generators $\mathbf{ZZI}, \mathbf{IZZ}$, and so we generate the set of eigenvalue assignments, for errors $\mathbf{E}_0 = \mathbf{III}, \mathbf{E}_1 = \mathbf{IIX}, \mathbf{E}_2 = \mathbf{IXI}, \mathbf{E}_3 = \mathbf{XII}$, itemized below.

Error	ZZI	IZZ
$\mathbf{E}_0$	+	+
$\mathbf{E}_1$	+	−
$\mathbf{E}_2$	−	−
$\mathbf{E}_3$	−	+

(9.35)

In other words, the action of error $\mathbf{E}$, an operator in $\mathscr{P}_3$, on the codespace leads to four distinct measurement possibilities. For example, $\mathbf{E}_1 = \mathbb{1} \otimes \mathbb{1} \otimes \mathbf{X}$ has the signature $+ -$ in (9.35). Not all operators that commute with all members of $\mathscr{S}$ are in $\mathscr{S}$. For example, $\mathbf{Z} \otimes \mathbf{Z} \otimes \mathbf{Z}, \mathbf{Z} \otimes \mathbf{I} \otimes \mathbf{I}, \mathbf{X} \otimes \mathbf{X} \otimes \mathbf{X}$, are just a few operators that commute with all members of $\mathscr{S}$ but are not in that group.

Definition 9.1 The normalizer of a stabilizer sub-group $\mathscr{S}$ of a Pauli group, is the set of elements in the Pauli group that commute with each element in $\mathscr{S}$. This set, labeled $N(\mathscr{S})$, is also a group.

The set $N(\mathscr{S}) - \mathscr{S}$ constitute all operators in the Pauli group that commute with elements of $\mathscr{S}$ but are not in $\mathscr{S}$. A QECC corrects errors that are not contained in $N(\mathscr{S}) - \mathscr{S}$, i.e. it corrects for elements either in $\mathscr{S}$, or that anti-commute with at least one member of $\mathscr{S}$ [4]. Without proof we offer the following theorem [1].

Theorem 9.1 *A code corrects the set of errors* $\mathbf{E}_0, \mathbf{E}_1, \mathbf{E}_2 \ldots$, *where* $\mathbf{E}_0$ *is the* $\mathscr{P}_n$ *identity operator, iff for every pair* $\mathbf{E}_i, \mathbf{E}_j$, *the product* $\mathbf{E}_i^\dagger \mathbf{E}_j$ *is not in* $N(\mathscr{S}) - \mathscr{S}$.

Lets take the set of possible errors $\mathbf{E}_0 = \mathbb{1} \otimes \mathbb{1} \otimes \mathbb{1}, \mathbf{E}_1 = \mathbb{1} \otimes \mathbb{1} \otimes \mathbf{X}, \mathbf{E}_2 = \mathbb{1} \otimes \mathbf{X} \otimes \mathbb{1}, \mathbf{E}_3 = \mathbf{X} \otimes \mathbb{1} \otimes \mathbb{1}$ and stabilizer $\mathscr{S}$. Now

$$\mathbf{E}_1^\dagger \mathbf{E}_0 = \mathbf{E}_0^\dagger \mathbf{E}_1 = \mathbb{1} \otimes \mathbb{1} \otimes \mathbf{X} \equiv \mathrm{IIX}$$
$$\mathbf{E}_2^\dagger \mathbf{E}_0 = \mathbf{E}_0^\dagger \mathbf{E}_2 = \mathbb{1} \otimes \mathbf{X} \otimes \mathbb{1} \equiv \mathrm{IXI}$$
$$\mathbf{E}_3^\dagger \mathbf{E}_0 = \mathbf{E}_0^\dagger \mathbf{E}_3 = \mathbf{X} \otimes \mathbb{1} \otimes \mathbb{1} \equiv \mathrm{XII}$$
$$\mathbf{E}_1^\dagger \mathbf{E}_2 = \mathbf{E}_2^\dagger \mathbf{E}_1 = \mathbb{1} \otimes \mathbf{X} \otimes \mathbf{X} \equiv \mathrm{IXX}$$
$$\mathbf{E}_1^\dagger \mathbf{E}_3 = \mathbf{E}_3^\dagger \mathbf{E}_1 = \mathbf{X} \otimes \mathbb{1} \otimes \mathbf{X} \equiv \mathrm{XIX}$$
$$\mathbf{E}_2^\dagger \mathbf{E}_3 = \mathbf{E}_3^\dagger \mathbf{E}_2 = \mathbf{X} \otimes \mathbf{X} \otimes \mathbb{1} = \mathrm{XXI}, \qquad (9.36)$$

and

$$\{\mathrm{IIX}, \mathrm{ZIZ}\}_+ = 0 \quad \{\mathrm{IXI}, \mathrm{ZZI}\}_+ = 0 \quad \{\mathrm{XII}, \mathrm{ZIZ}\}_+ = 0$$
$$\{\mathrm{IXX}, \mathrm{ZIZ}\}_+ = 0 \quad \{\mathrm{XIX}, \mathrm{IZZ}\}_+ = 0 \quad \{\mathrm{XXI}, \mathrm{IZZ}\}_+ = 0$$
$$\mathbf{E}_0^\dagger \mathbf{E}_0 = \mathbf{E}_1^\dagger \mathbf{E}_1 = \mathbf{E}_2^\dagger \mathbf{E}_2 = \mathbf{E}_3^\dagger \mathbf{E}_3 = \mathbf{E}_0$$

which implies that $\mathbf{E}_i^\dagger \mathbf{E}_j \notin N(\mathscr{S}) - \mathscr{S}$ for all i, j itemized above. Therefore, according to Theorem 9.1, the code-space for $\mathscr{S}$ allows an error correction protocol for $\mathbf{E}_0, \mathbf{E}_1, \mathbf{E}_2, \mathbf{E}_3$. As a counter example, suppose we wish to measure the phase error of qubit 1, and include $\mathbf{F} = \mathbb{1} \otimes \mathbb{1} \otimes \mathbf{Z}$ into a proposed error set $\{\mathbf{E}_0, \mathbf{F}\}$. But $\mathbf{F}^\dagger \mathbf{E}_0$ commutes with every element in $\mathscr{S}$ thus $\mathbf{F}^\dagger \mathbf{E}_0 \in N(\mathscr{S}) - \mathscr{S}$, violating the conditions of Theorem 9.1. Therefore, $\mathscr{S}$ does not allow a correction code for $\mathbf{E}_0, \mathbf{F}_1$.

9.4.2 Stabilizer Analysis of the Shor Code

We now apply our understanding of the stabilizer formalism in application to the nine-qubit Shor code. We know that the elements $\mathbf{M}_i$ are generators of a stabilizer group in $\mathscr{P}_9$. Vectors $|0\rangle_{9L}, |1\rangle_{9L}$ span the sub-space in the nine-qubit Hilbert space that is stabilized by members of $\mathscr{S}$. Let's take a set of all possible error operators acting only on a single qubit. First, the no-error operator is the identity

$$\mathbf{E}_0 = \mathbb{1} \otimes \mathbb{1} \otimes \mathbb{1} \otimes \mathbb{1} \otimes \mathbb{1} \otimes \mathbb{1} \otimes \mathbb{1} \otimes \mathbb{1} \otimes \mathbb{1},$$

and $\mathbf{X}_n$ correspond to single bit flip errors on the nth qubit, so that $\mathbf{X}_1 = \mathbb{1} \otimes \mathbb{1} \otimes \mathbb{1} \otimes \mathbb{1} \otimes \mathbb{1} \otimes \mathbb{1} \otimes \mathbb{1} \otimes \mathbb{1} \otimes \mathbf{X}$, $\mathbf{X}_9 = \mathbf{X} \otimes \mathbb{1} \otimes \mathbb{1} \otimes \mathbb{1} \otimes \mathbb{1} \otimes \mathbb{1} \otimes \mathbb{1} \otimes \mathbb{1} \otimes \mathbb{1}$, etc. $\mathbf{Z}_n$ represents a phase flip error in qubit n.

Let us investigate the action of each error on an arbitrary state, $|\psi\rangle = \alpha |+\rangle_{9L} + \beta |-\rangle_{9L}$ where $|\alpha|^2 + |\beta|^2 = 1$. If $\mathbf{M}_i$ commutes with $\mathbf{E}$, we know that $\mathbf{E}|\psi\rangle$ is an eigenstate of $\mathbf{M}_i$ with eigenvalue $(+1)$, but if $\mathbf{M}_i$ anti-commutes with $\mathbf{E}$, $\mathbf{E}|\psi\rangle$ is an

9.4 Stabilizers

eigenstate of the latter with eigenvalue (-1). In Table 9.3 we itemize all possible signatures (i.e., eigenvalues) corresponding to the eigenvalues of $\mathbf{M}_i$ for each $\mathbf{E}$. We note that the signature of every $\mathbf{X}_i$, $\mathbf{Y}_i$ error is distinct, and so that signature uniquely identifies that error. If the syndrome identifies the error $\mathbf{X}_j$, it is corrected by applying $\mathbf{X}_j^\dagger = \mathbf{X}_j$ on the nine-qubit Hilbert space. Note that the three $\mathbf{Z}$ errors have identical signatures, so at first sight, this seems to be a defect in the code. However, $\mathbf{Z}_1, \mathbf{Z}_2, \mathbf{Z}_3$ errors are indistinguishable in the subspace $|000\rangle + |111\rangle$ of the first three qubits. So, if the syndrome measurement results in $-++++++$, we know that there is a phase-flip error in at least one of the first three qubits. It does not matter which

Table 9.3 Shor code syndrome measurements

Error	$\mathbf{M}_1$	$\mathbf{M}_2$	$\mathbf{M}_3$	$\mathbf{M}_4$	$\mathbf{M}_5$	$\mathbf{M}_6$	$\mathbf{M}_7$	$\mathbf{M}_8$
$\mathbf{E}_0$	+	+	+	+	+	+	+	+
$\mathbf{X}_1$	+	+	−	+	+	+	+	+
$\mathbf{X}_2$	+	+	−	−	+	+	+	+
$\mathbf{X}_3$	+	+	+	−	+	+	+	+
$\mathbf{X}_4$	+	+	+	+	−	+	+	+
$\mathbf{X}_5$	+	+	+	+	−	−	+	+
$\mathbf{X}_6$	+	+	+	+	+	−	+	+
$\mathbf{X}_7$	+	+	+	+	+	+	−	+
$\mathbf{X}_8$	+	+	+	+	+	+	−	−
$\mathbf{X}_9$	+	+	+	+	+	+	+	−
$\mathbf{Y}_1$	−	+	−	+	+	+	+	+
$\mathbf{Y}_2$	−	+	−	−	+	+	+	+
$\mathbf{Y}_3$	−	+	+	−	+	+	+	+
$\mathbf{Y}_4$	−	−	+	+	−	+	+	+
$\mathbf{Y}_5$	−	−	+	+	−	−	+	+
$\mathbf{Y}_6$	−	−	+	+	+	−	+	+
$\mathbf{Y}_7$	+	−	+	+	+	+	−	+
$\mathbf{Y}_8$	+	−	+	+	+	+	−	−
$\mathbf{Y}_9$	+	−	+	+	+	+	+	−
$\mathbf{Z}_1$	−	+	+	+	+	+	+	+
$\mathbf{Z}_2$	−	+	+	+	+	+	+	+
$\mathbf{Z}_3$	−	+	+	+	+	+	+	+
$\mathbf{Z}_4$	−	−	+	+	+	+	+	+
$\mathbf{Z}_5$	−	−	+	+	+	+	+	+
$\mathbf{Z}_6$	−	−	+	+	+	+	+	+
$\mathbf{Z}_7$	+	−	+	+	+	+	+	+
$\mathbf{Z}_8$	+	−	+	+	+	+	+	+
$\mathbf{Z}_9$	+	−	+	+	+	+	+	+

qubit suffered the phase flip since, in the recovery operation, a phase flip in any of the first three qubits corrects a phase flip error. Similar considerations apply for qubits 4,5,6 and 7,8,9. This feature of the Shor code is common to *degenerate codes*. Measurements are typically stored in the space of the ancillary qubits. Suppose we label the ancillary-subspace qubits by the ket

$$|M_8\, M_7\, M_6\, M_5\, M_4\, M_3\, M_2\, M_1\rangle,$$

where M_i are the assignments tabulated in Table 9.3. If a syndrome measurements detect the $\mathbf{E}_0$ error (which, of course, is no error), the ancillary qubits are loaded to $|00000000\rangle$ corresponding to the first column in that Table. If an $\mathbf{X}_1$ error is detected, the ancillary qubits wind up, according to the second column in Table 9.3, in the state $|00000100\rangle$ and so on. Each syndrome measurement is unique, so every outcome has a unique binary assignment in the ancillary qubit Hilbert space. Because quantum mechanics is a linear theory, the Shor code will correct any single-qubit error that is a member of $\mathscr{P}_9$ [1].

9.5 Stabilizers II

In the previous section, we introduced the concept of a stabilizer and demonstrated its applications in error detection. Although designed primarily for quantum error correction, the stabilizer formalism has applications far beyond this initial purpose. One significant area is fault-tolerant quantum computers, where stabilizer codes form the backbone of protocols that enable quantum computers to operate reliably despite noise and imperfections. The stabilizer formalism is also crucial in constructing and analyzing topological quantum codes, such as surface codes, which are leading candidates for scalable quantum computation due to their robustness against local errors. The versatility of the stabilizer formalism in the description and manipulation of quantum states makes it an essential tool in the broader field of quantum information science.

9.5.1 Review of Coding Terminology

In this section, we provide a deeper dive into the stabilizer formalism and their application, but first review and introduce the following;

Definition 9.2 A physical qubit is the fundamental building block for quantum information processing; a vector describes the state of a physical qubit in a two-dimensional Hilbert space.

9.5 Stabilizers II

Definition 9.3 A quantum binary code C is a 2^k-dimensional subspace, called the codespace, of a 2^n dimensional Hilbert space. The length of a code is equal to the number of qubits n. A codeword is a vector in that codespace.

Definition 9.4 A Pauli operator $\mathbf{O}$ is a direct(tensor) product of n single-qubit Pauli operators. The weight of $\mathbf{O}$ is equal to the dimension of the Hilbert space n, minus the number of unit operators $\mathbf{I}$ contained in the expression for $\mathbf{O}$.

Definition 9.5 A stabilizer group $\mathscr{S}$ has as its elements a set of mutually commuting (Abelian) Pauli operators that do not contain the element $-\mathbf{I}$.

Definition 9.6 The eigenstates for each element in $\mathscr{S}$ with eigenvalues $(+1)$ of $\mathscr{S}$ form a vector sub-space, the codespace, of dimension $2^k < 2^n$.

Definition 9.7 The distance d of a code equals the minimum number of qubits that need to be altered to transform one codeword into another. A distance d stabilizer code can correct up to $t = (d-1)/2$ errors. An n-qubit, 2^k-dimensional code with distance d, is called a $[[n, k, d]]$ quantum code.

Theorem 9.2 [1] For the ℓ independent generators of a stabilizer group $\mathscr{S}$, $k = n - \ell$.

Definition 9.8 Symbols $|0\rangle_{nL}$, $|1\rangle_{nL}$, denote basis vectors in a two-dimensional, or $k = 1$, codespace. They constitute two orthogonal $(+1)$ eigenstates of $\ell = n - 1$ independent generators of a stabilizer group in a Hilbert space of n physical qubits. They serve as basis vectors for a logical qubit.

Definition 9.9 Given the state $|\Psi\rangle \equiv \alpha|0\rangle_{nL} + \beta|1\rangle_{nL}$, a logical gate $\mathbf{L}_n$ is a unitary operator that maps $|\Psi\rangle$ to another vector in this codespace.

To elucidate the importance of these definitions and theorem we will demonstrate their significance through a few examples given below.

9.5.2 Single-Qubit System

So, let's consider the Hilbert space of a single physical qubit. The Pauli group $\mathscr{P}_1$ consists of 16 elements, and we are looking for an Abelian subgroup for a stabilizer candidate. Many choices are available, one of which is $\mathbf{I}, \mathbf{X}$. Now $\mathbf{X}^2 = \mathbf{I}$ thus $\ell = 1$, as $\mathbf{X}$ is the sole independent generator for this stabilizer group. Because the number qubits $n = 1$, the dimension of the corresponding codespace $2^{(n-\ell)} = 2^{(1-1)} = 1$, and is spanned by the vector $|+\rangle = \frac{1}{\sqrt{2}}(|0\rangle + |1\rangle)$, as $\mathbf{X}|+\rangle = |+\rangle$. Alternatively, we could have chosen $\mathbf{Z}$ as a stabilizer generator with the corresponding codespace basis vector $|0\rangle$.

9.5.3 Two-Qubit System

Consider now a two-qubit system. One candidate for a set of stabilizer generators are $g_1 = \mathbf{I} \otimes \mathbf{X}$, $g_2 = \mathbf{X} \otimes \mathbf{I}$, as g_1, g_2 commute and are independent. (+1) eigenstates for g_1 are $\frac{1}{\sqrt{2}}|10\rangle + |11\rangle$ and $\frac{1}{\sqrt{2}}|00\rangle + |01\rangle$, whereas the (+1) eigenstates for g_2 are $\frac{1}{\sqrt{2}}|01\rangle + |11\rangle$ and $\frac{1}{\sqrt{2}}|00\rangle + |10\rangle$. We need to take a linear combination of these states to construct a simultaneous (+1) eigenstate for both g_1, g_2, given by

$$|\Psi\rangle = \frac{1}{2}(|00\rangle + |01\rangle + |10\rangle + |11\rangle).$$

Consider the stabilizer choice, $\mathbf{I} \otimes \mathbf{Z}$, $\mathbf{Z} \otimes \mathbf{I}$; these Pauli operators commute and a simultaneous (+1) eigenstate for this pair is given by state $|00\rangle$. Finally, consider the operators $\mathbf{Z} \otimes \mathbf{Z}$, $\mathbf{X} \otimes \mathbf{X}$, $-\mathbf{Y} \otimes \mathbf{Y}$, $\mathbf{I} \otimes \mathbf{I}$. These operators mutually commute and are derived from the generators $\mathbf{X} \otimes \mathbf{X}$, $\mathbf{Z} \otimes \mathbf{Z}$. The state

$$\frac{|00\rangle + |11\rangle}{\sqrt{2}} \tag{9.37}$$

is a (+1) simultaneous eigenstate for both generators. In all cases discussed above, the number of stabilizer generators $\ell = 2$, and since $n = 2$, the dimension of the corresponding codespace is $2^{n-\ell} = 1$. The three-qubit system allows a higher dimensional codespace. In particular, a two-dimensional code space that constitutes a logical qubit.

9.5.4 Three-Qubit System

Consider the following three-qubit operators $\mathbf{ZZI} \equiv \mathbf{Z} \otimes \mathbf{Z} \otimes \mathbf{I}$, $\mathbf{ZIZ} \equiv \mathbf{Z} \otimes \mathbf{I} \otimes \mathbf{Z}$, $\mathbf{IZZ} \equiv \mathbf{I} \otimes \mathbf{Z} \otimes \mathbf{Z}$, $\mathbf{III} \equiv \mathbf{I} \otimes \mathbf{I} \otimes \mathbf{I}$, they are members of an Abelian group, derived from the independent generators $g_1 \equiv \mathbf{ZZI}$, and $g_2 \equiv \mathbf{IZZ}$. g_1, g_2 are both stabilized by two orthogonal, (+1) eigenstates,

$$|0\rangle_{3L} = |000\rangle \quad |1\rangle_{3L} = |111\rangle \tag{9.38}$$

and which we recognize as the states, defined in Sect. 9.3, of a repetition code. Because $n = 3$ and $k = 2$, we obtain $2^{3-2} = 2$ basis vectors that span the codespace of a single logical qubit. Naively, we could define logical gates in this codespace by forming sums of outer products, e.g. $|0\rangle\langle 1|, |1\rangle\langle 1|$ etc. and making the substitutions $|0\rangle \to |0\rangle_{3L}$, $|1\rangle \to |1\rangle_{3L}$. For example, suppose we define a $\tilde{\mathbf{X}}_{3L}$ (NOT) gate

$$\tilde{\mathbf{X}}_{3L} \equiv |0\rangle_{3L}\langle 1|_{3L} + |1\rangle_{3L}\langle 0|_{3L}. \tag{9.39}$$

Given state $|\psi\rangle = \alpha|0\rangle_{3L} + \beta|1\rangle_{3L}$, gate $\tilde{\mathbf{X}}_{3L}$ acting on it generates the rotation

$$\tilde{\mathbf{X}}_{3L}|\psi\rangle = \alpha|1\rangle_{3L} + \beta|0\rangle_{3L}. \tag{9.40}$$

9.5 Stabilizers II

Thus $\tilde{X}_{3L}$ maps vectors in the codespace to vectors in the codespace, however, let's re-express this operator with the identity

$$\tilde{X}_{3L} = \mathbf{P}\,\mathbf{X}\,\mathbf{X}\,\mathbf{X}\,\mathbf{P} \tag{9.41}$$

where $\mathbf{P} = \mathbf{P}^\dagger$ is a projection operator. Because of the presence of projection operators $\tilde{X}_{3L}$ is not unitary in the Hilbert space of three qubits and, therefore, does not qualify as a unitary quantum gate. Instead, let's define the logical NOT gate

$$\mathbf{X}_{3L} \equiv \mathbf{X}\,\mathbf{X}\,\mathbf{X}. \tag{9.42}$$

Note $\mathbf{X}_{3L}|0\rangle_{3L} = |1\rangle_{3L}$, $\mathbf{X}_{3L}|1\rangle_{3L} = |0\rangle_{3L}$ and is unitary. Thus $\mathbf{X}_{3L}$ is a valid logical NOT gate. Because the commutators $[g_1, \mathbf{X}_{3L}] = [g_2, \mathbf{X}_{3L}] = 0$, $\mathbf{X}_{3L}$ is a member of the normalizer group $N(\mathscr{S})$, defined in Definition 9.1, of the stabilizer group. Consider an element of the normalizer, $\mathbf{N}_i \in N(\mathscr{S})$, and for any vector $|\psi\rangle$ in the codespace, define a vector $|\psi'\rangle \equiv \mathbf{N}_i|\psi\rangle$. Now,

$$\mathbf{g}_j|\psi'\rangle = \mathbf{g}_j\mathbf{N}_i|\psi\rangle = \mathbf{N}_i\mathbf{g}_j|\psi\rangle = |\psi'\rangle \tag{9.43}$$

Thus $|\psi'\rangle$ is also a $(+1)$ eigenstate for stabilizer $\mathbf{g}_j$, and by definition, belongs in this codespace. Thus members of the normalizer group for $\mathscr{S}$ map states in the codespace to states in the codespace; in other words, they act as logical gates.

Importantly, the vectors in this codespace allow for a syndrome-type diagnosis of certain errors. The measurement of generators g_1, g_2, is shown in Fig. 9.3. The first two control $\mathbf{Z}$ gates in that figure correspond to a measurement of the stabilizer generator $g_1 = \mathbf{Z} \otimes \mathbf{Z} \otimes \mathbb{1}$, whereas the second pair of Z gates acting on qubits 1 and 2, corresponds to a measurement of $g_2 = \mathbb{1} \otimes \mathbf{Z} \otimes \mathbf{Z}$. The resulting syndrome assignments for single-bit flip errors, $\mathbf{EX}_1 \equiv \mathbb{1} \otimes \mathbb{1} \otimes \mathbf{X}$, $\mathbf{EX}_2 \equiv \mathbb{1} \otimes \mathbf{X} \otimes \mathbb{1}$, $\mathbf{EX}_3 \equiv \mathbf{X} \otimes \mathbb{1} \otimes \mathbb{1}$ is tabulated in table (9.35). As shown in Sect. 9.2, phase shift errors within this code cannot be detected as they do not satisfy the conditions of Theorem 9.2. Instead, we introduce the stabilizer group with generators $\langle g_3 = \mathbf{XXI}, g_4 = \mathbf{IXX}\rangle$ whose stabilizer eigenstates $|+\rangle_{3L}, |-\rangle_{3L}$ are defined by Eq. (9.17), and whose syndrome measurement circuit is shown in Fig. 9.6.

In summary, a three-qubit quantum repetition code can either correct a bit-flip error with a stabilizer group $\langle g_1, g_2 \rangle$, or a phase flip errors using stabilizer group $\langle g_3, g_4 \rangle$ but not both.

From Definition 9.3 we can think of a code C as a linear mapping

$$C[|0\rangle] \to |0\rangle_L \quad C[|1\rangle] \to |1\rangle_L. \tag{9.44}$$

Suppose code C_1, maps $C_1[|0\rangle] \to |0\rangle_{3L}$, $C_1[|1\rangle] \to |1\rangle_{3L}$, and C_2, maps $C_2[|0\rangle] \to |+\rangle_{3L}$, $C_2[|1\rangle] \to |-\rangle_{3L}$. Since

$$C_2[C_1[|0\rangle]] = C_2[|0\rangle_{3L}]$$
$$C_2[C_1[|1\rangle]] = C_2[|1\rangle_{3L}] \tag{9.45}$$

we define,

$$C_2[|0\rangle_{3L}] \to |+\rangle_{3L} \otimes |+\rangle_{3L} \otimes |+\rangle_{3L} =$$
$$\frac{1}{\sqrt{8}}(|000\rangle + |111\rangle) \otimes (|000\rangle + |111\rangle) \otimes (|000\rangle + |111\rangle) \quad (9.46)$$

and which we recognize as codeword (9.19) for the Shor code. Similarly

$$C_2[C_1[|1\rangle]] = \frac{1}{\sqrt{8}}(|000\rangle - |111\rangle) \otimes (|000\rangle - |111\rangle) \otimes (|000\rangle - |111\rangle) \quad (9.47)$$

This method of combining codes C_1, C_2 is called *concatenation*, and, if we identify C_1 with the repetition code for bit flips and C_2 the repetition code for phase flips, we arrive at the Shor code, which, according to Table 9.3, allows syndrome assignments for either bit-flip or phase-flip errors. The Shor code is a 9 qubit code, encoding a single logical qubit and correcting for one error; thus, it is a [[9, 1, 3]] code. It is not the smallest $d = 3$ code; both the 5-qubit Laflamme code [5] and the 7-qubit Steane code [6], discussed below, also allow detection of bit-flip or phase-flip errors.

9.5.5 The Five-Qubit Code

The five-qubit, or Laflamme code, is the smallest code allowing the diagnosis of either a single bit-flip or a single phase-shift error in a logical qubit. The generators for this code are

$$\mathbf{M}_0 \equiv \mathbf{Z}_5 \otimes \mathbf{X}_4 \otimes \mathbf{X}_3 \otimes \mathbf{Z}_2 \otimes \mathbb{1}$$
$$\mathbf{M}_1 \equiv \mathbf{X}_5 \otimes \mathbf{X}_4 \otimes \mathbf{Z}_3 \otimes \mathbb{1} \otimes \mathbf{Z}_1$$
$$\mathbf{M}_2 \equiv \mathbf{X}_5 \otimes \mathbf{Z}_4 \otimes \mathbb{1} \otimes \mathbf{Z}_2 \otimes \mathbf{X}_1$$
$$\mathbf{M}_3 \equiv \mathbf{Z}_5 \otimes \mathbb{1} \otimes \mathbf{Z}_3 \otimes \mathbf{X}_2 \otimes \mathbf{X}_1 \quad (9.48)$$

where the subscripts identify qubits in the direct product space. This, five-qubit, code corrects $t = 1$ error with four stabilizer generators, so it is a [[$n = 5, k = (5-4), d = (2t+1)$]], or [[5, 1, 3]] quantum code. The stabilizers are fixed (stabilized) by the states,

$$|0\rangle_{5L} \equiv \mathbf{P}|00000\rangle \quad |1\rangle_{5L} \equiv \mathbf{P}|11111\rangle$$
$$\mathbf{P} \equiv \frac{1}{4}(\mathbb{1} + \mathbf{M}_0)(\mathbb{1} + \mathbf{M}_1)(\mathbb{1} + \mathbf{M}_2)(\mathbb{1} + \mathbf{M}_3). \quad (9.49)$$

This identity is proved by first evaluating $\mathbf{M}_0|0\rangle_{5L}$. Now, as $\mathbf{M}_0$ is a Pauli operator, commutes with all other stabilizer generators, and $\mathbf{M}_0\mathbf{M}_0 = \mathbb{1}$, $\mathbf{M}_0(\mathbb{1} + \mathbf{M}_0) = (\mathbb{1} + \mathbf{M}_0)$. Thus, if $\mathbf{P}|00000\rangle \neq 0$, $\mathbf{M}_0|0\rangle_{5l} = |0\rangle_{5L}$. An identical argument demonstrates that, for all i, $\mathbf{M}_i|0\rangle_{5L} = |0\rangle_{5L}$, as well as $\mathbf{M}_i|1\rangle_{5L} = |1\rangle_{5L}$. Also, $\langle 0|0\rangle_{5L} = \langle 1|1\rangle_{5L} = 1$, $\langle 0|1\rangle_{5L} = 0$. Thus $|0\rangle_{5L}, |1\rangle_{5L}$ are simultaneous (+1) eigenstates of (9.48) and serve as basis vectors for the codespace.

9.5 Stabilizers II

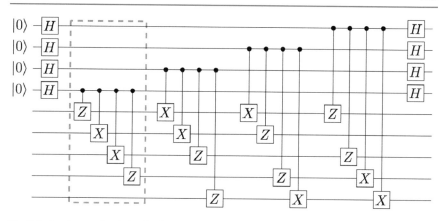

Fig. 9.9 Syndrome circuit for the five qubit code

We can expand Eq. (9.49) to express $|0\rangle_{5L}$ in terms of the five-qubit computational basis. For example, the factor $(\mathbb{1}+\mathbf{M}_3)|00000\rangle$ is equal to $|00000\rangle + |00011\rangle$, where we used the definition of $\mathbf{M}_3$ and the fact $\mathbf{Z}|0\rangle = |0\rangle$, and $\mathbf{X}|0\rangle = |1\rangle$. We repeat this procedure to show that

$$(\mathbb{1}+\mathbf{M}_2)(\mathbb{1}+\mathbf{M}_3)|00000\rangle = |00000\rangle + |10001\rangle + |00011\rangle - |10010\rangle. \tag{9.50}$$

Finishing this cycle, we finally obtain

$$\begin{aligned} 4|0\rangle_{5L} = &|00000\rangle + |10001\rangle + |00011\rangle - |10010\rangle + \\ &|11000\rangle - |01001\rangle - |11011\rangle - |01010\rangle - \\ &|10100\rangle - |00101\rangle - |10111\rangle + |00110\rangle + \\ &|01100\rangle - |11101\rangle - |01111\rangle - |11110\rangle. \end{aligned} \tag{9.51}$$

In Fig. 9.9, we illustrate the syndrome circuit for the five-qubit code.

Consider the block of gates enclosed by the dotted line. The input wires of the first five qubits contain the state $|\psi\rangle = \alpha|0\rangle_{5L} + \beta|1\rangle_{5L}$, and include a sixth syndrome qubit, initialized to $|0\rangle$. The remaining syndrome qubits are ignored as they are decoupled from this block. Exiting that block, we find the state

$$\frac{1}{\sqrt{2}}|0\rangle \otimes |\psi\rangle + \frac{1}{\sqrt{2}}|1\rangle \otimes \mathbf{M}_0|\psi\rangle = \frac{1}{\sqrt{2}}(|0\rangle + |1\rangle) \otimes |\psi\rangle \tag{9.52}$$

Now we can slide the second Hadamard gate on the sixth qubit inside that block to get

$$\left(\mathbf{H}\frac{1}{\sqrt{2}}(|0\rangle + |1\rangle)\right) \otimes |\psi\rangle = |0\rangle \otimes |\psi\rangle. \tag{9.53}$$

This procedure can be applied to every block in Fig. 9.9, resulting in the final state

$$|0000\rangle \otimes |\psi\rangle. \qquad (9.54)$$

Here, each of the syndrome qubits remains in its initial state $|0\rangle$. Suppose that now a bit flip occurred on the i'th qubit within that first block. We represent it by the vector $\mathbf{X}_i|\psi\rangle$, where identity operators on qubits $j \neq i$ are implicit. For the sake of illustration, let's take $i = 3$. Then Eq. (9.52) is replaced with

$$\frac{1}{\sqrt{2}}|0\rangle \otimes \mathbf{X}_3|\psi\rangle + \frac{1}{\sqrt{2}}|1\rangle \otimes \mathbf{M}_0\mathbf{X}_3|\psi\rangle = \frac{1}{\sqrt{2}}(|0\rangle + |1\rangle) \otimes \mathbf{X}_3|\psi\rangle \qquad (9.55)$$

where we used the fact that $\mathbf{M}_0$ commutes with $\mathbf{X}_3$. An additional Hadamard gate on the second qubit results in

$$|0\rangle \otimes \mathbf{X}_3|\psi\rangle \qquad (9.56)$$

As the output state of the first block. It is input to the second block, resulting in (ignoring the overall $\frac{1}{\sqrt{2}}$ factor).

$$|0\rangle \otimes |0\rangle \otimes \mathbf{X}_3|\psi\rangle + |1\rangle \otimes |0\rangle \otimes \mathbf{M}_1\mathbf{X}_3|\psi\rangle =$$
$$(|0\rangle - |1\rangle) \otimes |0\rangle \otimes \mathbf{X}_3|\psi\rangle \qquad (9.57)$$

where we used the fact that $\mathbf{M}_1$ anti-commutes with $\mathbf{X}_3$. The remaining Hadamard gate acting on state Eq. (9.57) results in

$$|1\rangle \otimes |0\rangle \otimes \mathbf{X}_3|\psi\rangle. \qquad (9.58)$$

Repeating this procedure for the third block in Fig. 9.9, we get,

$$|0\rangle \otimes |1\rangle \otimes |0\rangle \otimes \mathbf{X}_3|\psi\rangle + |1\rangle \otimes |1\rangle \otimes |0\rangle \otimes \mathbf{M}_2\mathbf{X}_3|\psi\rangle =$$
$$(|0\rangle + |1\rangle) \otimes |1\rangle \otimes |0\rangle \otimes \mathbf{X}_3|\psi\rangle, \qquad (9.59)$$

A final application of a Hadamard gate on the eighth syndrome qubit yields

$$|0\rangle \otimes |1\rangle \otimes |0\rangle \otimes \mathbf{X}_3|\psi\rangle. \qquad (9.60)$$

Proceeding, as above, for the remaining gate block $\mathbf{M}_3$, we obtain

$$|1010\rangle \otimes \mathbf{X}_3|\psi\rangle \qquad (9.61)$$

where $|1010\rangle$, is shorthand for the syndrome register. Consider now a $\mathbf{Z}_3$ error. Proceeding as above, we find that the input $|0000\rangle \otimes \mathbf{Z}_3|\psi\rangle$ is mapped to $|0001\rangle \otimes \mathbf{Z}_3|\psi\rangle$. Thus, a particular syndrome string heralds the presence of a distinct error, either a $\mathbf{X}_i$, or $\mathbf{Z}_i$ error. The syndrome assignment is unique for each error and identifies the qubit and the type of error inflicted on it. Having demonstrated a step-by-step circuit analysis to obtain syndrome assignments for errors $\mathbf{X}_3, \mathbf{Z}_3$, we obtain a quick mnemonic method for obtaining these assignments. Proceeding from left

9.5 Stabilizers II

Table 9.4 Five qubit code

Error	Syndrome register
X_1	$\lvert 0010\rangle$
X_2	$\lvert 0101\rangle$
X_3	$\lvert 1010\rangle$
X_4	$\lvert 0100\rangle$
X_5	$\lvert 1001\rangle$
Z_1	$\lvert 1100\rangle$
Z_2	$\lvert 1000\rangle$
Z_3	$\lvert 0001\rangle$
Z_4	$\lvert 0011\rangle$
Z_5	$\lvert 0110\rangle$

to right, the first syndrome entry is 0 if the error operator commutes with the qubit in the first block $\mathbf{M}_0$, and entry 1, if it anti-commutes. The second syndrome qubit entry is generated similarly but now with the stabilizer block $\mathbf{M}_1$. The process is repeated for the remaining stabilizer blocks; Table 9.4 summarizes the syndrome register assignments for all $\mathbf{X}$, or $\mathbf{Z}$ errors acting on the five-qubit circuit in Fig. 9.9.

> Mathematica Notebook 9.3: Five and Seven-qubit codes http://www.physics.unlv.edu/%7Ebernard/MATH_book/Chap9/chap9_link.html; See also https://bernardzygelman.github.io

9.5.6 The Seven-Qubit Steane Code

Consider $n = 7$ qubits and assemble the set of Pauli operators

$$\begin{aligned}
\mathbf{M}_0 &= \mathbf{Z}_7 \otimes \mathbf{Z}_6 \otimes \mathbf{Z}_5 \otimes \mathbf{I}_4 \otimes \mathbf{I}_3 \otimes \mathbf{I}_2 \otimes \mathbf{Z}_1 \\
\mathbf{M}_1 &= \mathbf{Z}_7 \otimes \mathbf{Z}_6 \otimes \mathbf{I}_5 \otimes \mathbf{Z}_4 \otimes \mathbf{I}_3 \otimes \mathbf{Z}_2 \otimes \mathbf{I}_1 \\
\mathbf{M}_2 &= \mathbf{Z}_7 \otimes \mathbf{I}_6 \otimes \mathbf{Z}_5 \otimes \mathbf{Z}_4 \otimes \mathbf{Z}_3 \otimes \mathbf{I}_2 \otimes \mathbf{I}_1 \\
\mathbf{M}_3 &= \mathbf{X}_7 \otimes \mathbf{X}_6 \otimes \mathbf{X}_5 \otimes \mathbf{I}_4 \otimes \mathbf{I}_3 \otimes \mathbf{I}_2 \otimes \mathbf{X}_1 \\
\mathbf{M}_4 &= \mathbf{X}_7 \otimes \mathbf{X}_6 \otimes \mathbf{I}_5 \otimes \mathbf{X}_4 \otimes \mathbf{I}_3 \otimes \mathbf{X}_2 \otimes \mathbf{I}_1 \\
\mathbf{M}_5 &= \mathbf{X}_7 \otimes \mathbf{I}_6 \otimes \mathbf{X}_5 \otimes \mathbf{X}_4 \otimes \mathbf{X}_3 \otimes \mathbf{I}_2 \otimes \mathbf{I}_1.
\end{aligned} \qquad (9.62)$$

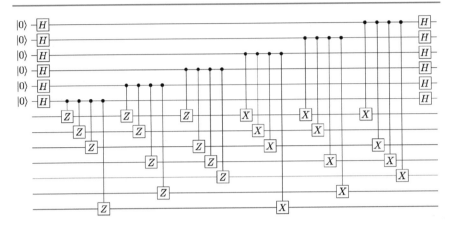

Fig. 9.10 Syndrome circuit for the Steane code

The set $\mathbf{M}_i$, itemized above, forms an Abelian group as all pairs commute and constitutes a set of stabilizer generators. They encode the logical qubits

$$|0\rangle_{7L} = \mathbf{P}_7 |0000000\rangle \quad |1\rangle_{7L} = \mathbf{P}_7 |1111111\rangle$$

$$\mathbf{P}_7 \equiv \frac{\sqrt{2}}{32} \prod_{i=0}^{5}(\mathbb{1} + \mathbf{M}_i) \tag{9.63}$$

whose syndrome detection circuit is shown in Fig. 9.10.

The Steane code has some interesting properties, for example, we can define logical gates acting on the codespace by expressing it as a direct product of the corresponding single qubit operator on each qubit. For example, consider the logical Hadamard gate $\mathbf{H}_{7L}$, which should have the property

$$\mathbf{H}_{7L}|0\rangle_{7L} = \frac{1}{\sqrt{2}}(|0\rangle_{7L} + |1\rangle_{7L})$$

$$\mathbf{H}_{7L}|1\rangle_{7L} = \frac{1}{\sqrt{2}}(|0\rangle_{7L} - |1\rangle_{7L}). \tag{9.64}$$

One finds that

$$\mathbf{H}_{7L} = \mathbf{H} \otimes \mathbf{H} \otimes \mathbf{H} \otimes \mathbf{H} \otimes \mathbf{H} \otimes \mathbf{H} \tag{9.65}$$

does indeed satisfy (9.64). We recognize that $\mathbf{H}_{7L}$ defined in (9.65) is not a Pauli operator, so it is not an element of the stabilizer group. $\mathbf{H}_{7L}$ does not commute with the stabilizer generators $\mathbf{M}_j \in \mathscr{S}$ defined in (9.62), therefore they are also not

9.5 Stabilizers II

elements in the normalizer group $N(\mathscr{S})$. However,

$$\mathbf{H}_{7L} \mathbf{M}_0 \mathbf{H}_{7L}^\dagger = \mathbf{M}_3$$
$$\mathbf{H}_{7L} \mathbf{M}_1 \mathbf{H}_{7L}^\dagger = \mathbf{M}_4$$
$$\mathbf{H}_{7L} \mathbf{M}_2 \mathbf{H}_{7L}^\dagger = \mathbf{M}_5$$
$$\mathbf{H}_{7L} \mathbf{M}_3 \mathbf{H}_{7L}^\dagger = \mathbf{M}_0$$
$$\mathbf{H}_{7L} \mathbf{M}_4 \mathbf{H}_{7L}^\dagger = \mathbf{M}_1$$
$$\mathbf{H}_{7L} \mathbf{M}_5 \mathbf{H}_{7L}^\dagger = \mathbf{M}_2. \tag{9.66}$$

and so the unitary operator $\mathbf{H}_{7L}$ maps any member of the stabilizer group $\mathscr{S}$, to $\mathscr{S}$. Note that $\mathbf{H}_{7L}$ is defined as a tensor product of single-qubit Hadamard gates. This property is not shared by, for example, the five-qubit code as $\mathbf{H}^{\otimes 5} = \mathbf{H} \otimes \mathbf{H} \otimes \mathbf{H} \otimes \mathbf{H} \otimes \mathbf{H}$, and

$$\mathbf{H}^{\otimes 5} |0\rangle_{5L} \neq \frac{1}{\sqrt{2}} (|0\rangle_{5L} + |1\rangle_{5L})$$
$$\mathbf{H}^{\otimes 5} |1\rangle_{5L} \neq \frac{1}{\sqrt{2}} (|0\rangle_{5L} - |1\rangle_{5L}). \tag{9.67}$$

Tensor products of single-qubit gates that form a logical gate, such as that shown by definition (9.65) is an example of a *transversal gate*. Transversal gates have the desirable property of suppressing the propagation and amplification of errors in a quantum circuit [1]. The Steane code also allows a transversal $\mathbf{CNOT}_L$ logical gate as well as the $\mathbf{S}_L$ logical phase-shift gate. As noted above, the five-qubit code allows transversal gates $\mathbf{X}_L$, but its logical Hadamard gate is not transversal.

9.5.7 Surface Codes

An additional feature of the Steane code is illustrated through a three-dimensional, space-time rendering of its syndrome measurement circuit. In the left panel of Fig. 9.11, time progresses along the vertical axis, and at any given moment, blocks of the $\mathbf{M}_0, \mathbf{M}_1, \mathbf{M}_2$ stabilizer generators intersect dashed lines that represent qubits situated on a 2D plane. Similar representations of the remaining $\mathbf{X}$ type stabilizers can be stacked above the $\mathbf{Z}$ stabilizers but are not shown in this figure. In the right panel, those blocks are projected onto a planar region defined by the labeled qubits. This projection describes a color code representation. A color code representation is a type of graph [7]; a mathematical structure consisting of a set of vertices (also called nodes) and a set of edges (also called links) that connect pairs of vertices. The Steane code features some of the properties of a color code. In the right hand side panel of Fig. 9.11, the labeled qubits define the vertices of the graph. An edge connects adjacent vertices. The edge between vertex 5 and 1 defines a (undirected) path. The sequence of paths $5 \to 1 \to 6 \to 7 \to 5$ constitutes a cycle. This cycle forms the boundary of a polygon, called a plaquette, whose face is colored (red). A

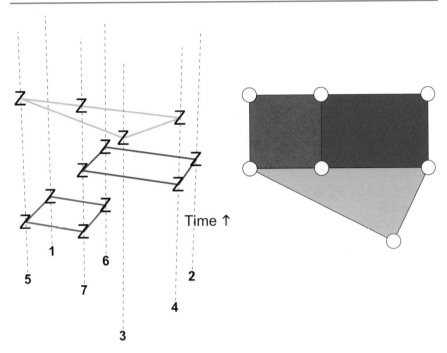

Fig. 9.11 The left panel is a 3D visualization of the syndrome circuit in Fig. 9.10, for the first three sub-blocks of **Z** measurements. The right hand panel is a projection of that diagram to a planar surface

color code generally describes a collection of plaquettes bounded by edges where each vertex is connected to three vertices(trivalent graph) and where the faces of adjacent plaquettes have one of three different colorings. In the Steane code, each stabilizer generator is associated with a plaquette. For example, generators $\mathbf{M}_0$, $\mathbf{M}_3$ are associated with the plaquette whose face is colored red. Similarly $\mathbf{M}_1$, $\mathbf{M}_4$ with the blue region, and $\mathbf{M}_2$, $\mathbf{M}_5$ with the green region.

Color codes represent a type of surface code, and it is widely believed that surface codes provide a route to achieving fault-tolerant quantum computing in the future.

In previous paragraphs, we expressed a stabilizer as a one-dimensional string of Pauli operators, their placement on the string based on the qubit index, for example, in the five-qubit code, stabilizer $\mathbf{M}_0 = \mathbf{Z}_5 \otimes \mathbf{X}_4 \otimes \mathbf{X}_3 \otimes \mathbf{Z}_2 \otimes \mathbb{1}$. The subscripts on the Pauli operators identify the qubit on which it is acting. So, in our convention, the first qubit is acted on by the unit operator, the second by a $\mathbf{Z}$ gate, the third by a $\mathbf{X}$ gate, and so on. Consider a system of twelve qubits, and instead of representing them graphically as the linear string shown on the top row of Fig. 9.12, we re-arrange the qubits labels on a lattice structure shown in that figure. We note that the numbered qubits reside on the edges of this graph. Let's define the following Pauli operator

$$\mathbf{g}_1 \equiv \mathbb{1} \otimes \mathbb{1} \otimes \mathbb{1} \otimes \mathbb{1} \otimes \mathbb{1} \otimes \mathbb{1} \otimes \mathbb{1} \otimes \mathbb{1} \otimes \mathbf{X}_4 \otimes \mathbf{X}_3 \otimes \mathbf{X}_2 \otimes \mathbf{X}_1. \quad (9.68)$$

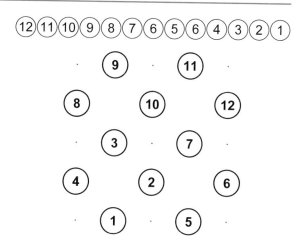

Fig. 9.12 Rearrangement of a linear array of qubits to a two-dimensional lattice array

Comparing it to the placement of qubits on the 2D grid in Fig. 9.12 we realize that g_1 contains a **X** operator on each edge of the cycle $1 \to 2 \to 3 \to 4 \to 1$. Alternatively, we can say that g_1 represent the set of operators on the face of a plaquette whose edges enclosing it form a boundary. With this convention, we express the cycle $2 \to 5 \to 6 \to 7 \to 2$, with the Pauli operator

$$\mathbf{g}_2 \equiv \mathbb{1} \otimes \mathbb{1} \otimes \mathbb{1} \otimes \mathbb{1} \otimes \mathbb{1} \otimes \mathbf{X}_7 \otimes \mathbf{X}_6 \otimes \mathbf{X}_5 \otimes \mathbb{1} \otimes \mathbb{1} \otimes \mathbf{X}_2 \otimes \mathbb{1}, \quad (9.69)$$

which corresponds to the face of the bottom right plaquettes on the lattice. Similarly, for the faces of the upper left and right plaquettes

$$\mathbf{g}_3 \equiv \mathbb{1} \otimes \mathbb{1} \otimes \mathbf{X}_{10} \otimes \mathbf{X}_9 \otimes \mathbf{X}_8 \otimes \mathbb{1} \otimes \mathbb{1} \otimes \mathbb{1} \otimes \mathbb{1} \otimes \mathbf{X}_3 \otimes \mathbb{1} \otimes \mathbb{1},$$
$$\mathbf{g}_4 \equiv \mathbf{X}_{12} \otimes \mathbf{X}_{11} \otimes \mathbf{X}_{10} \otimes \mathbb{1} \otimes \mathbb{1} \otimes \mathbf{X}_7 \otimes \mathbb{1} \otimes \mathbb{1} \otimes \mathbb{1} \otimes \mathbb{1} \otimes \mathbb{1} \otimes \mathbb{1} \quad (9.70)$$

respectively.

Consider the product of $\mathbf{g}_1\mathbf{g}_2$, using their string representations given above, and rules for multiplying operators, we find

$$\mathbf{g}_1\mathbf{g}_2 = \mathbb{1} \otimes \mathbb{1} \otimes \mathbb{1} \otimes \mathbb{1} \otimes \mathbb{1} \otimes \mathbf{X}_7 \otimes \mathbf{X}_6 \otimes \mathbf{X}_5 \otimes \mathbf{X}_4 \otimes \mathbf{X}_3 \otimes \mathbb{1} \otimes \mathbf{X}_1. \quad (9.71)$$

We observe that the collection of nontrivial operators **X** in $\mathbf{g}_1\mathbf{g}_2$, is the union of the **X** operators in $\mathbf{g}_1$ with the **X** operators in $\mathbf{g}_2$ minus the intersection of the latter, and generates the cycle $7 \to 3 \to 4 \to 1 \to 5 \to 6 \to 7$. That cycle forms the boundary, shown in the panel labeled g_1g_2 in Fig. 9.13, of a larger polygon. In the same way, we can graphically represent each member of this stabilizer group shown in Fig. 9.13 by one or more polygons.

Introduced as an illustrative example, this stabilizer code is not very useful. It does not correct **X** or bit-flip errors. With $n = 12$ and four stabilizers, the code encodes eight qubits.

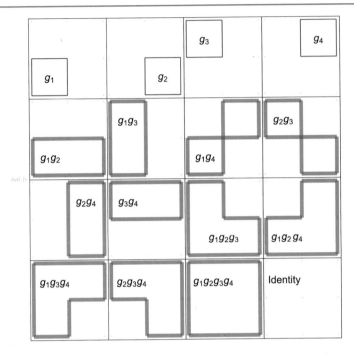

Fig. 9.13 Top row provides a graphical representation of stabilizer generators $\mathbf{g}_i$, for $i = 1, \cdots 4$. Remaining panels provide a graphical representation, formed by the solid outlines of polygons, for each member of the stabilizer group expressed as products of the generators $\mathbf{g}_i$. Each of the 16 larger squares represents the qubit array shown in Fig. 9.12

The first surface code, the toric code, was introduced by Kitaev [8] and is illustrated in Fig. 9.14. In that figure, qubits are located on the edges of a lattice structure. The solid red lines surround the face of a plaquette, and each edge hosts a Pauli **Z** operator. The blue solid lines intersect at a vertex, and the qubits on the edges are associated with the endpoints of the horizontal and vertical lines. In the toric code, these endpoints are associated with the Pauli **X** operators labeled by the corresponding qubit index. As plaquette operators **Z** only intersect with vertex operators **X** by sharing a pair of qubits at a time, as shown in the first panel of Fig. 9.14, all face and vertex operators covering the lattice commute and, therefore, form an acceptable stabilizer generator set. The products of these generators are members of the stabilizer group. The lattice of qubits in the first panel is a local section of a larger lattice, as depicted in Fig. 9.14. The global lattice exists on the surface of a torus and does not have a boundary, meaning that it does not have a distinct beginning or end.

In contrast to the toric code, surface codes with boundaries are known as planar surface codes and were initially proposed by Kitaev and Bravyi [9]. In Fig. 9.15 we illustrate the smallest surface code [10], consisting of nine qubits residing on the vertices of the lattice. In that figure, the upper left plaquette the **Z** operators form a cycle $3 \to 6 \to 5 \to 2 \to 3$ of vertices. In our convention, that set represents the

9.5 Stabilizers II

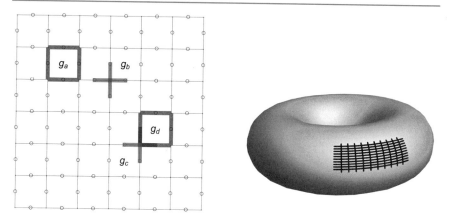

Fig. 9.14 The toric code is illustrated, with the left panel displaying a local section of a global lattice on a torus surface. In this depiction, stabilizer generators g_a, g_d consist of a sequence of **Z** operators forming a loop of edges marked by red borders. Similarly, blue lines centered on the lattice vertices indicate a sequence of **X** operators

Fig. 9.15 The smallest surface code [10] with a boundary

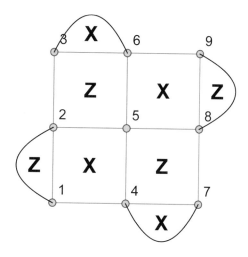

generator

$$\mathbb{1} \otimes \mathbb{1} \otimes \mathbb{1} \otimes \mathbb{1} \otimes \mathbb{1} \otimes \mathbb{1} \otimes Z_6 \otimes Z_5 \otimes \mathbb{1} \otimes Z_3 \otimes Z_2 \otimes \mathbb{1} \quad (9.72)$$

whereas the generator associated with the boundary loop connecting vertex 5 with vertex 6 is given by.

$$\mathbb{1} \otimes \mathbb{1} \otimes \mathbb{1} \otimes \mathbb{1} \otimes \mathbb{1} \otimes \mathbb{1} \otimes X_6 \otimes \mathbb{1} \otimes \mathbb{1} \otimes X_3 \otimes \mathbb{1} \otimes \mathbb{1} \quad (9.73)$$

Because each **X** and **Z** stabilizer generator shares an even number of vertices, all listed generators in Fig. 9.15 commute. This figure contains nine qubits and eight generators therefore, this code encodes one qubit.

These examples show that graphical representations of a stabilizer code offer valuable insights; however, the true strength of these codes extends far beyond merely serving as a visual aid. Surface codes possess profound mathematical structures that are revealed through topology. A thorough investigation of this topic is beyond the scope of this monograph; however, utilizing the stabilizer formalism, an interested reader can delve deeper through more advanced literature [3].

9.5.8 Fault Tolerant Computing and the Threshold Theorem

In the previous section, we discussed an important property of the Steane code. Several key single and two-qubit gates introduced in Chap. 2, such as the Hadamard gate, the S-phase gate, and the two-qubit CNOT gate, can be expressed as transversal logical gates that act on the codespace spanned by the basis $|0\rangle_{7L}$ and $|1\rangle_{7L}$. According to DiVincenzo's criteria (iv), discussed in Sect. 7.1.1, an operational quantum computer requires a set of universal quantum gates, that is, a set of gates from which any unitary operator in the Hilbert space of n qubits can be constructed. A popular universal set includes [1], the Hadamard, CNOT, and the S and T phase gates, defined in Eq. (3.44). The Steane code can implement three transversal gates, i.e., the logical Hadamard, CNOT, and S gates. The latter form a set of gates called Clifford gates.

Definition 9.10 Clifford gates map Pauli operators into Pauli operators under conjugation.

That is, if **C** represents a Clifford gate and **P** is a member of the Pauli group, we say that **P** is mapped by conjugation with **C** following the operation.

$$\mathbf{C P C}^{\dagger}. \quad (9.74)$$

As **C** is a Clifford gate, we can be assured that $\mathbf{C P C}^{\dagger}$ is also a member of the Pauli group. For example, (9.66) demonstrates how the Steane code generators are mapped by conjugation with logical Hadamard operators to other generators. Interestingly, there is a very important theorem [1], the Gottesman-Knill theorem [1] concerning Clifford gates, which, without proof, we paraphrase below.

Theorem 9.3 *[1] Quantum computations only involving specific operations, such as preparing qubits in computational basis states, applying quantum gates from the Clifford group, and measuring qubits in the computational basis, can be efficiently simulated on a classical computer.*

The theorem implies that, despite being quantum operations, a circuit consisting of only Clifford gates does not offer a quantum advantage over a classical simulation of that circuit. However, as pointed out above, Clifford gates do not form a universal gate set and, to allow quantum speedup, a quantum circuit containing the Clifford

9.5 Stabilizers II

Fig. 9.16 Concatenation of a simple circuit

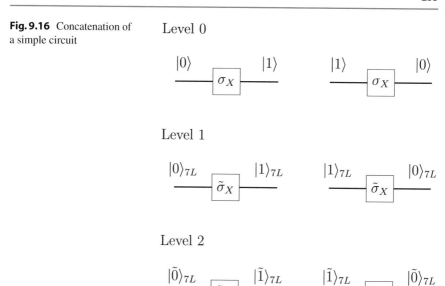

set of Hadamard, CNOT, and S gates must be supplemented by a non-Clifford gate, typically a T gate. However, these gates are challenging to implement with high fidelity. They introduce complexity and error rates that are difficult to manage directly. One proposal for introducing T gates into a circuit is a method called magic-state distillation [8].

Given that we have established an appropriate error correction code, we must be assured that the code delivers an acceptable answer within a certain allowed error probability $p_\varepsilon \ll 1$.

First, let's consider a simple calculation, whose circuit is shown in Fig. 9.16. In the upper panel of the figure, labeled Level 0, a single qubit travels along a quantum channel (denoted by a wire) from left to right; it enters a Pauli-X gate, whence it exits and embarks on an additional quantum channel before finding its final destination. In an ideal setting with the absence of noise, the information is unscathed as it traverses the three channels. But there is always noise, and we isolate three of its sources. The qubit could become corrupted as it is shuttled along the quantum channel represented by the left-hand side wire. A faulty gate is also the source of error. Finally, the qubit travels along the remaining noisy quantum channel, the r.h.s wire. Let's assume that all three error probabilities add to have gate failure probability p. We also know that our algorithm requires N gates to arrive at an answer. In Fig. 9.17 we present a simulation for the circuit error probability as a function of the number N gates. In this simulation, we choose the three values $p = 0.05, 0.025, 0.01$. From that figure, we note that for small N, the circuit failure probability scales as Np before saturating.

Suppose we are constrained with physical qubits that allow a minimum gate error probability $p = 0.01$. Let's assume, for the purpose of illustration, that we require $N = 20$ level zero gates for the algorithm to terminate with an answer. Suppose

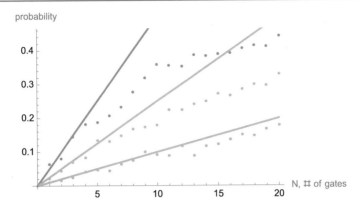

Fig. 9.17 Icons represent data points for circuit failure probability given as a function of circuit gate number N for a given gate failure probability p. Blue corresponds to $p = 0.05$, orange to $p = 0.025$, and green to 0.01. The solid lines are linear fits Np, for the corresponding gate error probabilities p

we accept the answers that satisfy the maximum error bound $p_\varepsilon < 0.01$. Clearly, according to Fig. 9.17 the Level 0 circuit is not acceptable as the circuit failure probability $p_C \approx 0.17$ at $N = 20$ gates is over an order of magnitude larger than p_ε.

Thus, we need an error-correcting scheme by introducing the seven-qubit logical codewords $|0\rangle_{7L}, |1\rangle_{7L}$ of the Steane code (see Mathematica Notebook 9.3) to mitigate the error rate in the first channel, retrieve a corrected state, and feed it into a physical Pauli-X gate. But that defeats the goal of fault-tolerant computing, as the gate introduces additional errors. Instead, it is desirable to introduce a new gate $\tilde{\sigma}_X$, defined by the following

$$\tilde{\sigma}_X |0\rangle_{7L} = |1\rangle_{7L}$$
$$\tilde{\sigma}_X |1\rangle_{7L} = |0\rangle_{7L}. \quad (9.75)$$

This gate behaves like a Pauli-X gate on the codewords $|0\rangle_{7L}, |1\rangle_{7L}$, and the stabilizer formalism provides a roadmap for constructing it from physical qubits. In the Steane code, this logical gate is transversal and given by

$$\tilde{\sigma}_X = \mathbf{XXXXXXX}. \quad (9.76)$$

Gate $\tilde{\sigma}_X$ is called a fault-tolerant gate if it restricts the probability of error to an upper bound value of cp^2, where p is the probability of failure of the gate Level 0 and c a constant. This merit figure improves the fidelity of the circuit provided $cp^2 < p$ or $p < p_{th}$ where $p_{th} \equiv 1/c$ is the *threshold probability*. For the sake of simplicity, we assume that the circuit in the panel labeled Level 1 in Fig. 9.16 suffers a gate error cp^2. For N independent gates, the circuit error probability is shown in Fig. 9.18.

Note that whereas the Level 0 circuit requires a total of $N = 20$ qubits for the algorithm to halt, we now require more resources as a logical qubit is constituted from a number $d > 1$ physical qubits. For the Steane code, we assume $d \approx 7$, and so

9.5 Stabilizers II

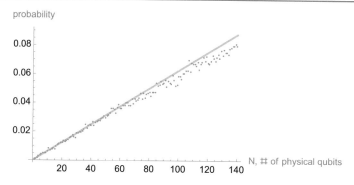

Fig. 9.18 Circuit failure probability p_C for Level 1 gates, where $p = 0.01$, $c = 6.25$ as a function of the number of physical qubits

we require $d\,20 \approx 140$ qubits for the algorithm to halt. Though, from Fig. 9.18, we find a circuit probability $p_C \approx 0.07$, an improvement from the Level 0 $p_C \approx 0.17$, it does not achieve our prescribed error bound $p_\varepsilon = 0.01$. We can improve the fidelity of the circuit By repeating this process, using the method of concatenation, lower error bounds are enforced. We know that the codewords

$$|0\rangle_{7L} = \sum c_{(0)}(k_1, k_2 \ldots k_7) |k_7\rangle \otimes |k_6\rangle \otimes \cdots \otimes |k_1\rangle$$
$$|1\rangle_{7L} = \sum c_{(1)}(k_1, k_2 \ldots k_7) |k_7\rangle \otimes |k_6\rangle \otimes \cdots \otimes |k_1\rangle \quad (9.77)$$

where $c_{(0)}(k_1, k_2 \ldots k_7)$, $c_{(1)}(k_1, k_2 \ldots k_7)$ are expansion coefficients for the Steane code, and the sum is over all values $k_1, k_2 \ldots k_7$, where $k_i \in \{0, 1\}$. Let's define new code words $\left|\tilde{0}\right\rangle_{7L}$, $\left|\tilde{1}\right\rangle_{7L}$

$$\left|\tilde{0}\right\rangle_{7L} = \sum c_{(0)}(k_1, k_2 \ldots k_7)|k_7\rangle_{7L} \otimes |k_6\rangle_{7L} \otimes \cdots \otimes |k_1\rangle_{7L}$$
$$\left|\tilde{1}\right\rangle_{7L} = \sum c_{(1)}(k_1, k_2 \ldots k_7)|k_7\rangle_{7L} \otimes |k_6\rangle_{7L} \otimes \cdots \otimes |k_1\rangle_{7L}, \quad (9.78)$$

for $k_i \in \{0, 1\}$. The codewords $\left|\tilde{0}\right\rangle_{7L}$, $\left|\tilde{1}\right\rangle_{7L}$ are defined by the Steane code, except that they are expressed as a linear combination of the logical states $|0\rangle_{7L}, |1\rangle_{7L}$. In addition we define a new operator

$$\tilde{\tilde{\sigma}}_X \left|\tilde{0}\right\rangle_{7L} = \left|\tilde{1}\right\rangle_{7L}$$
$$\tilde{\tilde{\sigma}}_X \left|\tilde{1}\right\rangle_{7L} = \left|\tilde{0}\right\rangle_{7L}. \quad (9.79)$$

which acts like a Pauli-X gate for codewords $\left|\tilde{0}\right\rangle_{7L}$, $\left|\tilde{1}\right\rangle_{7L}$. The last panel, labeled Level 2, in Fig. 9.16 represents a concatenated version of the Level 1 diagram. With this circuit, the probability of error is reduced by another factor to order

$d^2 N c(cp^2)^2 = Nd^2 c^3 p^4$. According to the concatenation cycle

$$\begin{array}{ccccccc} k=0 & k=1 & k=2 & k=3 & \cdots & k \\ p \to & dcp^2 \to & d^2 c(cp^2)^2 \to & d^3 c(c(cp^2)^2)^2 & & \\ \hline p & dcp^2 & d^2 c^3 p^4 & d^3 c^7 p^8 & \cdots & d^k \frac{(cp)^{2^k}}{c} \end{array} \qquad (9.80)$$

$$p_C = N d^k (cp)^{2^k}/c. \qquad (9.81)$$

If $p < 1/c$, we can attain a prescribed error probability bound after k concatenations provided that,

$$N d^k (cp)^{2^k}/c \le \varepsilon. \qquad (9.82)$$

Dividing by N, multiplying by c, and taking the base-2 logarithm on both sides we get

$$\log_2 d^k + 2^k \log_2(cp) \le \log_2 \frac{\varepsilon c}{N} \qquad (9.83)$$

since the second term on the left hands side of Eq. (9.83) dominates for large k, we replace that equation with

$$2^k \log_2(cp) < \log_2 \frac{\varepsilon c}{N} \qquad (9.84)$$

or

$$2^k < \frac{\log_2(\varepsilon c/N)}{\log_2(pc)} = \frac{\log_2(N/\varepsilon c)}{\log_2(1/pc)}. \qquad (9.85)$$

Solving for k by taking the logarithm of both sides and exponentiating d with it, we get

$$d^k < \left(\frac{\log_2(N/\varepsilon c)}{\log_2(1/pc)} \right)^{\log_2 d} \qquad (9.86)$$

where we used the identity $x^{\log y} = y^{\log x}$. In other words, the total number of resources required to obtain the desired error bound ε is proportional to N, multiplied by the factor (9.86), a polynomial of a logarithmic argument of N. Because of this bound, we can be assured that our circuit will not grow in an unbounded manner, as we require more levels of concatenations to meet the maximum allowed error margin.

Mathematica Notebook 9.4: Fault tolerance http://www.physics.unlv.edu/%7Ebernard/MATH_book/Chap9/chap9_link.html; See also https://bernardzygelman.github.io

Problems

Mathematica Notebook 9.5: Problems and exercise http://www.physics.unlv.edu/%7Ebernard/MATH_book/Chap9/chap9_link.html; See also https://bernardzygelman.github.io

References

1. M.E. Nielsen, I.L. Chuang, *Quantum Computation and Quantum Information* (Cambridge University Press, 2011)
2. A. Zee, *Group Theory in a Nutshell for Physicists* (Princeton University Press, 2016)
3. K. Fujii, *Quantum Computation with Topological Codes*, Springer Briefs in Mathematical Physics (2015)
4. D. Gottsman, *Stabilizer Codes and Quantum Error Correction*, arxiv:quant-ph/9705052v1 (1997)
5. R. Laflamme, C. Miquel, J.P. Paz, W.H. Zurek, *Perfect Quantum Error Correction Code* (1996). arXiv:quant-ph/9602019
6. A. Steane, Multiple-particle interference and quantum error correction. Proc. Roy. Soc. Lond. A **452**, 2551 (1996)
7. https://en.wikipedia.org/wiki/Graph_(discrete_mathematics)
8. A.Yu. Kitaev, *Fault-tolerant quantum computation by anyons*, https://arxiv.org/pdf/quant-ph/9707021 (1997); Ann. Phys. **303**, 2 (2003)
9. S.B. Bravyi, A.Yu. Kitaev, *Quantum codes on a lattice with boundary*, arXiv:quant-ph/9811052 (1998)
10. A.G. Fowler, D. Sank, J. Kelly, R. Barends, J.M. Martinis, *Scalable extraction of error models from the output of error detection circuits*, arXiv:quant-ph/1405.1454 (2014)

Adiabatic Quantum Computing   10

Abstract

This chapter introduces Adiabatic Quantum Computing (AQC) and highlights its potential as an alternative to the circuit model. We introduce, via numerical demonstration, the quantum adiabatic theorem, which asserts that a system will remain in its ground state provided the Hamiltonian changes slowly enough. This principle forms the basis of AQC's approach, making it capable of addressing complex problems like determining the ground states of quantum many-body systems, or the extremum of a cost function. The chapter covers the technique of embedding a time-independent Hamiltonian in a time-dependent one, facilitating a transition from a simple, solvable system to a more complex target problem. By slowly evolving the system, the initial Hamiltonian's ground state becomes the target Hamiltonian's ground state. Additionally, the text explores its application in areas traditionally linked to the circuit model, like Grover's algorithm. Theoretical concepts are demonstrated with practical examples. Lastly, the chapter examines geometric and topological phases generated by adiabatic evolution and explores their potential as a resource.

10.1 Introduction

In addition to the circuit model, adiabatic quantum computing (AQC) shows great promise as a computing platform. AQC does not necessarily require a description based on gates and circuit components. However, the circuit model and AQC share common features. The equivalence of the circuit model and the AQC paradigm was demonstrated in a definition of the latter provided in [1]. The historical roots of AQC are related to a classical algorithm known as *simulated annealing*. The goal of simulated annealing is to find the extremum (usually the minimum) of a multivariate function $f(x_1 x_2...x_n)$, which is sometimes referred to as a *cost function*. This is achieved by simulating a physical system that encapsulates this function and using statistical and thermodynamic principles to reach a configuration where the function reaches its extremum at the point $x_1^* x_2^*...x_n^*$.

Typically, AQC aims to find the ground state (energy eigenstate with the lowest energy) of a quantum system by simulating the continuous Hamiltonian evolution of the latter. Problems of this type are essential in many applications. For example, finding the ground state of large molecules can require tremendous time and memory resources on standard digital computers. This problem has many applications, including drug design and understanding of protein folding. AQC also has potential applications in other fields, such as machine learning, which is still an emerging area. In addition, algorithms typically associated with the circuit model, such as the Grover algorithm, can be implemented in AQC. The AQC paradigm, as defined in [1,2], is believed to be universal [2].

Our first task is to find the ground state of a Hamiltonian **H**. To achieve the desired goal, one takes advantage of the quantum adiabatic theorem, which we paraphrase (for the particular case of the ground state) below.

10.1.1 The Quantum Adiabatic Theorem

Consider a quantum system described by a Hamiltonian $\mathbf{H}(t)$ that evolves slowly (i.e., adiabatically) in time t. If this system possesses a non-degenerate ground state $|u(t_0)\rangle$ at $t = t_0$, then $\exp(i\gamma(t))|u(t)\rangle$ is a solution to the Schrödinger equation corresponding to this Hamiltonian provided that $|u(t)\rangle$ is an instantaneous eigenstate of $\mathbf{H}(t)$, at time t [3,4], i.e.

$$\mathbf{H}(t)|u(t)\rangle = E(t)|u(t)\rangle \tag{10.1}$$

where $E(t)$ is an instantaneous eigenvalue of $\mathbf{H}(t)$. The phase

$$\exp(i\gamma(t)) = \frac{\langle u(t)|\mathbf{U}(t,t_0)|u(t_0)\rangle}{|\langle u(t)|\mathbf{U}(t,t_0)|u(t_0)\rangle|} \tag{10.2}$$

where $\mathbf{U}(t,t_0)$ is the time development operator for $\mathbf{H}(t)$. Note that, in general, $|\psi(t)\rangle \equiv \mathbf{U}(t,t_0)|u(t_0)\rangle \neq |u(t)\rangle$.

How slowly must $\mathbf{H}(t)$ evolve for this theorem to hold? We will only partly address that question in this discussion. Instead, we will experiment with a model system to illustrate this feature of the AQC platform.

10.1 Introduction

10.1.2 The AQS Strategy

Our goal is to find the ground state, or eigenstate with the smallest eigenvalue, of a time-independent Hamiltonian **H**. If **H** is embedded in a Hilbert space spanned by n qubits, a direct method would require the construction of a matrix representation of **H** in that Hilbert space and the solution of an eigenvalue equation for the smallest eigenvalue of that matrix. In physical systems of n qubits, it requires the manipulation of a $2^n \times 2^n$ matrix. Even with a modest number of qubits, say $n = 50$, this task becomes intractable for standard digital computing machines. Here, we demonstrate how the quantum adiabatic theorem can be exploited to find a ground state.

The strategy involves embedding **H** in a time-dependent Hamiltonian $\mathbf{h}(t)$ in such a way that the time-independent Hamiltonian $\mathbf{H}_0 \equiv \mathbf{h}(t_0)$ has a simple form, allowing us to efficiently determine its eigenstates and eigenvalues. However, at some large value $t = T$, $\mathbf{h}(t)$ merges into $\mathbf{h}(T) = \mathbf{H}$, the "hard problem". The adiabatic theorem tells us that if this evolution is carried out sufficiently slowly, the eigenstate at $t = t_0$ will evolve into a solution to the Schrödinger equation at $t = T$ an eigenstate of the "hard" Hamiltonian **H**.

Consider the following parameterization for $\mathbf{h}(t)$

$$\mathbf{h}(t) \equiv \mathbf{H}_0(1 - \frac{t}{T}) + \frac{t}{T}\mathbf{H}. \tag{10.3}$$

Note that at $t = t_0 = 0$, $\mathbf{h}(t_0) = \mathbf{H}_0$ and at $t = T$, $\mathbf{h}(T) = \mathbf{H}$. Consider an array of qubits as shown in Fig. 10.1 that allows quantum control of qubit-qubit interactions and single qubit interactions with an external agent. Given an initial state $|u(t_0)\rangle$ for this set of qubits, the quantum state at a later time $t > t_0$ evolves according to the Schrödinger equation (3.36), with the generator of that evolution being the Hamiltonian operator $\mathbf{h}(t)$. The time dependence of $\mathbf{h}(t)$ has the form shown in Eq. (10.3), and is induced by an external agent (i.e. depending on the qubit architecture, laser, microwave, optical, etc. fields). The initial state is taken to be a ground state of $\mathbf{h}(t_0 = 0) = \mathbf{H}_0$, chosen judiciously so that its eigenstates can be easily obtained using standard methods. The quantum adiabatic theorem guarantees that if the time rate of change in $\mathbf{h}(t)$ is small, $|u(t_0)\rangle$ evolves into an eigenstate of **H**, at $t = T$. Now

$$\frac{d\,\mathbf{h}(t)}{dt} = -\frac{\mathbf{H}_0}{T} + \frac{\mathbf{H}}{T} \tag{10.4}$$

suggesting that as $T \to \infty$, the conditions required to invoke the quantum adiabatic theorem are satisfied.

Fig. 10.1 Differences between the circuit model and AQC. In both upper and lower panels, amplitude $|\psi_{in}\rangle$ at initial time t_0, evolves into $|\psi_{out}\rangle = \mathbf{U}(t_f, t_0)|\psi_{in}\rangle$. In the upper panel, the time evolution operator is digitized so that $\mathbf{U} = \mathbf{U}_n \ldots \mathbf{U}_3\mathbf{U}_2\mathbf{U}_1$. In the second panel $\mathbf{U}(t) = T \exp\left(-i \int_{t_0}^{t} \mathbf{h}(\tau) d\tau\right)$ is parameterized by a continuous variable t. $\mathbf{H}_0(t)$ describes single-qubit interactions with an external time dependent field. $\mathbf{H}_{int}(t)$ describes qubit-qubit interactions and $\mathbf{h}(t) = \mathbf{H}_0(t) + \mathbf{H}_{int}(t)$

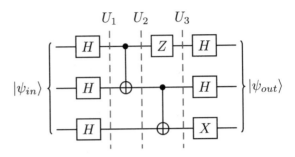

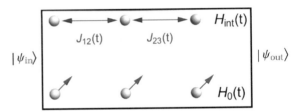

10.1.3 Model System

To demonstrate some of the features of the AQC paradigm, we introduce the following time-independent Hamiltonian for a three-qubit system.

$$\mathbf{H} = \hbar\omega \left(\sigma_X \otimes \mathbb{1} \otimes \mathbb{1} - \mathbb{1} \otimes \sigma_X \otimes \mathbb{1} + \mathbb{1} \otimes \mathbb{1} \otimes \sigma_x\right) + \hbar\Omega \left(\sigma_X \otimes \sigma_X \otimes \mathbb{1} + \sigma_X \otimes \mathbb{1} \otimes \sigma_X + \mathbb{1} \otimes \sigma_X \otimes \sigma_X\right). \quad (10.5)$$

Its matrix representation with respect to the computational basis is

$$\underline{\mathbf{H}} = \begin{pmatrix} 0 & \hbar\omega & -\hbar\omega & \hbar\Omega & \hbar\omega & \hbar\Omega & \hbar\Omega & 0 \\ \hbar\omega & 0 & \hbar\Omega & -\hbar\omega & \hbar\Omega & \hbar\omega & 0 & \hbar\Omega \\ -\hbar\omega & \hbar\Omega & 0 & \hbar\omega & \hbar\Omega & 0 & \hbar\omega & \hbar\Omega \\ \hbar\Omega & -\hbar\omega & \hbar\omega & 0 & 0 & \hbar\Omega & \hbar\Omega & \hbar\omega \\ \hbar\omega & \hbar\Omega & \hbar\Omega & 0 & 0 & \hbar\omega & -\hbar\omega & \hbar\Omega \\ \hbar\Omega & \hbar\omega & 0 & \hbar\Omega & \hbar\omega & 0 & \hbar\Omega & -\hbar\omega \\ \hbar\Omega & 0 & \hbar\omega & \hbar\Omega & -\hbar\omega & \hbar\Omega & 0 & \hbar\omega \\ 0 & \hbar\Omega & \hbar\Omega & \hbar\omega & \hbar\Omega & -\hbar\omega & \hbar\omega & 0 \end{pmatrix}. \quad (10.6)$$

Calculating its eigenvalues typically involves finding the roots of a high-degree polynomial, in this case, a $2^3 = 8$ degree polynomial involving two parameters ω, Ω. This task is not problematic with standard numerical eigenvalue solvers on an average laptop. However, for a modest $n = 50$ qubit system, the cited classical algorithm leads to a 2^{50} degree polynomial.

10.1 Introduction

In the quantum adiabatic algorithm (QAA), we first define an "easy" Hamiltonian

$$\mathbf{H}_0 = \hbar\omega(\sigma_Z \otimes \mathbb{1} \otimes \mathbb{1} + \mathbb{1} \otimes \sigma_Z \otimes \mathbb{1} + \mathbb{1} \otimes \mathbb{1} \otimes \sigma_Z) \tag{10.7}$$

whose matrix representation,

$$\begin{pmatrix} 3\hbar\omega & 0 & 0 & 0 & 0 & 0 & 0 & 0 \\ 0 & \hbar\omega & 0 & 0 & 0 & 0 & 0 & 0 \\ 0 & 0 & \hbar\omega & 0 & 0 & 0 & 0 & 0 \\ 0 & 0 & 0 & -\hbar\omega & 0 & 0 & 0 & 0 \\ 0 & 0 & 0 & 0 & \hbar\omega & 0 & 0 & 0 \\ 0 & 0 & 0 & 0 & 0 & -\hbar\omega & 0 & 0 \\ 0 & 0 & 0 & 0 & 0 & 0 & -\hbar\omega & 0 \\ 0 & 0 & 0 & 0 & 0 & 0 & 0 & -3\hbar\omega \end{pmatrix} \tag{10.8}$$

allows us to read off its eigenvalues on the diagonal. The state with the lowest energy, $-3\hbar\omega$, is given by ket $|111\rangle$ or $|7\rangle_3$ (Remember that $\sigma_Z|1\rangle = -|1\rangle$.) Thus, we construct the time-dependent system Hamiltonian

$$\mathbf{h}(t) = \mathbf{H}_0(1 - \frac{t}{T}) + \mathbf{H}\frac{t}{T} \tag{10.9}$$

and initialize the system into the quantum state $|u(t_0)\rangle = |111\rangle$, the ground state of $\mathbf{H}_0$. The system evolves according to the Schrödinger equation

$$i\hbar\frac{d|\psi(t)\rangle}{dt} = \mathbf{h}(t)|\psi(t)\rangle \tag{10.10}$$

where $|\psi(t_0)\rangle = |u(t_0)\rangle$. We let the system evolve until time T to the final state $|\psi(T)\rangle$. Under the adiabatic assumption $|\psi(T)\rangle$ is equal, up to a phase, to $|u(T)\rangle$, the ground state of $\mathbf{H}$. Below, we simulate quantum evolution using a classical algorithm.

10.1.3.1 Solution to the Schrödinger Equation

Mathematica Notebook 10.1: An Introduction to Adiabatic Quantum Computing http://www.physics.unlv.edu/%7Ebernard/MATH_book/Chap7/chap7_link.html; see also https://bernardzygelman.github.io

We are dealing with a three-qubit Hilbert space that is spanned by the vectors $|0\rangle_3, |1\rangle_3, |2\rangle_3, \ldots |7\rangle_3$. Thus a solution to the Schrödinger Eq. (10.10) can be expressed as

$$|\psi(t)\rangle = \sum_{i=0}^{i=7} c_i(t)|i\rangle_3 \tag{10.11}$$

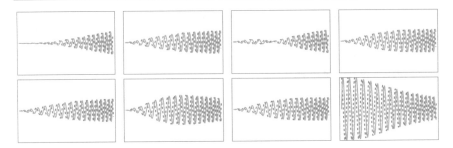

Fig. 10.2 Plot of the real (blue dotted line) and imaginary (orange solid line) of the amplitudes $c_i(t)$ as a function of t (horizontal axis). The $i = 0$ amplitude is shown in the upper left-hand panel, followed going left to right by the amplitudes $i = 1, 2 \ldots 7$

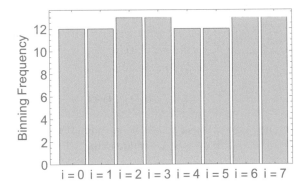

Fig. 10.3 Plot of measurement statistics for observation of final state $|\psi(T)\rangle$, with the computational basis. For example, out of 100 measurements (on the same $|\psi(T)\rangle$), 12 instances reveal the configuration 000, corresponding to $i = 0$

where we require the initial value condition $c_i(t_0) = 0$ for $i \neq 7$, $c_7(t_0) = 1$. Inserting expression (10.11) into Eq. (10.10), we obtain a set of first-order coupled differential equations for coefficients $c_i(t)$. In an AQC experiment, protected from and free of decoherence, Hamiltonian evolution is guaranteed by the laws of nature, but here we must simulate this evolution using a classical algorithm. We set $\Omega = 5\omega$, $\omega = 1$, $\hbar = 1$ in our simulation.

Solutions to the Schrödinger are shown in Fig. 10.2, note that at $t = 0$ all amplitudes save the real part of $c_7(0)$ vanish. A later time, $t = T \gg t_0$ we measure, as shown in Fig. 10.3, the qubit configurations in the computational basis. The probability for observing a specific configuration is given by the Born rule $p_i = |\langle\psi(T)|i\rangle_3|^2$. We require multiple runs and measurements with identical initial conditions to obtain good statistics. For example, in Fig. 10.3, a bar chart illustrates the binning occupancy for each state i, at time $t = T$ after $N = 100$ runs.

To estimate the ground state energy of $\mathbf{H}$, one needs to evaluate the matrix element $\langle\mathbf{H}\rangle \equiv \langle\psi(T)|\mathbf{H}|\psi(T)\rangle$. Though we have access to $|\psi(T)\rangle$, measurement in the computational basis does not provide the phase information needed to evaluate $\langle\mathbf{H}\rangle$. The process of reconstructing a quantum state through repeated measurements is known as *quantum state tomography* [5], and a detailed discussion of this topic is beyond the scope of this monograph. However, a method for

10.1 Introduction

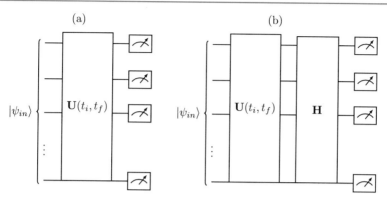

Fig. 10.4 **a** Measurement, in the computational basis, of the bit string configuration for state $|\psi(T)\rangle \sim \mathbf{U}(T, t_0)|u(t_0)\rangle$. **b** Measurement, in the computational basis, of $\mathbf{H}|\psi(T)\rangle \approx E_g|\psi(T)\rangle$, in the adiabatic approximation

evaluating $\langle \mathbf{H} \rangle$ is available [6] if the operator $\mathbf{H}$ can be represented as a sum of tensor products of Pauli operators. Instead, here we offer a different approach. First, we measure the system in state $|\psi(T)\rangle$ with repeated measurements in the computational basis, as shown in panel (a) of Fig. 10.4. We estimate the value $|c_i(T)|^2$, shown in Fig. 10.3, for a given i from that data. Next, we include an additional gate $\mathbf{H}$ as shown in panel (b) of Fig. 10.4, acting on the output $|\psi(T)\rangle$. Its output is $|\psi'(T)\rangle = \mathbf{H}|\psi(T)\rangle$. Repeated measurements, in the computational basis, allow us to estimate $|c_i'(T)|^2 = |\langle i|\psi'(T)\rangle|^2 = |\langle i|\mathbf{H}|\psi(T)\rangle|^2 = E_g^2|c_i(T)|^2$, where we used, according to the adiabatic condition, the fact that $\mathbf{H}|\psi(T)\rangle \approx E_g|\psi(T)\rangle$ and E_g is the ground state eigenvalue of $\mathbf{H}$. Thus

$$|E_g(i)| = \sqrt{\frac{|c_i'(T)|^2}{|c_i(T)|^2}}. \tag{10.12}$$

In Fig. 10.5, we plot the observed values, from our simulation, the ratios Eq. (10.12) for each measurement output configuration.

We observe variation in the estimation of E_g as a function of the bit string configuration index, i, with the average value $\langle E_g \rangle = 5.00$. The correct sign for E_g, can be gleaned from the structure of matrix $\mathbf{H}$, or using the variational principle $\langle \phi|\mathbf{H}|\phi\rangle \geq E_g$, for an arbitrary state $|\phi\rangle$. Exact diagonalization of $\mathbf{H}$ reveals that $E_g = -5.00$, in harmony with results posted in Table 10.1, in which we itemize $\langle \psi(T)|\mathbf{H}|\psi(T)\rangle$, for values of the adiabatic parameter T in the range $T = 2$ to $T = 32$.

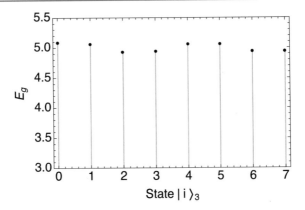

Fig. 10.5 The graph shows the ratio $|E_g|$ for each configuration index i of the observed bit string. For instance, when $i = 0$, it corresponds to the bit string 000, and when $i = 7$, it corresponds to 111. The average value for all observed bit string values is $\langle |E_g| \rangle = 5.00$

Table 10.1 Estimate of $\langle \psi(T)|\mathbf{H}|\psi(T)\rangle$, as a function of the adiabaticity parameter T

T	$E_g = \langle \psi(T)|\mathbf{H}|\psi(T)\rangle$
2.0	-3.07
4.0	-4.58
8.0	-4.96
16.0	-4.98
32.0	-5.00

The fast convergence for the estimate E_g is a consequence of the fact that the ground adiabatic state of $\mathbf{h}(t)$, shown in Fig. 10.6, is far removed from the excited states of $\mathbf{h}(t)$ for all values of t. This is not usually the case; for many practical Hamiltonians, the ground state can suffer avoided crossings, in which the ground state energy eigenvalue almost merges with an excited state eigenvalue for some short time interval Δt. In such cases, we need to conduct a more thorough analysis.

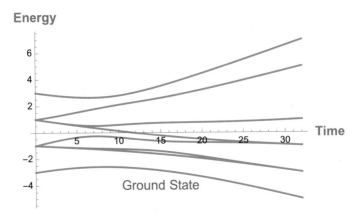

Fig. 10.6 The instantaneous adiabatic energies, as a function of t, for the Hamiltonian $\mathbf{h}(t)$

10.1.4 Avoided Crossings

In many instances, the curves corresponding to the adiabatic energies exhibit an avoided crossing at some particular time t_c. For example, consider the single-qubit Hamiltonian

$$\mathbf{h}(t) = \mathbf{H}_0(1 - \frac{t}{T}) + \frac{t}{T}\mathbf{H}$$

$$\mathbf{H}_0 = \begin{pmatrix} \Delta & 0 \\ 0 & -\Delta \end{pmatrix} \quad \mathbf{H} = \Delta \begin{pmatrix} -1 & \xi \\ \xi & 1 \end{pmatrix} \tag{10.13}$$

where Δ, ξ are real parameters. The instantaneous eigenstates of $\mathbf{h}(t)$ are illustrated in Fig. 10.7.

Figure 10.7 illustrates an avoided crossing at $s = 1/2$ as a function of parameter ξ. It is evident that the energy difference $\Delta E(t_c) \to 0$ at the instance of closest approach as $\xi \to 0$. As this is a single-qubit system, the eigenvalues of $\mathbf{H}$ are easily calculated to give

$$E_g = -\Delta(1 + \xi^2). \tag{10.14}$$

Thus $\langle \mathbf{H} \rangle \equiv \langle \psi | \mathbf{H} | \psi \rangle$ where $|\psi(t)\rangle$ is a solution to the Schrödinger equation for the state that "grows" out of the ground state of $\mathbf{H}_0$ at $t = 0$ should approach E_g in the adiabatic limit. In Table 10.2, we itemize the calculated values for $\langle \mathbf{H} \rangle$, for various values of T, and ξ. The data in this Table demonstrates that the convergence of $\langle \mathbf{H} \rangle$ to its adiabatic limit is highly dependent on the parameter ξ, which controls the value of the energy gap $\Delta E(t_c)$ at the point of avoided crossing. For instance, when $\xi = 1.0$, $\Delta E(t_c) = 1$, and when $\xi = 0.5$, $\Delta E(t_c) = 0.5$, the estimate of $\langle \mathbf{H} \rangle$

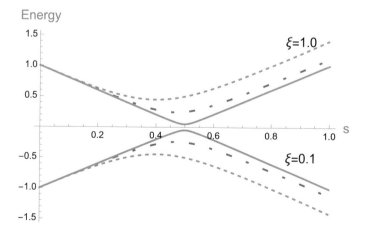

Fig. 10.7 The instantaneous adiabatic energies, as a function of $s \equiv t/T$, for the Hamiltonian $\mathbf{h}(t)$ given in Eq. (10.13). The dotted line corresponds to the eigenvalue for $\xi = 1.0$, the dashed-dotted line for $\xi = 0.5$, and the solid line for $\xi = 0.1$. The parameter $\Delta = 1$ for all three cases

Table 10.2 Estimate of $\langle H \rangle$, as a function of the adiabaticity parameter T. The Table illustrates the convergence of $\langle H \rangle$ toward the exact values predicted by Eq. (10.14). The parameter $\Delta = 1$

T	$\xi = 1.0$	$\xi = 0.5$	$\xi = 0.1$
10.0	−1.276	−0.215	0.925
20.0	−1.406	−0.751	0.851
100.0	−1.414	−1.118	0.355
500.0	−1.414	−1.118	−0.721
1×10^3	−1.414	−1.118	−0.965
3×10^3	−1.414	−1.118	−1.005
5×10^3	−1.414	−1.118	−1.005

converged to the correct value by $T = 100$. However, for $\xi = 0.1$, $\Delta E(t_c) = 0.1$, convergence is only achieved at $T = 3 \times 10^3$. To further explore this behavior, we appeal to Landau-Zener theory [7,8].

10.1.4.1 Landau-Zener Transitions, and the Scheduling Function

The theory that describes processes in which a dynamical system navigates a region in which adiabatic energies suffer an avoided crossing is called Landau-Zener theory, after L. Landau and C. Zener. An application of the latter is in the theory of slow atomic collisions [9].

When an atom slowly approaches an ion, one can calculate the adiabatic energies of the system at any given time or particle separation. The latter is governed by the Hamiltonian that describes all Coulomb interactions between nuclei and electrons. A diagram that illustrates the behavior of the adiabatic energies as a function of particle separation is called a potential energy diagram. It turns out [9] that the appearance of an avoided crossing of the ground state with an excited state is typical in such diagrams. If the system starts in the ground state, at some later time it will probably experience an avoided crossing at some critical inter-atom distance or some instance of time t_c. Landau-Zener theory states that at the avoided crossing the system suffers a transition from an incoming, initial, adiabatic state $|i\rangle$, to excited adiabatic state $|f\rangle$, with probability

$$P_{i \to f} \sim \exp(-\frac{\Gamma}{v}) \qquad (10.15)$$

where Γ is a parameter that is a function of the avoided crossing energy gap, that is, $\Delta(t_c)$, at the crossing time t_c, and v is the relative velocity of the collision partners at the instance of crossing. This expression informs us that, for a given Γ, the probability of a transition increases exponentially for larger values of v. Alternatively, small values of v exponentially decrease the likelihood of a transition and enhance enforcement of adiabaticity.

This suggests that we can improve the convergence property of the system described above if we can control the time or speed at which we pass through an

10.1 Introduction

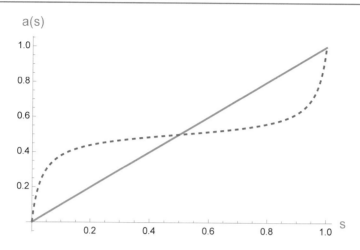

Fig. 10.8 Scheduling function $a_1(s) = s$, shown by the solid line. The dashed line represents the scheduling function $a_2(s)$

avoided crossing. This is indeed the case, but a scheduling function must be introduced. Consider again the Schrödinger equation,

$$i \frac{d|\psi(t)\rangle}{dt} = \mathbf{h}(t)|\psi(t)\rangle \tag{10.16}$$

and let us define a new variable $s \equiv t/T$, so that Eq. (10.16) becomes

$$\frac{i}{T} \frac{|d\psi(s)\rangle}{ds} = \mathbf{h}(s)|\psi(s)\rangle \tag{10.17}$$

where as t ranges from $t = 0$ to $t = T$, $0 \le s \le 1$. According to this parametrization

$$\mathbf{h}(s) = \mathbf{H}_0(1 - a(s)) + a(s)\mathbf{H} \quad a(s) = s \equiv \frac{t}{T}. \tag{10.18}$$

$a(s)$ is called the scheduling function defined so that $a(0) = 0$ and $a(1) = 1$. The linear function $a(s) = s$, used in our discussions so far, is shown in Fig. 10.8 by the solid line. We note that at the instance of an avoided crossing $s_c = 1/2$, the derivative $a'(s) \equiv da/ds = 1$ is non-zero. According to Landau-Zener theory, it would be preferable to have $a'(s = 1/2) \ll 1$ to avoid a so-called nonadiabatic transition to an excited state. The dashed line in Fig. 10.8 is a candidate for such a scheduling function.

In Table 10.3, we tabulate the inner product $\gamma \equiv |\langle \phi | \psi(T) \rangle|$ where $|\psi(T)\rangle$ is the solution of the Schrödinger equation discussed above, and $|\phi\rangle$ is the exact ground eigenstate for Hamiltonian $\mathbf{H}$. A value $\gamma < 1$ indicates failure of the adiabatic approximation as the system has made a Landau-Zener transition into an excited state. In that Table, we calculate γ using two different functional forms, illustrated in Fig. 10.8, for the scheduling function $a(s)$. For this calculation, we use the value

Table 10.3 Estimates for the overlap γ as a function of the adiabaticity parameter T. The table illustrates the convergence of γ toward unity, the adiabatic limit

T	$a_1(s)$	$a_2(s)$
10.0	0.199	0.405
20.0	0.277	0.659
100.0	0.569	0.972
300.0	0.831	1.000
400.0	0.889	1.000
500.0	0.927	1.000
1×10^3	0.990	1.000
2×10^3	1.000	1.000

$\xi = 0.1$. The first column of that Table includes the entries corresponding to the scheduling function $a_1(s) = s$, whereas the second column corresponds to $a_2(s)$, the dashed line shown in Fig. 10.8. It is evident, from this data, that the use of the scheduling function a_s greatly enhances the convergence to the ground state of $\mathbf{H}$.

10.1.5 The Grover Algorithm

In the previous discussion, we focused on minimizing the cost function $\langle \psi | \mathbf{H} | \psi \rangle$. However, the AQC algorithm allows a myriad of applications [2], some touching on problems associated with circuit-based algorithms; for example, the Grover algorithm, studied in Chap. 4, which seeks to identify a target state $|\xi\rangle$ in an n-qubit system, under a series of oracle queries. In the gate model version of the Grover algorithm, a n-qubit system is initialized to the superposition state $|\phi\rangle \equiv \mathbf{H}^{\otimes n}|0\rangle_n$. A query of $|\phi\rangle$ provides the correct item $|\xi\rangle$ with probability $1/N$, where $N = 2^n$, the classical figure of merit for an unsorted search. The Grover circuit requires approximately $\sqrt{N}$ gates to drive the success probability of finding $|\xi\rangle$ to unity.

In the AQC paradigm, we define an initial Hamiltonian $\mathbf{H}_0$ whose ground state is given by $|\phi\rangle$. This Hamiltonian is embedded in a time-dependent Hamiltonian $\mathbf{h}(t)$, so that the ground state of $\mathbf{H}_0$ is given by $|\phi\rangle$. Thus, consider the Hamiltonian,

$$\mathbf{H}_0 \equiv \mathbb{1} - |\phi\rangle\langle\phi| \qquad (10.19)$$

where

$$|\phi\rangle \equiv \frac{1}{\sqrt{N}} \sum_{i=0}^{N-1} |i\rangle_n \qquad (10.20)$$

in an $N = 2^n$ dimensional Hilbert space spanned by basis vectors $|i\rangle_n$. From its definition, the ground state of $\mathbf{H}_0$ is given by the vector $|\phi\rangle$, and since $\mathbf{H}_0|\phi\rangle = 0$, its ground state energy $E_g = 0$. The remaining $N - 1$ vectors orthogonal to $|\phi\rangle$, are degenerate eigenstates of $\mathbf{H}_0$, with eigenvalue $(+1)$. We also define

$$\mathbf{H} \equiv \mathbb{1} - |\xi\rangle\langle\xi| \qquad (10.21)$$

10.1 Introduction

where ξ is a label for some arbitrary basis vector $|i\rangle_3$. $|\xi\rangle$ represents the ground state, with eigenvalue 0, of $\mathbf{H}$. Basis vector $|i\rangle_n \neq |\xi\rangle$, is a degenerate excited eigenstate of $\mathbf{H}$ with eigenvalue $(+1)$.

We initialize the system to the linear superposition $|\phi\rangle$ and seek to identify the vector $|\xi\rangle$ following a measurement with respect to the computational basis. According to Born's rule, the probability of identification of the sought item is given by $|\langle\xi|\phi\rangle|^2 = 1/N$, the classical figure of merit. Our goal is to enhance this probability by employing Hamiltonian evolution from the initial state $|\phi\rangle$ to a final state that approaches $|\xi\rangle$ within a time interval T. As $|\phi\rangle$ is the ground state of $\mathbf{H}_0$, we anticipate that it adiabatically evolves into the ground state of $\mathbf{H}$, the sought for item $|\xi\rangle$.

To that end, we introduce the Hamiltonian

$$\mathbf{h}(s) = \mathbf{H}_0(1 - a(s)) + a(s)\mathbf{H}, \tag{10.22}$$

the generator for the time development operator for $|\phi\rangle$. Following our discussion in Chap. 4 we introduce the vector

$$|\eta\rangle = \sqrt{\frac{N}{N-1}}|\phi\rangle - \frac{1}{\sqrt{N-1}}|\xi\rangle \tag{10.23}$$

and note that $\langle\xi|\eta\rangle = 0$, $\langle\eta|\eta\rangle = 1$. We also find, using the fact $\langle\xi|\phi\rangle = \frac{1}{\sqrt{N}}$ and expressing $|\phi\rangle$ as a linear combination of $|\eta\rangle$ and $|\xi\rangle$, that

$$\mathbf{H}_0|\eta\rangle = \frac{1}{N}|\eta\rangle - \frac{\sqrt{N-1}}{N}|\xi\rangle$$

$$\mathbf{H}_0|\xi\rangle = \frac{N-1}{N}|\xi\rangle - \frac{\sqrt{N-1}}{N}|\eta\rangle. \tag{10.24}$$

Likewise

$$\mathbf{H}|\eta\rangle = |\eta\rangle$$

$$\mathbf{H}|\xi\rangle = 0, \tag{10.25}$$

and so $\mathbf{h}(s)$ induces rotations in the subspace spanned by $|\eta\rangle$, $|\xi\rangle$. Thus we diagonalize $\mathbf{h}(s)$ by restricting the matrix representation of $\mathbf{h}(s)$ to this subspace. We ignore all other eigenstates outside this manifold, as they do not mix with $|\xi\rangle$, $|\eta\rangle$.

Evaluating

$$\begin{pmatrix} \langle\eta|\mathbf{h}(s)|\eta\rangle & \langle\eta|\mathbf{h}(s)|\xi\rangle \\ \langle\xi|\mathbf{h}(s)|\eta\rangle & \langle\xi|\mathbf{h}(s)|\xi\rangle \end{pmatrix} \tag{10.26}$$

we find the matrix representation

$$\underline{\mathbf{h}}(s) = \begin{pmatrix} 1/2 & 0 \\ 0 & 1/2 \end{pmatrix} - \frac{1}{2}\begin{pmatrix} \frac{2(a(s)-1)}{N} - 2a(s) + 1 & \frac{2(1-a(s))\sqrt{N-1}}{N} \\ \frac{2(1-a(s))\sqrt{N-1}}{N} & \frac{2-2a(s)}{N} + 2a(s) - 1 \end{pmatrix} \tag{10.27}$$

or,

$$\mathbf{h}(s) = \begin{pmatrix} 1/2 & 0 \\ 0 & 1/2 \end{pmatrix} - \frac{\Delta(s)}{2} \begin{pmatrix} \cos\theta(s) & \sin\theta(s) \\ \sin\theta(s) & -\cos\theta(s) \end{pmatrix} \quad (10.28)$$

where

$$\Delta(s) = \sqrt{\frac{4(a(s)-1)a(s)(N-1)}{N} + 1}$$

$$\cos\theta(s) = \frac{1 - 2a(s) + 2(a(s)-1)/N}{\Delta(s)}$$

$$\sin\theta(s) = 2\frac{(1-a(s))}{\Delta(s)}\frac{\sqrt{N-1}}{N}. \quad (10.29)$$

We recognize that

$$\mathbf{h}(s) = \mathbf{U}(s)\mathbf{h}_d(s)\mathbf{U}^\dagger(s) \quad (10.30)$$

where $\mathbf{h}_d(s)$ is a diagonal matrix

$$\mathbf{h}_d(s) = \frac{1}{2}(\mathbb{1} - \Delta(s)\sigma_Z) \quad (10.31)$$

and σ_Z is the Pauli-Z matrix. The instantaneous adiabatic eigenvalues can be read off immediately from Eq. (10.31) so that

$$E_{ad}(s) = 1/2 \pm \Delta(s), \quad (10.32)$$

where the $+$ sign gives the excited state energy, and the $(-)$ sign the ground state energy. The unitary operator

$$\mathbf{U}(s) = \exp(-i\,\theta(s)\,\sigma_Y/2) = \begin{pmatrix} \cos\frac{\theta(s)}{2} & -\sin\frac{\theta(s)}{2} \\ \sin\frac{\theta(s)}{2} & \cos\frac{\theta(s)}{2} \end{pmatrix} \quad (10.33)$$

where σ_Y is the Pauli-Y matrix. Therefore, the ground adiabatic state is given by

$$|u(s)\rangle_g = \cos\frac{\theta(s)}{2}|\eta\rangle + \sin\frac{\theta(s)}{2}|\xi\rangle. \quad (10.34)$$

Noting that for $s = 0$, $\cos\theta(0) = \frac{N-2}{N}$, or $\cos\frac{\theta(0)}{2} = \sqrt{\frac{N-1}{N}}$, and $\sin\frac{\theta(0)}{2} = \sqrt{\frac{1}{N}}$, we find

$$|u(0)\rangle_g = \sqrt{\frac{N-1}{N}}|\eta\rangle + \sqrt{\frac{1}{N}}|\xi\rangle = |\phi\rangle \quad (10.35)$$

10.1 Introduction

in harmony with definition (10.19). Likewise for $s = 1, a(s) = 1, \cos(\theta(1)) = 0, \cos(\frac{\theta}{2}) = 1$

$$|u(1)\rangle_g = |\xi\rangle. \quad (10.36)$$

For a linear schedule function $a(s) = s = 1/2$, the adiabatic energy surfaces suffer an avoided crossing as the energy defect parameter $\Delta(s)$ reaches its minimum

$$\Delta(1/2) = \sqrt{\frac{1}{N}} \quad (10.37)$$

Thus, with large values of $N = 2^n$ and smaller energy gaps, we anticipate poor convergence to the adiabatic limit. In the previous sections, we qualitatively illustrated the relationship between the energy gap at the avoided crossing and the period of time T required to achieve adiabaticity. A more quantitative relationship that provides bounds [2] on the critical value of T is beyond the scope of this monograph. But a detailed analysis [2] shows that for a linear schedule function, a solution to the Schrödinger equation approaches the adiabatic limit when

$$T \gg 3N. \quad (10.38)$$

That is, for large N, the time T for completion of the evolution scales linearly in N. This scaling behavior is similar to the classical algorithm for an unsorted search. Given N candidates, repeated queries to the oracle deliver the correct answer after an average of $N/2$ trials. Thus, the bound (10.38) does not appear to achieve a quantum advantage over the classical figure of merit. However, a judicious choice for the scheduling function $a(s)$ may provide algorithmic speed-up. This is indeed the case, choosing [2]

$$a(s) = \frac{1}{2} + \frac{1}{2\sqrt{N-1}} \tan\left((2s-1)\arctan\sqrt{N-1}\right) \quad (10.39)$$

one finds [2]

$$T \propto \sqrt{N}, \quad (10.40)$$

a quadratic speed-up.

> Mathematica Notebook 10.2: The Grover Algorithm Redux; Pick a Card, Any Card: https://bernardzygelman.github.io

10.1.5.1 Stoquastic Versus Non-stoquastic

Hamiltonians are called *stoquastic* if all off-diagonal elements in their matrix representation are real and nonpositive. The Hamiltonian presented in Eq. (10.6) includes both negative and positive entries in the computational basis. If there exists no efficient algorithm to find a basis in which the Hamiltonian contains non-positive off-diagonal elements, the Hamiltonian is referred to as *non-stoquastic*. This distinction is crucial because stoquastic Hamiltonians are usually tackled using classical Monte Carlo methods [2], and it remains uncertain whether a quantum advantage can be achieved with these systems. Non-stoquastic Hamiltonians frequently lead to the sign problem [2], making classical Monte Carlo simulations either impractical or significantly more challenging. Non-stoquastic Hamiltonians are considered prime candidates for realizing quantum advantage, as their non-stoquastic nature renders certain problems exponentially difficult for classical algorithms but potentially more manageable for quantum algorithms.

10.1.6 Summary

Quantum adiabatic computing holds promise for solving specific optimization problems more efficiently than classical algorithms. The concept is to encode the solution to a problem in the ground state of a Hamiltonian and to evolve the system adiabatically from an easily prepared initial ground state to the desired ground state. Despite these implementations, there are substantial challenges. Many practical problems are not efficiently solvable due to issues such as thermal noise, decoherence, and the difficulty of maintaining adiabatic conditions. Empirical studies and benchmarking against classical algorithms have produced mixed results. For some problems, quantum annealers have outperformed classical algorithms, but for others, they have not demonstrated a clear advantage. In conclusion, while quantum adiabatic computing shows potential, especially for certain optimization problems, it faces significant challenges and is currently regarded as one part of the broader quantum computing landscape. Research in QAC has led to the development of new hybrid algorithms, such as the Quantum Approximate Optimization Algorithm (QAOA) [11], which shows promise as a variational quantum algorithm designed to solve combinatorial optimization problems. The field is evolving and ongoing research is likely to further elucidate and enhance its role in the future of computation.

10.2 What About the Phase? Exploring Holonomic Quantum Computing

In Sect. 10.1.1, we studied the adiabatic evolution of a quantum state controlled by a time-dependent Hamiltonian $\mathbf{h}(t)$. If a system is initially in the ground eigenstate of $\mathbf{h}(0) = \mathbf{H}_0$, $|u_g(0)\rangle$, the adiabatic theorem states that the probability of finding the system in the instantaneous ground eigenstate of $\mathbf{h}(T) = \mathbf{H}$, approaches unity in the adiabatic limit. In that discussion, we focus on the probability of finding the system in state $|u_g(T)\rangle$. Because the probability is not affected by an overall phase factor, we did not draw attention to the accompanying phase factor. In this section, we examine the properties of this phase factor and explore its potential as a resource.

10.2 What About the Phase? Exploring Holonomic Quantum Computing

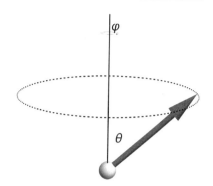

Fig. 10.9 A spin-1/2 particle with a magnetic moment subjected to a time-dependent magnetic field, represented by a vector, as it moves in a circuit on the dashed line at a constant angle θ along the vertical axis

Hamiltonians can be parameterized by variables in addition to time. For example, consider a single qubit, such as an electron, exposed to a magnetic field that changes its orientation in space. In Fig. 10.9, we depict a spin-1/2 qubit with a magnetic moment subjected to a magnetic field that makes an adiabatic circuit along the dashed line in that figure. The polar angle θ is constant, but its orientation along that circuit is labeled by the parameter φ, an azimuthal angle. The Hamiltonian for this system is given in Eq. (3.41)

$$\mathbf{H}(t) = \Delta \begin{pmatrix} \cos\theta & \sin\theta \exp(-i\varphi(t)) \\ \sin\theta \exp(i\varphi(t)) & -\cos\theta \end{pmatrix} \tag{10.41}$$

where Δ is a constant proportional to the coupling strength of the magnetic moment in the magnetic field.

In the setup described in Fig. 10.9, φ is a function of time such that $\varphi(t=0) = 0$ and $\varphi(t=T) = 2\pi$, and $\mathbf{H}(0) = \mathbf{H}(T)$. Hamiltonians possessing this property are called cyclic Hamiltonians. Thus, according to the adiabatic theorem, an eigenstate of $\mathbf{H}(0)$, $|u(0)\rangle$, returns to $|u(0)\rangle$ at $t = T$ with the addition of a phase factor, provided that the adiabatic condition is satisfied.

Following the discussion of the previous section, we consider the Hamiltonian,

$$\mathbf{h}(t) = \mathbf{H}_0(1 - t/T) + t/T\,\mathbf{H}. \tag{10.42}$$

This Hamiltonian is not cyclic, but we can make it cyclic if we set $\mathbf{H} = \mathbf{H}_0$, as then $\mathbf{h}(0) = \mathbf{h}(T)$. For the latter, an analytic solution to the Schrödinger equation for the ground state amplitude is given by

$$|\psi_g(t)\rangle = |u_g\rangle \exp\left(-i \int_0^t d\tau\, E_g(\tau)\right) = |u_g\rangle \exp(-i\,E_g\,t)$$
$$\mathbf{H}_0\,|u_g\rangle = E_g\,|u_g\rangle, \tag{10.43}$$

where we used the fact that the instantaneous eigenvalue $E_g(t)$ is constant. As promised by the adiabatic theorem, $|u_g(0)\rangle = |u_g\rangle$ evolves[1] into $|u_g\rangle$ at $t = T$, with the addition of a phase factor. In general, the quantity

$$\gamma_D \equiv \int_0^T d\tau \, E_g(\tau), \tag{10.44}$$

is called the dynamic phase, as it is dependent on the history of the instantaneous energy eigenvalue $E_g(t)$ during adiabatic evolution. This observation raises the question: for any time-dependent Hamiltonian $\mathbf{h}(t)$ does the adiabatic theorem allow for an additional phase factor that differs in nature from the dynamic phase? i.e.

$$|\psi(T)\rangle \stackrel{?}{=} \exp(-i\gamma_D) \exp(i\gamma_G) |u_g(T)\rangle. \tag{10.45}$$

For over half a century, a generation of textbooks argued for the case $\gamma_G = 0$; that is, one can always find a representation where any nonvanishing γ_G can effectively be "gauged" away, thus leaving only a dynamic phase factor. In a seminal paper [4], M. V. Berry demonstrated that γ_G cannot be eliminated under certain circumstances. Importantly, Berry's, or the geometric, phase has geometric and topological significance that has spurred explorations of possible application in quantum information processing [12]. This line of investigation is sometimes referred to as geometric phase or holonomic quantum computing [12]. Here, we explore aspects of this feature by analyzing the adiabatic evolution of a single qubit governed by the Hamiltonian (10.41).

First, we note

$$\mathbf{H}(t) = \mathbf{U}(t) \, \mathbf{H}_{BO} \, \mathbf{U}^\dagger(t)$$
$$\mathbf{U}(t) = \exp(-i\sigma_z \varphi(t)/2) \exp(-i\sigma_y \theta/2)) \exp(i\sigma_z \varphi(t)/2)$$
$$\mathbf{H}_{BO} = \Delta \sigma_z = \begin{pmatrix} \Delta & 0 \\ 0 & -\Delta \end{pmatrix} \tag{10.46}$$

where σ_z, σ_y are Pauli matrices, and from which we can find the instantaneous adiabatic eigenfunctions and eigenvalues, viz.

$$|u_e(t)\rangle = \mathbf{U}(t)|0\rangle = \begin{pmatrix} \cos(\theta/2) \\ \exp(i\varphi(t)) \sin(\theta/2) \end{pmatrix}$$
$$|u_g(t)\rangle = \mathbf{U}(t)|1\rangle = \begin{pmatrix} -\exp(-i\varphi(t)) \sin(\theta/2) \\ \cos(\theta/2) \end{pmatrix} \tag{10.47}$$

where

$$|0\rangle \equiv \begin{pmatrix} 1 \\ 0 \end{pmatrix} \quad |1\rangle \equiv \begin{pmatrix} 0 \\ 1 \end{pmatrix}. \tag{10.48}$$

---

[1] Since the Hamiltonian is constant, this example is a trivial demonstration of the adiabatic theorem.

10.2 What About the Phase? Exploring Holonomic Quantum Computing

To prove this assertion, we use expression (10.47) and Eq. (10.46) to find

$$\mathbf{H}(t)|u_e(t)\rangle = \Delta|u_e(t)\rangle$$
$$\mathbf{H}(t)|u_g(t)\rangle = -\Delta|u_g(t)\rangle. \quad (10.49)$$

In other words, they are adiabatic eigenstates with instantaneous energy eigenvalues $\pm\Delta$, respectively.

It is useful to visualize these adiabatic states on the Bloch sphere. Using the Hopf map defined in Eq. (2.16) we find that $|u_e(t)\rangle$ follows the trajectory

$$x(t) = \cos\varphi(t)\sin\theta$$
$$y(t) = \sin\varphi(t)\sin\theta$$
$$z(t) = \cos\theta \quad (10.50)$$

on the Bloch sphere, which we recognize as that taken by the magnetic field in Fig. 10.9. Thus $|u_e(t)\rangle$ follows the external magnetic field on the Bloch sphere and similarly one finds that the ground state $|u_g(t)\rangle$ is anti-aligned to $|u_e(t)\rangle$ on the Bloch sphere.

In the adiabatic regime, so that $d\varphi/dt \ll 1$, solutions to the Schrödinger equation (setting $\hbar = 1$)

$$i\frac{|d\psi(t)\rangle}{dt} = \mathbf{H}(t)|\psi(t)\rangle \quad (10.51)$$

with initial conditions $|\psi(0)\rangle = |u_g(0)\rangle$, or $|\psi(0)\rangle = |u_e(0)\rangle$, carry out the Bloch sphere trajectories cited above. We also know that those solutions possess an extra phase factor, not recorded in the Bloch sphere representation, one of which is the dynamic phase factor corresponding to $\gamma_D = \pm\Delta t$, for the excited and ground states, respectively.

> Mathematica Notebook 10.3: Adiabatic Evolution in the Rabi Model: https://bernardzygelman.github.io

10.2.1 The Rabi Model

We need a solution to Eq. (10.51) to identify γ_G for that evolution. Fortunately, analytical solutions are available for the case where $\varphi(t) = \omega t$, a magnetic field traversing the path at constant angular velocity ω. The Hamiltonian corresponding to that choice is called the Rabi model after I. Rabi, a mid-20th century atomic physics pioneer. Solutions to the Rabi model are well known and can be found in several textbook treatments [10]. Here, we express those solutions in the form

$$|\psi(t)\rangle = \mathbf{U}(t)\exp\left(-i\,\sigma_z\,(\Delta + \omega\,\sin^2\frac{\theta}{2})t\right)\mathbf{W}(t)\mathbf{U}^\dagger(0)|\psi(0)\rangle \quad (10.52)$$

where $|\psi(0)\rangle$ is an arbitrary qubit amplitude at $t = 0$, and $\mathbf{W}(t, 0)$, obeys the following

$$i\frac{d\mathbf{W}(t)}{dt} = \mathbf{V}(t)\,\mathbf{W}(\mathbf{t}) \quad \mathbf{W}(0) = \mathbb{1}$$
$$\mathbf{V}(t) = \frac{\omega \sin\theta}{2}\begin{pmatrix} 0 & v(t) \\ v^*(t) & 0 \end{pmatrix}$$
$$v(t) = \exp(i(2\,\Delta - \omega\cos\theta)\,t). \tag{10.53}$$

To demonstrate (see Mathematica Notebook 10.3) that Eq. (10.52) is a solution to Eq. (10.51), we substitute the former for the latter and use the definition of $\mathbf{W}(t)$. An analytic expression for $\mathbf{W}(t)$ is available, but we are only interested in the adiabatic limit in which the ratio $\omega/\Delta \to 0$. In that case,

$$\mathbf{W}(t) \to \mathbb{1} + \mathcal{O}\left(\frac{\omega}{\Delta}\right). \tag{10.54}$$

Thus, keeping only the first term in this limit, we find

$$|\psi(t)\rangle \to \mathbf{U}(t)\exp\left(-i\,\sigma_z\,(\Delta + \omega\,\sin^2\frac{\theta}{2})\,t\right)\mathbf{U}^\dagger(0)\,|\psi(0)\rangle. \tag{10.55}$$

If the initial state $|\psi_g(0)\rangle = |u_g(0)\rangle = \mathbf{U}(0)|1\rangle$, we find that after a completion of one circuit at $t = T = 2\pi/\omega$,

$$|\psi_g(T)\rangle = \exp(i\,\Delta\,T)\exp\left(i\,2\pi\,sin^2\frac{\theta}{2}\right)|u_g(0)\rangle \tag{10.56}$$

in harmony with the adiabatic theorem. In deriving (10.56) we used the fact that $\mathbf{U}(\frac{2\pi}{\omega}) = \mathbf{U}(0)$, $\sigma_z|1\rangle = -|1\rangle$. Similarly, if the initial state is $|u_e(0)\rangle$,

$$|\psi_e(T)\rangle = \exp(-i\,\Delta\,T)\exp\left(-i\,2\pi\,sin^2\frac{\theta}{2}\right)|u_e(0)\rangle. \tag{10.57}$$

10.2.2 Geometric Phase as a Holonomy

When first introduced to vectors, you probably learned that the magnitude and orientation of a vector uniquely define it. However, for vector fields, it is also crucial to identify the point in space where a vector is located. We need to be able to compare a vector positioned at point A with one at point B. How can we assess the difference or similarity between the two? To address this question, we must introduce the concept of *parallel transport* [13]. Parallel transport is a prescription that informs us how to "move" a vector from point $\mathbf{a}$ to a neighboring point $\mathbf{a} + \Delta\mathbf{a}$ without changing its length and direction with respect to the local geometry. We illustrate this concept in Fig. 10.10. In the left panel of that figure, we depict a vector at point A and parallel transport it to different points along the sides of a triangle ABC on the surface of a flat plane. At each point, the vector assumes the orientation and length of the original

10.2 What About the Phase? Exploring Holonomic Quantum Computing

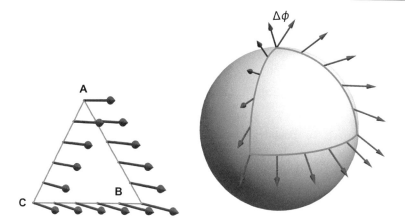

Fig. 10.10 (left panel) Parallel transport of a vector on triangle ABC. (right panel) Illustration of the angle shift $\Delta\phi$ as a vector is parallel transported about a triangular path on the surface of a sphere

vector at point A. This behavior aligns with our intuitive notion of parallelism. Now, we assume that there is a clock attached to the vector. As we transport the vector around the circuit, it arrives at the same point in space (i.e., point A), but the clock has registered a time interval. We repeat this construct on a triangular path on a curved surface, as shown in the second panel of Fig. 10.10. Since we are performing parallel transport on a curved surface (i.e., the surface of a sphere), the vector being transported will not return to point A with the same orientation as it had at the start of its journey. Indeed, differential geometry [13] tells us that there is a change $\Delta\phi$ in the relative orientation between the initial and final vectors. That difference is a function of the surface's curvature and the circuit's enclosed area. In differential geometry, this feature is called a *holonomy* and signals a failure of a parallel transported vector along geodesics to return to its original orientation.

We now return to Eq. (10.56), which describes Schrödinger evolution, in the adiabatic limit, of a state initially at a point on the Bloch sphere specified by $|u_g(0)\rangle$. At time $T = 2\pi/\omega$, it returns to that same point on the Bloch sphere, with a change of a phase factor. The dynamic phase γ_D is proportional to T, the time interval it takes to make one complete circuit, and so serves as an effective "clock." Thus, the dynamic phase is analogous to the clock in our example of parallel transport of the vectors in Fig. 10.10. But Eq. (10.56) features an additional phase given by

$$\gamma_G = 2\pi \sin^2 \frac{\theta}{2}. \qquad (10.58)$$

Is this an analog of the holonomy for parallel-transported vectors? Is there a curvature associated with this holonomy? Remarkably, the answer is yes.

10.2.3 Berry's Phase

With a similar but more general Hamiltonian compared to the one defined by Eq. (10.41), Berry demonstrated that the quantity γ_G can be represented as a surface integral.

$$\gamma_G = \int_S d\mathbf{a} \cdot \mathbf{B}. \tag{10.59}$$

where the surface S is bounded by a closed path $\mathbf{R}(t)$ and $\mathbf{B}$ is a vector field. For Berry's Hamiltonian

$$\mathbf{H} = \frac{1}{2}\sigma \cdot \mathbf{R} \quad \mathbf{R} = X\hat{\mathbf{i}} + Y\hat{\mathbf{j}} + Z\hat{\mathbf{k}} \tag{10.60}$$

in a spherical coordinate system ($R\,\theta\,\phi$), one obtains

$$\mathbf{B} = \pm\frac{\hat{R}}{2R^2}. \tag{10.61}$$

Below we restrict our discussion to the positive sign in the expression (10.61), and observe that this field has a monopole-like structure illustrated in Fig. 10.11. In mathematics-oriented literature, $\mathbf{B}$ is sometimes called a curvature. It can also be interpreted as an effective magnetic field of a monopole. Consider integral (10.59) over a surface whose boundary is enclosed by the rim of the cone-like structure shown in Fig. 10.11. It represents a path $\mathbf{R}(t)$ that parameterizes Hamiltonian $\mathbf{H} = \frac{1}{2}\sigma \cdot \mathbf{R}(t)$. Comparing with Hamiltonian (10.41), we have

$$\mathbf{R}(t) = 2\Delta\left(\sin\theta\cos\varphi(t)\,\hat{\mathbf{i}} + \sin\theta\sin\varphi(t)\,\hat{\mathbf{j}} + \cos\theta\,\hat{\mathbf{k}}\right)$$
$$\varphi(t) = \omega t \tag{10.62}$$

where θ represents the angle of a cone with respect to the vertical axis. Thus, using Eq. (10.61) in expression (10.59) and spherical coordinates,

$$\gamma_G = \int_0^{2\pi} d\phi \int_0^{\theta} d\Omega \sin\Omega \,\frac{1}{2} = \pi(1 - \cos\theta) = 2\pi \sin^2\frac{\theta}{2}. \tag{10.63}$$

Typically, a holonomy is expressed as a line integral of a vector potential or *connection*. Indeed, using *Stokes theorem*

$$\int_S d\mathbf{a} \cdot \mathbf{B} = \int_S d\mathbf{a} \cdot (\nabla \times \mathbf{A}) = \oint_C d\mathbf{R} \cdot \mathbf{A}$$
$$\mathbf{A} = \hat{\phi}\,\frac{(1 - \cos\theta)}{2R\sin\theta} \tag{10.64}$$

where C is a circuit that is a boundary of the surface S, and the connection $\mathbf{A}$ is expressed in a spherical coordinate system. Thus, the phase γ_G observed in the adiabatic evolution of solutions corresponding to Hamiltonian (10.41) can be interpreted in the language of differential geometry as a holonomy.

10.2 What About the Phase? Exploring Holonomic Quantum Computing

Fig. 10.11 Field lines of the monopole (10.61), the cone has an opening angle θ and the path $\mathbf{R}(t)$ follows the rim of the cone

In Eq. (10.64), we relate the phase $\gamma_G = \pi(1 - \cos\theta)$ to a line integral of the connection given therein

$$\gamma_G = \oint_C d\mathbf{R} \cdot \mathbf{A} = \int_0^T dt \frac{d\mathbf{R}}{dt} \cdot \hat{\boldsymbol{\phi}} \frac{(1 - \cos\theta)}{2R\sin\theta} = \pi(1 - \cos\theta). \quad (10.65)$$

In the derivation of Eq. (10.64) we used

$$\frac{d\mathbf{R}}{dt} = 2\Delta\,\dot{\varphi}(t)(-\hat{\mathbf{i}}\sin\varphi(t) + \hat{\mathbf{j}}\cos\varphi(t)) \quad (10.66)$$

and $\phi(0) = 0$, $\phi(T) = 2\pi$. Now, there is a complementary way to obtain that expression for γ_G. For the specific path shown in Fig. 10.9 θ, $R = 2\Delta\cos\theta$ are constant, so we can also express $\mathbf{R}(t)$ with cylindrical coordinates (ρ, φ, z) viz.

$$\mathbf{R}(t) = \rho\cos\varphi(t)\,\hat{\mathbf{i}} + \rho\sin\varphi(t)\,\hat{\mathbf{j}} + z\,\hat{k}. \quad (10.67)$$

We want to express γ_G as

$$\int_{t=0}^T dt \frac{d\mathbf{R}}{dt} \cdot \mathbf{A}' = \int_{t=0}^T dt\,\rho\left(-\hat{\mathbf{i}}\sin\varphi(t) + \hat{\mathbf{j}}\cos\varphi(t)\right) \cdot \mathbf{A}' \quad (10.68)$$

Thus, we require

$$\mathbf{A}' = \hat{\varphi}\frac{(1-\cos\theta)}{2\rho} \tag{10.69}$$

in order to recover the correct expression $\gamma_G = \pi(1 - \cos\theta)$.

10.2.4 Non-Abelian Phases

In the previous sections, we focused on the scalar function $\exp(i\gamma_G)$ accompanying an adiabatic amplitude as it returns to its initial position on the Bloch sphere. Because the Schrödinger equation is linear, a superposition of solutions (10.56) and (10.57) is also a valid solution to the Schrödinger equation in the adiabatic limit. Equivalently, that solution is given by Eq. (10.55), where $|u(0)\rangle$ is a linear superposition of $|u_g(0)\rangle, |u_e(0)\rangle$. Now at $t = T = 2\pi/\omega$

$$|\psi(T)\rangle = \mathbf{U}(T)\exp\left(-i\,\sigma_z\,(\Delta + \omega\sin^2\frac{\theta}{2})t\right)\mathbf{U}^\dagger(0)\,|u(0)\rangle =$$
$$\exp\left(-i\,\hat{\mathbf{n}}_\theta\cdot\sigma\,(\Delta T + 2\pi\sin^2\frac{\theta}{2})\right)|u(0)\rangle. \tag{10.70}$$

In the derivation of Eq. (10.70) we used $\mathbf{U}(T) = \mathbf{U}(0) = \exp(-i\,\sigma_y\theta/2)$, $\mathbf{U}\sigma_z\mathbf{U}^\dagger = \hat{\mathbf{n}}_\theta\cdot\sigma$, $\hat{\mathbf{n}}_\theta = \sin\theta\,\hat{\mathbf{i}} + \cos\theta\,\hat{\mathbf{k}}$. Suppose we choose the energy defect Δ so that $\Delta T = 2\pi\,n$, where n is an integer, we then obtain

$$|\psi(T)\rangle = \exp(-i\pi\,\hat{\mathbf{n}}_\theta\cdot\sigma\,(1 - \cos\theta))|u(0)\rangle. \tag{10.71}$$

Note that $|\psi(T)\rangle \neq \exp(i\gamma)|u(0)\rangle$, i.e., the traditional adiabatic theorem is no longer valid. Indeed, Eq. (10.71) describes a rotation, about the $\hat{\mathbf{n}}_\theta$-axis through the angle $\pi(1 - \cos\theta)$ on the Bloch sphere. Alternatively, we can think of this rotation as a matrix-valued or non-Abelian geometric phase factor. It is non-Abelian in the sense that if we take two rotations $\mathbf{R}(\theta) = \exp(-i\pi\,\hat{\mathbf{n}}_\theta\cdot\sigma\,(1 - \cos\theta))$, corresponding to parameters θ_1, θ_2, then

$$\mathbf{R}(\theta_1)\mathbf{R}(\theta_2) - \mathbf{R}(\theta_2)\mathbf{R}(\theta_1) = 2i\,\sin(\theta_1 - \theta_2)\sin(\pi\cos\theta_1)\sin(\pi\cos\theta_2)\,\sigma_2 \tag{10.72}$$

and generally does not vanish.

10.2.5 From Phases to Forces

Hamiltonian (10.41) resides in a 2-dimensional Hilbert space usually denoted by the symbol $\mathbb{C}^2$. However, that Hilbert space is parameterized by a classical variable $\varphi(t)$, and in some applications, it is desirable to elevate $\varphi(t)$ to a quantum variable

10.2 What About the Phase? Exploring Holonomic Quantum Computing

(i.e., a Hilbert space operator), donated by the symbol $\boldsymbol{\varphi}$. In other words

$$\mathbf{H}(\varphi(t)) \to \mathcal{H}_{ad}$$

$$\mathcal{H}_{ad} \equiv \Delta \begin{pmatrix} \cos\theta & \sin\theta \exp(-i\boldsymbol{\varphi}) \\ \sin\theta \exp(i\boldsymbol{\varphi}) & -\cos\theta \end{pmatrix}. \tag{10.73}$$

Note that $\mathcal{H}_{ad}$ lives in a direct product Hilbert space, $\mathcal{H}_\varphi \otimes \mathbb{C}^2$, where $\mathcal{H}_\varphi$ is an infinite-dimensional Hilbert space whose basis vectors are square-integrable functions of the eigenvalues of $\boldsymbol{\varphi}$, simply labeled φ. The eigenstates of $\mathcal{H}_{ad}$ are called adiabatic eigenstates but, unlike in the semiclassical realm, they are time-independent in the Schrödinger picture.

Dynamics is generated by the presence of a kinetic energy term $\mathcal{H}_{KE} = -\hbar^2 \nabla^2/2m$ added to $\mathcal{H}_{ad}$. So, for the case of 3D dynamics, one considers a total Hamiltonian

$$\mathcal{H}_{total} = -\frac{\hbar^2}{2m}\nabla^2 + \mathcal{H}_{ad}(\varphi) \tag{10.74}$$

where m is a constant and we replaced operator $\boldsymbol{\varphi}$ with its eigenvalue. Here θ is a constant parameter and should not be treated as a quantum variable in $\mathcal{H}_{ad}$.

Consider an amplitude $u_g(\varphi)$ that is a ground eigenstate of $\mathcal{H}_{ad}$. Does this scenario allow for an analog of the adiabatic theorem, namely, under what conditions does the system remain in the ground state as it evolves in time? Studies show that the probability of remaining in the ground state approaches unity, provided that there exists, throughout the evolution, a sufficiently large energy gap, 2Δ, between the ground and first excited state eigenvalues of $\mathcal{H}_{ad}$ and the expectation value $\langle \mathcal{H}_{KE} \rangle \ll \Delta$. This observation forms the basis for the Born-Oppenheimer (BO) approximation in which the system amplitude $u(\varphi, \chi)$ is expressed as a product of the ground adiabatic state with an effective amplitude that is a function of φ as well as other quantum variables χ not included in $\mathcal{H}_{ad}$, namely

$$u(\varphi, \chi) = F(\varphi, \chi) u_g(\varphi). \tag{10.75}$$

In a Born-Oppenheimer treatment for the system described by an adiabatic Hamiltonian similar to (10.73) we find [28] the following effective Schrödinger equation,[2] expressed with cylindrical coordinates, for amplitude F

$$-\frac{\hbar^2}{2m}(\nabla - i\mathbf{A})^2 F + V_{BO} F = i\hbar \frac{\partial F}{\partial t}$$

$$\mathbf{A} = \hat{\varphi}\frac{(1-\cos\theta)}{2\rho} \quad V_{BO} = -\Delta. \tag{10.76}$$

---

[2] Here we ignore a higher order gauge correction V_{AD} [22].

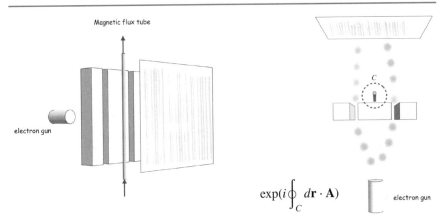

Fig. 10.12 Illustration of an AB effect experiment. A beam of electrons passes through a double-slit barrier in which a thin, inaccessible, magnetic flux tube is placed. A detection screen is positioned in front of the barrier as shown. The right panel is a top-down view of the setup

This effective equation highlights the substitution of the gradient operator ∇ with a covariant derivative, i.e.

$$\nabla \to \nabla - i\mathbf{A} \tag{10.77}$$

where $\mathbf{A}$, is the connection introduced in Eq. (10.69). Solutions to Eq. (10.76) exhibit a phenomenon called the Aharonov-Bohm (AB) effect [14], illustrated in Fig. 10.12. The AB effect is an iconic template for the role of topology in quantum physics. Topology is the study of the global properties of a manifold. For example, the topology of a sphere is the same as that of a solid cube because the cube can be smoothly reshaped into a sphere (think of shaping a ball of clay into a cubic shape). However, the topology of a coffee cup with a single handle differs from that of a sphere because it has a hole, and turning it into a sphere would require tearing the handle. However, you could smoothly reshape a coffee cup into a donut while maintaining a single hole's presence. This last example has led to the humorous catchphrase, "A topologist can't tell the difference between a coffee cup and a donut."

The AB effect involves an experimental setup shown in Fig. 10.12 in which a beam of electrons is projected onto a wall with two slits and detected on a screen behind that wall. The detection pattern on the screen is similar to that observed in Young's double-slit experiment for light [15]. That experiment in the early 19th century confirmed the wave nature of light. Similarly, the quantum double-slit experiment highlights the dual wave-particle nature of electrons. In addition to the double slit, the AB (gedanken) experiment introduces a thin solenoid that traps a magnetic field (also called a magnetic flux tube) along its axis. Because the magnetic field is localized, it cannot exert a force on the electrons. The electrons pass through the slits without experiencing any force, as if no magnetic flux tube were present. However, the electron detection screen registers the presence of the flux tube because changing the

10.2 What About the Phase? Exploring Holonomic Quantum Computing

magnetic flux within the tube results in a shift in the screen pattern. In some way, the electrons "know" that there is a "hole" (the flux tube) in the space through which they are traveling.

Quantitatively, this phenomenon is described by positing a Schrödinger equation that is coupled to a vector potential that has a form given in Eq. (10.76), specifically, in a cylindrical coordinate system,

$$A = \frac{e}{\hbar c}\frac{\Phi}{2\rho}\hat{\varphi} \qquad (10.78)$$

where e is the charge of the electron, c is the speed of light and Φ is the magnetic flux trapped in the solenoid. Furthermore, using Feynman's many paths formulation of quantum mechanics [27], the pattern on the screen arises due to contributions coming from a sum of terms that involve the quantity

$$\exp(i\oint_C d\mathbf{r}\cdot\mathbf{A}). \qquad (10.79)$$

C is an arbitrary closed loop in the electron configuration space. In the Feynman formulation, there are two classes of contributions, one that involves loops C_1 that contain the flux tube and others that do not. The former determines the nature of the fringe shifts on the detection screen. To a topologist, this feature points to a "hole" in the manifold. The phase factor (10.79) indicates the location of that "hole" since smoothly contracting C_1 will eventually "lasso" the flux tube and reveal its location in the manifold. Note the similarity between this phase factor and $\exp(i\gamma_G)$ where γ_G is obtained using the connection defined in Eq. (10.69).

The Aharonov-Bohm (AB) effect was originally introduced within the realm of electromagnetism, requiring the coupling of a charged particle with a gauge potential (10.78). However, in addition to an AB-like effect described by Eq. (10.76), which does not necessitate interaction with an electric charge, the first suggestion of an AB-like effect stemming from a nonelectromagnetic source came from Alden Mead [16], and is now known as the molecular Aharonov-Bohm (MAB) effect. Research on this topic [17, 18] continues to be an active area of investigation in theoretical chemistry.

Equation (10.76) includes the presence of a nontrivial connection **A** derived within a Born-Oppenheimer approximation of the system described by Hamiltonian (10.73). The latter contains no overt gauge couplings; could the appearance of **A** be an artifact with no predictive value of the BO approximation? Despite arguments raised in some literature, the answer is no! In a numerical simulation, the wave packets propagated through a double slit wall using a Hamiltonian similar to that given by Eq. (10.73), and which did not invoke a BO approximation, we were able to reproduce [28] an AB fringe pattern on a detection screen. That AB fringe pattern was evident when our simulation guaranteed a large energy defect $2\Delta \gg \langle \mathcal{H}_{KE}\rangle$ between the ground and excited eigenstates of $\mathcal{H}_{ad}$. At smaller defects, the fringe pattern reverted to that produced by a double-slit configuration without the presence of a flux tube.

So far, we have considered connections that do not generate gauge forces. To explore the possibility of the latter, consider the following

$$\mathbf{H}(\phi(t), \theta(t)) \to \mathcal{H}_{ad}$$

$$\mathcal{H}_{ad} \equiv \Delta \begin{pmatrix} \cos\theta & \sin\theta \exp(-i\phi) \\ \sin\theta \exp(i\phi) & -\cos\theta \end{pmatrix}. \tag{10.80}$$

That is, elevate both parameters[3] θ, ϕ into quantum variables, $\boldsymbol{\theta}, \boldsymbol{\phi}$. Including the term $\mathcal{H}_{KE}$, and using the BO approximation, once again arrives at the effective Schrödinger equation given by Eq. (10.76) but with the connection in that equation replaced with, in spherical coordinates,

$$\mathbf{A} = \hat{\boldsymbol{\phi}} \frac{(1 - \cos\theta)}{2R \sin\theta}. \tag{10.81}$$

Such effective gauge potentials generate holonomies in toy and diatom systems [19,20]. Importantly, as first demonstrated in [21], they can generate, unlike the case for the AB-like gauge potential, effective Lorentz-like gauge forces. For example in the slow approach of two atoms whose relative motion, denoted by vector $\mathbf{R}$, is coupled to gauge connection similar to that of (10.81), but with integer "charge", an application of the Eherenfest theorem [27] suggests that the collision partners experience a mutual force, described by the following

$$\frac{d\langle\mathbf{R}\rangle}{dt} = -\frac{i}{m}\langle\nabla - i\mathbf{A}\rangle \equiv \langle\mathbf{v}\rangle$$

$$m\frac{d\langle\mathbf{v}\rangle}{dt} = \frac{1}{2}\langle\mathbf{v} \times \mathbf{B} - \mathbf{B} \times \mathbf{v}\rangle + \mathbf{F}_{grad}$$

$$\mathbf{F}_{grad} = -\langle\nabla V_{BO}(R)\rangle - \langle\nabla V_{AD}(R)\rangle. \tag{10.82}$$

$\mathbf{B} = \nabla \times \mathbf{A}$, is an effective "magnetic" field, and $\mathbf{F}_{grad}$ are, "electric"-like gradient forces.

$\mathbf{F}_{grad}$ has two components; the first is obtained by taking the gradient along the interatomic axis of the Born-Oppenheimer potential $V_{BO}(R)$ and is sometimes called the Feynman-Hellmann force. The second component originates from a higher-order scalar gauge correction V_{AD} [21–23].

In this application, the effective $\mathbf{B}$ field has a monopole-like structure similar to that given by the Berry curvature Eq. (10.61). In Fig. 10.13 we illustrate the dramatic differences between the scattering behavior resulting from a gradient force and that of a magnetic monopole Lorentz force. In addition to the monopole, one can generate different configurations for the effective curvature $\mathbf{B}$ by constructing, using external fields, alternative adiabatic Hamiltonians. For example, one can "engineer" $\mathcal{H}_{ad}$ in such a way to generate effective $\mathbf{B}$ fields that can guide, via effective Lorentz forces, neutral matter. Indeed, simulations show how effective gauge forces can be used to

---

[3] In general, one should also include the radial coordinate R as a quantum variable. See the discussion in [21].

10.2 What About the Phase? Exploring Holonomic Quantum Computing

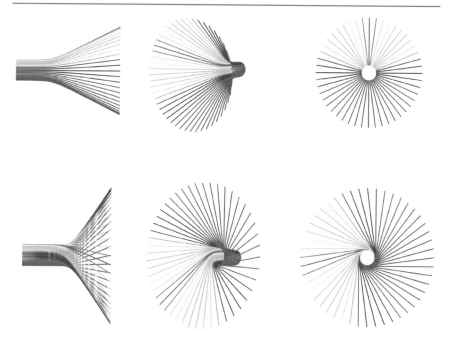

Fig. 10.13 Top panel shows trajectories, from three different perspectives, of a particle scattered by a gradient force e.g. a Coulomb potential. The lower panel illustrates trajectories of a particle, with identical initial conditions to those on the upper panel, by a magnetic monopole field

construct an effective magnetic lens for neutral matter [24]. This phenomenon has alternatively been called geometric [24], artificial [25], or synthetic magnetism [26].

10.2.6 A Topological Rabi Model

The Rabi model discussed in Sect. 10.2.1 describes adiabatic circuits of constant radius about the origin of a 2D plane. We seek to study adiabatic behavior as a system navigates a wider array of loops shown in Fig. 10.14. We consider planar loops, so the z coordinate is taken to be constant. In addition, we consider the displacement parameter x_0, which allows loops to be displaced in the xy plane. Thus we introduce

$$\mathbf{H}_{ad}(t) = B_1 \frac{\sigma_x \, x(t) + \sigma_y \, y(t)}{\sqrt{x(t)^2 + y(t)^2}} + B_0 \, \sigma_3$$
$$x(t) = a(t) \cos \omega t + x_0$$
$$y(t) = a(t) \sin \omega t$$
$$B_1 = \Delta \sin \theta \quad B_0 = \Delta \cos \theta. \tag{10.83}$$

The adiabatic Hamiltonian generated by these loops is parameterized not only by an azimuthal angle, and x_0, but also by the parameter $a(t)$. For the special case $x_0 = 0$

Fig. 10.14 Various loops C_1, C_2, C_3, C_4, associated with circuits generated by different parameters in Eq. (10.83). There are two classes of loops: those that encircle a tube along the z-axis and those that do not

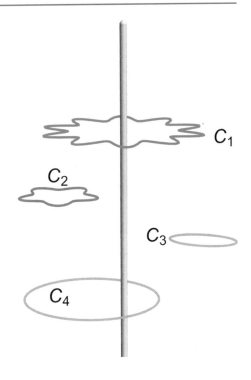

and $a(t) = 1$, we recover the standard Rabi Hamiltonian. We are not aware of analytic solutions to the Schrödinger equation corresponding to Hamiltonian (10.83), and use numerical methods (see Mathematica Notebook 10.4) to propagate the Schrödinger solution of the ground adiabatic eigenstate $u_g(0)$, for $\mathbf{H}_{ad}(0)$. Our condition for adiabaticity is, as before, $\omega/\Delta \ll 1$. We obtain the numerical solution $u(t)$ at $t = T = 2\pi$, and compare it the prediction of the adiabatic theorem

$$|\psi(T)\rangle = \exp(-i\gamma_D) \exp(i\gamma_G) |u_g(0)\rangle. \tag{10.84}$$

Furthermore, the adiabatic energy eigenstates are the same as before, i.e. the constants $\pm\Delta$ for excited and ground adiabatic states, respectively. Thus, calculating the overlap

$$\Omega \equiv \exp(i\gamma_D) \langle u_g(0)|\psi(T)\rangle \tag{10.85}$$

we can assess whether the conditions required for the quantum adiabatic theorem are satisfied. If $|\Omega| = 1$ then $|\psi(T)\rangle$ is proportional, up to a phase, to $|u_g(0)\rangle$. If that condition is met we obtain the equality

$$\Omega = \exp(i\gamma_G) \tag{10.86}$$

as we canceled the dynamic phase in expression (10.84).

10.2 What About the Phase? Exploring Holonomic Quantum Computing

Table 10.4 Calculated values for the probability $p \equiv |\langle u_g(0)|u(T)\rangle|^2$, and γ_G for the loops shown in Fig. 10.14

	C_1	C_2	C_3	C_4
p	1.00	1.00	1.00	1.00
γ_G	1.57	0.00	0.00	1.57

In Table 10.4 we show the results for the calculation of $|\langle u_g(0)|\psi(T)\rangle|^2$, and γ_G for each of the loops shown in Fig. 10.14. In these calculations, we set $\Delta = 1.0$, $\omega = 0.01$, and $\theta = \pi/3$. The parameters $a(t)$, x_0 were adjusted accordingly for each of the loops shown in that figure. The first row of Table 10.4 confirms that the conditions for adiabaticity are satisfied. The second column demonstrates that γ_G vanishes for loops C_2 and C_3 and values $\gamma_G \approx \pi/2$, for loops C_1, C_4. Thus, regardless of shape and location, loops that encircle the z-axis, illustrated by the tube in Fig. 10.14, have a non-vanishing and identical geometric phase, consistent with the theoretical value predicted from an AB-like gauge potential Eq. (10.46), i.e. in a cylindrical coordinate system,

$$\gamma_G = \oint_C d\mathbf{R} \cdot \mathbf{A} = \pi(1 - \cos\theta)$$
$$\mathbf{A} = \hat{\varphi} \frac{1 - \cos\theta}{2\rho}. \tag{10.87}$$

> Mathematica Notebook 10.4: Numerical Solutions to a Topological Rabi Model: https://bernardzygelman.github.io

Thus Hamiltonian (10.83) exhibits topological behavior and can be interpreted as a semi-classical analog of the scattering of a charged particle about an AB flux tube of "charge" $(1 - \cos\theta)$. This interpretation is consistent with other values chosen for θ (See Mathematica Notebook 10.4), that are not tabulated in Table 10.4.

10.2.7 Outlook

In Sect. 10.1.1 we focused on the probability that a quantum system remains in a ground state as it evolves adiabatically. Since that probability is independent of a global phase factor, the latter is usually ignored in applications of AQC. In this section, we reconsidered that phase and demonstrated its deep geometric and topological underpinning. Topology deals with questions that concern the global properties of a manifold. Therefore, the study of the latter can inform us how a system responds to perturbations impressed on a system. A detailed discussion of adiabatic and other topological phases and their possible application as a resource in quantum information science is available in [12].

Problems

> Mathematica Notebook 10.5: Problems and exercises https://bernardzygelman.github.io

References

1. D. Aharonov, W. van Dam, J. Kempe, Z. Landau, S. Loyd, O. Regev, *Adiabatic Quantum Computation is Equivalent to Standard Quantum Computation*, arXiv:quant-ph/0405098 (2005)
2. T. Albash, D. Lidar, Adiabatic quantum computing. Rev. Mod. Phys. **90**, 015002 (2018)
3. M. Born, V.A. Fock, Beweis des Adiabatensatzes. Zeitschrift fur Physik A **51**, 165 (1928)
4. M.V. Berry, Quantal phase factors accompanying adiabatic changes. Proc. R. Soc. A **392**, 45 (1984)
5. https://en.wikipedia.org/wiki/Quantum_tomography
6. A. Peruzzo, J. McClean, P. Shadbolt, M.-H. Yung, X.-Q. Zhou, P.J. Love, A. Aspuru-Guzik, J.L. O'Brien, A variational eigenvalue solver on a photonic quantum processor. Nat. Commun. **5**, 4213 (2014)
7. C. Zener, Non-adiabatic crossing of energy levels. Proc. R. Soc. Lond. A **174**, 696 (1932)
8. L. Landau, Zur Theorie der Energiubertagung II. Physikalishe Zeitschrift der Sowjetunion **2**, 46 (1932)
9. B. Zygelman, P.C. Stancil, M. Gargaud, R. McCarrol, *Ion-Atom Charge Transfer Reactions at Low Energies*, Springer Handbook of Atomic, Molecular, and Optical Physics Springer Handbooks, 805 (2023)
10. C. Cohen-Tannoudji, B. Dieu, F. Laloe, *Quantum Mechanics*, vol. I (Wiley, 1991)
11. E. Farhi, J. Goldstone, S. Gutmann, L. Zhou, The quantum approximate optimization algorithm and the Sherrington-Kirkpatrick model at infinite size. Quantum **6**, 759 (2022)
12. J. Zhang, T.H. Kyaw, S. Filipp, L. Kwek, E. Sjöqvist, D. Tong, Geometric and holonomic quantum computation. Phys. Rep. **1027**, 1 (2023)
13. M. Stone, P. Goldbart, *Mathematics for Physics: A Guided Tour for Graduate Students* (Cambridge University Press, 2009)
14. A. Aharanov, D. Bohm, Significance of electromagnetic potentials in quantum theory. Phys. Rev. **115**, 485 (1959); W. Ehrenberg, R.E. Siday, The refractive index in electron optics and the principles of dynamics. Proc. Phys. Soc. B **62**, 8 (1949)
15. R. Feynman, R. Leighton, M. Sands, *The Feynman Lectures on Physics Vol I-III, The New Millennium Edition* (Basics Books, New York, 2011)
16. C. Alden Mead, The molecular Aharonov-Bohm effect in bound states. Chem. Phys. **49**, 23 (1980)
17. B.K. Kendrick, Geometric phase effects in the $H + D_2 \to HD + D$ reaction. J. Chem. Phys. **112**, 5679 (2000)
18. B. Zygelman, The molecular Aharonov-Bohm effect redux. J. Phys. B: At. Mol. Phys. **50**, 025102 (2016)

19. M. Stone, Born-Oppenheimer approximation and the origin of Wess-Zumino terms: some quantum mechanical examples. Phys. Rev. D **33**, 1191 (1986)
20. J. Moody, A. Shapere, F. Wilczek, Realizations of magnetic monopole gauge fields: diatoms and spin precession. Phys. Rev. Lett. **56**, 893 (1986)
21. B. Zygelman, Appearance of gauge potentials in atomic collision physics. Phys. Lett. A **125**, 476 (1987)
22. B. Zygelman, A. Dalgarno, Direct charge transfer of He^+ in neon. Phys. Rev. A **33**, 3853 (1986)
23. B. Zygelman, Non-Abelian geometric phase and long-range atomic forces. Phys. Rev. Lett. **64**, 256 (1990)
24. B. Zygelman, Geometric-phase atom optics and interferometry. Phys. Rev. A **92**, 043620 (2015)
25. J. Dalibard, F. Gerbier, G. Juzeliunas, P. Öhberg, Artificial gauge potentials for neutral atoms. Rev. Mod. Phys. **83**, 1523 (2011)
26. Y.J. Lin, R.L. Compton, K. Jimenez-Garcia, J.V. Porto, I.B. Spielman, Synthetic magnetic fields for ultracold neutral atoms. Nature **462**, 628 (2009)
27. J.J. Sakurai, J. Napolitano, *Modern Quantum Mechanics*, 3rd edn. (Cambridge University Press, 2021)
28. B. Zygelman, Topological behavior of a neutral spin-1/2 particle in a background magnetic field. Phys. Rev. A **103**, 042212 (2021); Erratum: Phys. Rev. A **107**, 029901 (2023)

Index

A
Abelian group, 216
Adiabatic Quantum Computing, 177
Adjoint, 12
Aharonov, Y., 266
Alan Turing, 56
Albert Einstein, 123
Algorithm, 2
Alonzo Church, 56
Ancillary qubit, 207
Anharmonicity, 197
Anti-correlated, 141
Anti-Jaynes-Cummings, 174
Anton Zeilinger, 144
Arthur Ekart, 138
Aspect, A., 144

B
Balanced function, 68
Bandwidth limited, 79
Bardeen, J., 195
Basis, 7, 109
$BB84$, 123
BB84 protocol, 138
BCS theory, 195
Bell Laboratories, 147
Bell pairs, 142
Bell states, 127
Bell's Theorem, 136
Bernstein-Vazirani problem, 71
Berry, M.V., 258
Bertlmann's socks, 133
Bi-chromatic Raman beams, 175
Bit, 1
Bit-flip error, 203
Black-box quantization, 199

Bloch sphere, 25
Blue-sideband transitions, 174
Bohm, D., 266
Bohr magneton, 73
Boltzmann constant, 193
Boolean logic, 147
Boolean logic adder circuit, 147
Boolean logic gates, 52
Boris Podolsky, 123
Born rule, 5, 107
Boundary conditions, 182
Brian Josephson, 194

C
Capacitor, 190
Carrier frequency, 174
Carrier transition, 174
Cartesian product, 13
CERN, 124
Channels, 203
Charles Bennet, 138
CHSH inequalities, 133, 143
Cipher, 93
Cirac-Zoller gate, 164
Cirac-Zoller mechanism, 160, 171
Circuit impedance, 193
Circuit model, 51
Circuit model of quantum computing, 177
Classical bits, 142
Classical dynamical systems, 148
Classical error correction, 203
Claude Shannon, 147
Clauser-Horne-Shimony-Holt inequalities, 123
Clauser, J., 133
Closed system, 108

Closure, 109
CNOT gate, 60, 128
Code, 204
Codeword, 204
Coherent, 105
Coherent state, 108
Collapse hypothesis, 108
Commutators, 176
Completeness, 19
Computational basis, 18, 106, 139, 169
Computational complexity, 55
Concatenation, 224
Conjugate momenta, 149
Connection, 262
Constant function, 68
Control bit, 61
Control phase gate, 169
Control qubit, 128
Cooley, J., 80
Cooper, F., 195
Cooper pairs, 195
Correlation expectation value, 131
Cost function, 242
Coulomb force, 149
Coulomb repulsion, 170
Coupling constant, 159

D
Daniel Greenberger, 144
Data compression, 142
David Bohm, 124
David Deutsch, 63
David Wineland, 189
Degenerate, 12
Degenerate codes, 220
Degrees of freedom, 149
Density matrix, 109, 117, 131
Density operator, 106, 108
Detuning, 174
Deutsch-Josza algorithm, 69
Deutsch-Josza problem, 71
Deutsch's algorithm, 63
DFT, 80
Diffusion gate, 102
Direct, or tensor, product, 13
Direct product, 110
DiVincenzo criteria, 148
Dual space, 6
Dyadic, 9

E
Effective time-independent Hamiltonian, 177
Eigenstates, 152
Eigenvalue, 12, 107
Eigenvalue equation, 152
Eigenvector, 12
Einstein energy relation, 171
Ekert protocol, 123
Electric field, 155
Electromagnetic spectrum, 171
Electron, 73, 154
Electronic digital computing machine, 147
Encoding, 204
Energy spectrum, 153
Entangled states, 111, 120
Entanglement, 39
EPR, 123
Error correction codes, 141
Error recovery, 204
Error syndrome, 205
Eugene Wigner, 103
Euler formula, 78
Expectation value, 131, 136

F
Fan-out, 53
Farad, 190
Faraday induction law, 191
Fast Fourier Transform, 80
Fault-tolerant computing, 205
FFT, 80
Flying qubits, 140
Foundational postulates, 108
Fourier series, 78
Frequency interpretation of probability, 10

G
Gedanken experiment, 9, 123
Generalized coordinates, 149
Generators, 33
George Boole, 51
GHZ, 143
GHZ states, 123
Gilles Brassard, 138
Greatest Common Divisor (GCD), 94
Group, 25
Group generators, 216
Grover iteration, 98
Grover operator, 97, 99
Grover quantum algorithm, 77

Index

H
Hadamard, 59
Hadamard gate, 143, 155, 158
Hamiltonian, 73, 119, 149, 155, 163
Hamiltonian equations, 149
Hamiltonian functional, 149
Hamiltonian interaction, 177
Hamiltonian, R.W., 149
Harmonics, 183
Heat equation, 78
Heike Kamerlingh Onnes, 193
Heisenberg spin model, 177
Henry, unit of inductance, 191
Hermitian, 12
Hermitian operator, 107
Hidden variables, 136
High-Q, 189
Hilbert space, 5, 105
Holonomy, 261
Holt, M., 144
Holt, R., 133
Hopf map, 29
Horn, M., 133
Hydrogen atom, 149
Hyperfine interactions, 171
Hyperfine qubits, 172

I
Inductance, 191
Inductor, 190
Inner product, 6
Interacting qubits, 159
Interacting rotors, 163
Interaction Hamiltonian, 166, 167
Interaction picture, 156, 167, 169
Interference, 70
Ion crystals, 170
Ion-phonon coupling, 171
Ion-phonon operator, 176
Ion-qubits, 170
Ion-traps, 170
Irreversible, 56
Isomorphism, 25

J
James Clerk Maxwell, 35
Jaynes-Cummings Hamiltonian, 188
Jean-Baptiste Fourier, 78
John Atanasoff, 147
John Bell, 124, 127, 133
John von Neumann, 103
Jones vector, 36
Joseph Henry, 191
Josephson junctions, 194

K
Kinetic energy, 150, 170
Kirchoff loop law, 191
Konrad Zuse, 147
Kronecker delta, 7
Kronecker product, 16

L
Lagrangian, 149
Lagrangian formulation, 149
Lamb-Dicke limit, 175
Lamb-Dicke parameter, 174
Laser, 166
Laser beat frequency, 174
Laser-cooled ions, 170
Laser cooling, 170
Laser field, 155
Làszlò Ràtz, 103
Lev Grover, 95
Levi-Civita symbol, 33
Linearly independent, 5
Linear vector space, 3
Local realism, 136
Logical bits, 204
Louis de-Broglie, 123
Lowering operator, 163
Ludwig Boltzmann, 118

M
Magnetic dipole radiation, 172
Magnetic field, 171
Magnetic interactions, 171
Magnetic moment, 40
Majority vote, 205
Man-in-the-Middle (MITM) attack, 141
Matrix equation, 154
Matrix representation, 165
Max Born, 5
Maxwell's equations, 165
Measurement operator, 13
Measurement operators, 107
Meissner effect, 193
Metastable states, 170
Michael Faraday, 190
Microprocessor chip, 147
Microwave radiation, 172
Miller's algorithm, 94

Mixed ensemble, 110
Modes, 183
Mølmer-Sørenson coupling, 174
Multi-qubit gates, 171

N
Nathan Rosen, 123
Negative correlation, 132
Niels Bohr, 123
Nine-qubit code, 206
No-cloning theorem, 125
Non-commutative, 32
Non-demolition measurements, 189
Nondegenerate eigenvalue, 108
Non-stoquastic, 256
Normalized, 7
Normalizer of a group, 217
Nucleus, 171
Number operator, 163
Nyquist frequency, 79
Nyquist-Shannon sampling theorem, 79

O
One-time pad, 138
Operator, 6
Optical qubit, 170, 171
Oracle, 95
Order of a group, 216
Orthogonal, 7
Orthonormal, 8, 110
Outer product, 6

P
Parallel transport, 260
Parity, 205
Partial differential equation, 152
Partial trace, 110
Paul Dirac, 6
Pauli group, 214
Pauli matrices, 73
Pauli-X gate, 165
Pauli-X operator, 166
Pauli-Z gate, 172
Peter Shor, 71, 81
Phase gate, 57, 158
Phase shift gate, 73
Phonon, 163, 166
Phonon bus state, 175
Phonon Hilbert space, 163
Photon, 39, 123, 128, 170
Planck's constant, 41, 151, 171

Plane polarized light, 36
Positive correlation, 132
Power spectrum, 83
Poynting vector, 182
Prime numbers, 93
Private key, 93, 138
Private key distribution problem, 138
Private key encryption, 138
Probability measure, 135
Projection operator, 17
Proton, 149
Public key encryption, 93
Pure ensemble, 110
Pure state, 105
Purification, 116

Q
QFT, 83
Quantum Dense Coding, 142
Quantum Electrodynamics (QED), 39, 184
Quantum Error Correcting Codes (QECC), 206
Quantum interference, 63
Quantum Key Distribution (QKD), 123, 138
Quantum logic gates, 147
Quantum parallelism, 63, 70
Quantum state tomography, 246
Quantum statistical physics, 177
Qubit, 1

R
Rabi coupling, 173
Rabi flopping, 172
Rabi-frequency, 173
Rabi, I., 259
Rabi transitions, 174
Raising operators, 163
Raman laser beams, 173
Raman-laser pair, 175
Random variable, 136
Read-out, 171
Red-detuned pulse, 175
Red-sideband transitions, 174
Register, 2
Reinhold Bertlmann, 133
Relativity, 123
Representation, 26
Resonance, 170
Reversible, 56
Ripple circuit, 54

Index 279

Roger Penrose, 144
Rotating wave approximation (RWA), 157
Rotor qubits, 158
RSA encryption protocol, 93
RSA public key encryption protocol, 138
RWA approximation, 167, 174
Rydberg atom, 189

S
Sample space, 135
Sampling rate, 79
Schmidt number, 114, 116
Schrieffer, J.R., 195
Schrödinger equation, 155
Schroedinger equation, 72, 167
Second order differential equation, 149
Semi-prime numbers, 93
Serge Haroche, 189
Shannon information entropy, 117
Shimony, A., 133
Shor algorithm, 77, 138
Shor code, 211
Shor's algorithm, 71
Simon's problem, 71
Simple harmonic motion, 161
Simulated annealing, 241
Single-mode, 163
Singular-value decomposition, 115
SI units, 190
Space-like interval, 143
Special relativity, 40
2-sphere, 30
3-sphere, 29
Spin, 25, 39, 40, 171
Spin measurements, 141
Spontaneous emission, 170, 172
Stabilizer group, 216
Stabilizers, 214
Standing wave, 183
State, 1
Stephen Hawking, 136
Stephen Wiesner, 138
Stern-Gerlach device, 40
Stimulated Raman Excitation (SRE), 172
Stokes theorem, 262
Stoquastic, 255
Superconductor, 192

Superposition principle, 4
Superposition state, 175
Syndrome decoding, 205

T
Target bit, 61
Target qubit, 128
The method of separation of variables, 152
Threshold probability, 236
Time dependent expectation value, 120
Time development operator, 159, 169
Translating Rotor, 150
Transpose, 31
Transversal gate, 229
Trap loss, 170
Trapping potential, 170
Truth tables, 51
Tukey, J., 80
Tunneling, 195
Turing model, 56

U
Uncertainty, 43
Uncertainty principle, 45
Unitary, 12
Unitary evolution, 167
Unitary time evolution operator, 158
Universal gates, 170

V
Vacuum energy, 162
Vacuum state, 162
Vibrational qubit, 166
Volterra-Dyson series, 72
Von Neumann, 108
Von Neumann entropy, 117

W
Wavenumber, 183
Willard Gibbs, 118

Y
Yb^+, 172

Z
Zeeman Hamiltonian, 154

Printed in the United States
by Baker & Taylor Publisher Services